Systems Biology of Cell Signaling

Systems Biology of Cell Signaling

Recurring Themes and Quantitative Models

James E. Ferrell, Jr.

CRC Press is an imprint of the
Taylor & Francis Group, an **informa** business
A GARLAND SCIENCE BOOK

Mathematica® and the *Mathematica* logo are registered trademarks of Wolfram Research, Inc. (WRI - www.wolfram.com http://www.wolfram.com/) and are used herein with WRI's permission. WRI did not participate in the creation of this work beyond the inclusion of the accompanying software, and offers it no endorsement beyond the inclusion of the accompanying software.

First edition published 2022
by CRC Press
6000 Broken Sound Parkway NW, Suite 300, Boca Raton, FL 33487-2742

and by CRC Press
2 Park Square, Milton Park, Abingdon, Oxon, OX14 4RN

© 2022 Taylor & Francis Group, LLC

CRC Press is an imprint of Taylor & Francis Group, LLC

Reasonable efforts have been made to publish reliable data and information, but the author and publisher cannot assume responsibility for the validity of all materials or the consequences of their use. The authors and publishers have attempted to trace the copyright holders of all material reproduced in this publication and apologize to copyright holders if permission to publish in this form has not been obtained. If any copyright material has not been acknowledged please write and let us know so we may rectify in any future reprint.

Except as permitted under U.S. Copyright Law, no part of this book may be reprinted, reproduced, transmitted, or utilized in any form by any electronic, mechanical, or other means, now known or hereafter invented, including photocopying, microfilming, and recording, or in any information storage or retrieval system, without written permission from the publishers.

For permission to photocopy or use material electronically from this work, access www.copyright.com or contact the Copyright Clearance Center, Inc. (CCC), 222 Rosewood Drive, Danvers, MA 01923, 978-750-8400. For works that are not available on CCC please contact mpkbookspermissions@tandf.co.uk

Trademark notice: Product or corporate names may be trademarks or registered trademarks and are used only for identification and explanation without intent to infringe.

Library of Congress Cataloging-in-Publication Data
Names: Ferrell, James Ellsworth, author.
Title: Understanding cell signaling : motifs, recurring themes, and the theory of nonlinear dynamics / James Ferrell.
Description: Boca Raton : Taylor & Francis, 2021. | Includes bibliographical references and index. | Summary: "All living cells continually detect and respond to external signals. This is true of prokaryotes, whether they are living alone or in biofilms, and it is even more manifestly true in multicellular eukaryotes, where communication between cells and coordination of the cells' behavior enables the organism to function as a unified whole. In large multicellular organisms like us humans, cells receive signals from their immediate neighbors through short-range signals like neurotransmitters and cell-surface molecules. They receive signals from more distant neighbors via longer-range diffusible molecules like morphogens, and from still-more distant neighbors by means of hormones that flow through the circulatory system. And they receive signals from the outside world via sense organs. Cells also monitor their own internal status, and there is a great deal of overlap between the cellular components involved in cell-cell communication and internal monitoring"— Provided by publisher.
Identifiers: LCCN 2021007837 (print) | LCCN 2021007838 (ebook) | ISBN 9780367643836 (hardback) | ISBN 9780815346036 (paperback) | ISBN 9781003124269 (ebook)
Subjects: LCSH: Cellular signal transduction.
Classification: LCC QP517.C45 F47 2021 (print) | LCC QP517.C45 (ebook) | DDC 571.7/4--dc23
LC record available at https://lccn.loc.gov/2021007837
LC ebook record available at https://lccn.loc.gov/2021007838

ISBN: 978-0-367-64383-6 (hbk)
ISBN: 978-0-8153-4603-6 (pbk)
ISBN: 978-1-003-12426-9 (ebk)

DOI: 10.1201/9781003124269

Typeset in ITC Leawood Std Book
by codeMantra

Instructors can access the Figure Slides from the Instructor Hub upon registering. To register, please visit https://routledgetextbooks.com/textbooks/instructor_downloads/

CONTENTS

DETAILED CONTENTS

PREFACE

We biologists study life, and the processes of life are endlessly fascinating. Cannonball trajectories and masses on springs are fine, but they cannot hold a candle to cell division.

This may be why we see so many physicists turning their sights on biology, trying to get to the bottom of its wonderful phenomenology, and making good headway too—some of the deepest biology papers of the past decades have been written by physicists-turned biologists.

One of the secrets of their success is mathematics, approaches and tools borrowed from celestial mechanics, chemical kinetic theory, and control theory. You can have a terrific, satisfying career in biology without ever making use of the quadratic formula, let alone bifurcation theory, but for those willing to give it a go, the rewards can be great. It gives you the chance to go beyond just describing fascinating phenomena to really understanding how and why they happen.

This book is an attempt to introduce biologists to some powerful mathematical approaches from the theory of dynamical systems. The biological focus is on cell signaling, the interplay between a cell and the outside world that allows it to "know" things, respond to them, adjust to them, and remember them. Cell signaling is a big, complicated field, but it turns out that evolution has come up with the same handful of tricks over and over again to build reliable signaling systems. There are commonalities to seemingly disparate cell signaling phenomena that become apparent once you decide to take the plunge and apply a little math.

This book arose out of a course I teach on systems biology and mathematical modeling. It is a 30-h course for Ph.D. students, and I devote 10 h to lectures and 20 h to hands-on modeling. Occasionally a student in the course will come in with some knowledge of biology plus a good background in matrix algebra or dynamical systems theory, but for most of the students the math is completely new, or the biology is completely new, or both. The 17 chapters that make up this book include too much material for my course—maybe they will be just right for your course, or you could do what I do and pick and choose among the offerings.

There are written problem sets to accompany Chapters 2–16. They make use of *Mathematica*®, and so the problem sets attempt to both solidify the concepts introduced in the book chapters and gently introduce the students to writing simple *Mathematica*® code. I have also tried doing the course in MATLAB®, but I have had better luck with *Mathematica*®, especially for students who have no previous exposure to either. The problem sets, with answers, are available to instructors. Please go to the book's product page (https://www.routledge.com/Systems-Biolo-gy-of-Cell-Signaling-Recurring-Themes-and-Quantitative/Ferrell/p/book/9780815346036), and then register for the "Instructor Hub" by clicking on the relevant link. Some of the problem sets are based on famous systems biology papers. This shows that a diligent student can (usually) run published models for themselves, and, in the space of an hour or two, get a much better idea of what the modeling shows and means than could be obtained by just reading a paper.

This book does not attempt to give a comprehensive overview of cell signaling—there are just too many genes, proteins, and pathways for that. Instead it repeatedly calls upon a group of interesting, important, and reasonably well-understood signaling archetypes, including two-component signaling in bacteria, G-protein-coupled receptors, receptor tyrosine kinase signaling, and oscillators from the embryonic cell cycle and neurobiology. My strong impression has been that both the physicists and biologists in my course benefit from this restricted biological focus.

James E. Ferrell, Jr.

Palo Alto

ACKNOWLEDGEMENTS

Thanks so much to my colleagues who encouraged me to write something like this way back when—especially David Morgan, Mike Tyers, Andreas Meyerhans, Steve Martin, Jeremy Thorner, and Ted Weinert.

Thanks to the many colleagues who taught me, through discussions and through their published work, so much of the stuff I am relaying here. Especially Eduardo Sontag, Dennis Thron, and Steven Strogatz, but also Uri Alon, Naama Barkai, Michael Elowitz, Daniel Fisher, Erwin Frey, Albert Goldbeter, Jeff Hasty, Mogens Jensen, Boris Kholodenko, Arthur Lander, Stan Leibler, Wendell Lim, Tobias Meyer, Béla Novák, Rama Ranganathan, Galit Lahav, Stas Shvartsman, Eric Siggia, Gürol Süel, Peter Swain, Chao Tang, John Tyson, and many others.

Many thanks to the publishers and editors who helped along the way—Miranda Robertson, Summers Scholl, Chuck Crumly, and Jordan Wearing.

Big thank-yous to my students and postdocs, especially those who helped with revisions and proofreading, including Oshri Afanzar, Yuxin Chen, Yuping Cheng, Xianrui Cheng, Jo-Hsi Huang, Lendert Gelens, William Huang, Julia Kamenz, Zhengda Li, Shixuan Liu, and Connie Phong. Thanks to Ferrell lab alumni whose photos and data were used in the figures, including Oshri Afanzar, Graham Anderson, Xianrui Cheng, Sang Hoon Ha, Eric Machleder, Joe Pomerening, Tony Tsai, and Wen Xiong. And to many colleagues, including John Albeck, Uri Alon, Brian Kobilka, Shinya Kuroda, Mark Lemmon, Stephen Smith, Matthew Spade, Jane Stout, Nick Tonks, and Claire Walzcak, for allowing me to use their published figures and photos.

And the biggest thanks to Britta Erickson for every big and little thing.

This book is dedicated to my parents James and Sara Ferrell.

AUTHOR

James E. Ferrell, Jr., PhD, studied physics, mathematics, and chemistry at Williams College and graduated in 1976. He earned doctoral degrees in chemistry (1984) and medicine (1986) from Stanford, and studied cell signaling at UC Berkeley as a postdoctoral fellow from 1986 to 1990. He joined the faculty at the University of Wisconsin in 1990 and moved back to Stanford in 1992. He was the first chair of the Department of Chemical and Systems Biology and is currently professor of Chemical and Systems Biology and of Biochemistry.

INTRODUCTION

1

IN THIS CHAPTER . . .

DOI: 10.1201/9781003124269-1

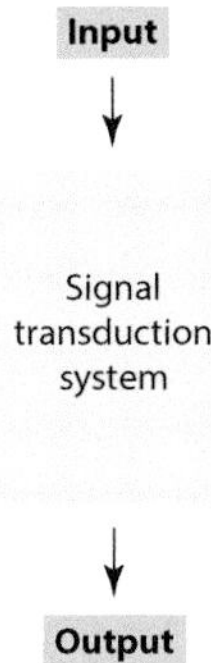

Figure 1.1 A schematic view of a generic signal transduction process.

SIGNAL TRANSDUCTION COMPONENTS AND SYSTEMS

All living cells continually detect and respond to external signals. This is true for prokaryotes, whether they are living alone or in biofilms, and it is even more manifestly true in multicellular eukaryotes, where communication between cells and coordination of the cells' behavior enables the organism to function as a unified whole. In large multicellular organisms like us humans, cells receive signals from their immediate neighbors through short-range signals like neurotransmitters and cell-surface molecules. They receive signals from more distant neighbors via longer range diffusible molecules such as morphogens and from still-more distant neighbors by means of hormones that flow through the circulatory system. They receive signals from the outside world via sense organs. Cells also monitor their own internal status, and there is a great deal of overlap between the cellular components involved in cell–cell communication and internal monitoring. Ultimately a cell processes input signals through a process termed **signal transduction**, shown schematically in **Figure 1.1**.

Signal transduction allows us to see, hear, taste, smell, and feel. It allows us to think, remember, and move. Signaling determines if and when a cell grows and divides and often determines if and when it dies. Signaling drives differentiation, enables the formation of all of our tissues and organs during development, and maintains them after they have formed. It allows our blood to clot and our immune system to fight infection. Signaling allows us to heal our wounds and to adapt to the unpredictable world around us. Signaling proteins are the targets of six of the ten most widely prescribed drugs in the United States (**TABLE 1.1**) and are the targets of probably all recreational drugs.

Thus, signal transduction is of special importance to neurobiologists, cell biologists, developmental biologists, hematologists, immunologists, and pharmacologists. Increasingly, it has been attracting the

TABLE 1.1 **Most Widely Prescribed Drugs in the United States**

				U.S. Prescriptions (millions)				
	Drug	Indications	Mechanism of Action	2014	2015	2016	2017	2018
1	Atorvastatin	High cholesterol	Inhibits cholesterol synthesis	74	94	97	105	112
2	*Levothyroxine*	Hypothyroidism	Activates thyroid hormone receptors	100	113	114	102	105
3	*Lisinopril*	Hypertension	Inhibits the last step in the production of the hormone angiotensin II	114	110	109	104	97
4	Metformin	Type II diabetes mellitus	Inhibits mitochondrial respiratory-chain complex 1	85	83	80	78	84
5	*Amlodipine*	Hypertension, angina pectoris	Inhibits voltage-gated calcium channels	63	71	75	73	76
6	*Metoprolol*	Hypertension, angina pectoris, and myocardial infarction	Inhibits β_1-adrenergic receptors	71	69	73	67	71
7	*Albuterol*	Asthma and chronic obstructive pulmonary disease	Activates β_2-adrenergic receptors	48	50	47	50	61
8	Omeprazole	Gastroesophageal reflux disease and gastric ulcers	Inhibits H^+/K^+ ATPase	71	71	70	58	58
9	*Losartan*	Hypertension	Inhibits angiotensin II receptors	37	47	49	52	51
10	Simvastatin	High cholesterol	Inhibits cholesterol synthesis	97	89	80	73	66

Six of the ten drugs on this list work by activating or inhibiting signaling proteins, or by inhibiting the production of a hormone. These are highlighted in italic. Source: Agency for Healthcare Research and Quality. Total purchases in by prescribed drug, United States, 1996–2018. Medical Expenditure Panel Survey. Generated interactively: Wed Jan 20 2021.

attention of physicists, control theorists, and electrical engineers—scientists who want to use the tools of their fields to deepen our understanding of this fascinating but highly complicated aspect of life.

SIGNAL TRANSDUCTION IS CARRIED OUT BY SYSTEMS OF VARYING COMPLEXITY

1.1 SIGNAL TRANSDUCERS ARE CELLULAR COMPONENTS THAT ACT MAINLY BY REGULATING OTHER CELLULAR COMPONENTS

Deciding which components of a cell count as signal transducers, and which do not, is not a trivial task. Often signal transducers are proteins or protein complexes, but they can also be RNAs, small molecules, or ions. Perhaps a few examples will help us sharpen our ideas of what is and what is not a signal transducer.

Receptors, protein kinases, and small G-proteins are signaling proteins, but glycolytic enzymes and motor proteins are not. MicroRNAs are signal transducers—they regulate mRNA stability and translation—but mRNAs, tRNAs, and rRNAs are not. Calcium ions are signal transducers—they regulate protein kinases, phosphoprotein phosphatases, motor proteins, and many other proteins—but magnesium ions are not. The membrane lipids PIP_3 and diacylglycerol are signal transducers—they both regulate particular protein kinases—but phosphatidylcholine is not; it (mainly) acts as a structural component of membranes. And the nucleotide cAMP is a signal transducer, allosterically regulating a subunit of protein kinase A, but its relative ADP is not; it is a metabolic intermediate. In general, signal transducers are cell components that vary dynamically in abundance or activity and affect a cell's function by regulating something else; they are more like managers than workers.

Some consider transcription factors—DNA-binding proteins that regulate the transcription of specific genes—to be terminal effectors of signal transduction systems rather than being signal transducers themselves. Here the main distinction is time scales; transcription is often slower than signal transduction processes like ion fluxes or protein phosphorylation. In other respects, though, transcription factors are just like other signal transduction proteins, relaying signals from upstream inputs (often protein kinases) to downstream targets (the genes whose transcription they regulate).

1.2 THE SIGNAL TRANSDUCTION PARTS LIST IS LONG

We now have a close-to-comprehensive parts list for the signaling proteins and other signaling molecules from all of the widely studied model organisms (e.g., humans, mice, *Drosophila melanogaster*, *Caenorhabditis elegans*, *Saccharomyces cerevisiae*, and *Escherichia coli*). The simplest of these model organisms in terms of the length of the parts list is, by a wide margin, the prokaryote *E. coli*. Many of its signaling pathways make use of one of 29 histidine-specific protein kinases, and these kinases phosphorylate about an equal number of downstream substrate proteins. Compared to the numbers of protein kinases and kinase substrates involved in mammalian signaling, these numbers are small, but still, these are the two largest families of paralogous genes in *E. coli*. In any case, with this limited cast of components, it is possible for a diligent student of cell signaling to acquire a reasonably comprehensive understanding of *E. coli* signal transduction.

The situation is much more complicated in human cells; the proteins involved in cell signaling are more numerous and more varied in their structures and functions than the bacterial proteins are. There are more than a dozen families of receptors in human cells, including **G-protein-coupled receptors** or GPCRs (more than 800 in humans), receptor tyrosine kinases or RTKs (49 in humans), tyrosine kinase-associated cytokine receptors, integrins, receptor serine/threonine kinases, receptor phosphatases, receptor guanylyl cyclases, the Hedgehog receptor Patched, the Wnt co-receptor LRP, cadherins, Toll-like receptors, ligand-gated ion channels, and steroid/retinoid receptors. These receptors act through at least as many classes of downstream signaling proteins: adaptor proteins, GTP-binding G-proteins, non-receptor protein kinases and phosphatases, methylases and demethylases, acetylases and deacetylases, **second messengers** (small molecules and ions), translational regulators, degradation regulators, and many classes of transcription factors and chromatin regulators, to name only a few. In total there are thousands of genes for signaling proteins in the human genome. The large number attests to the importance of cell signaling in complex organisms. However, it also presents a formidable challenge to students of cell signaling; there is a lot to learn.

1.3 SIGNAL TRANSDUCTION IN BACTERIA IS ACCOMPLISHED BY SHORT, (MOSTLY) LINEAR, (MOSTLY) NON-INTERCONNECTED PATHWAYS

Once an organism's signaling parts list is completed, the next step toward understanding how a signal transduction process works is to figure out how the relevant parts are wired together into pathways, circuits, and networks. All of these not-quite-synonymous terms can be used to describe systems of signaling proteins, with the particular term used depending on the size and topology of the system: the term "networks" tends to be used for big systems with all sorts of complications, "pathways" are generally smaller systems with little or no feedback, and "circuits" are often intermediate in size and complexity. Obtaining a reliable systems map or circuit diagram is much more difficult than obtaining a list of components. Nevertheless, we have a good understanding of a handful of signaling pathways in a variety of model organisms, from *E. coli* through humans.

As mentioned above, in *E. coli*, signal transduction is often mediated by the so-called two-component systems, with the two components of the **two-component system** being a histidine-specific protein kinase and a kinase substrate called a **response regulator**. There are other important types of signaling system in *E. coli* as well; for example, the bacterium responds to changes in the availability of lactose and glucose through the binding of lactose to a transcriptional repressor and the binding of cAMP, a surrogate for low glucose levels, to a transcriptional activator, without the intermediacy of a kinase or response regulator. But two-component systems account for a good share of signal transduction in *E. coli* and other prokaryotes.

Three specific examples of two-component systems are shown in **Figure 1.2**. The simplest of these is the quorum-sensing (Qse)

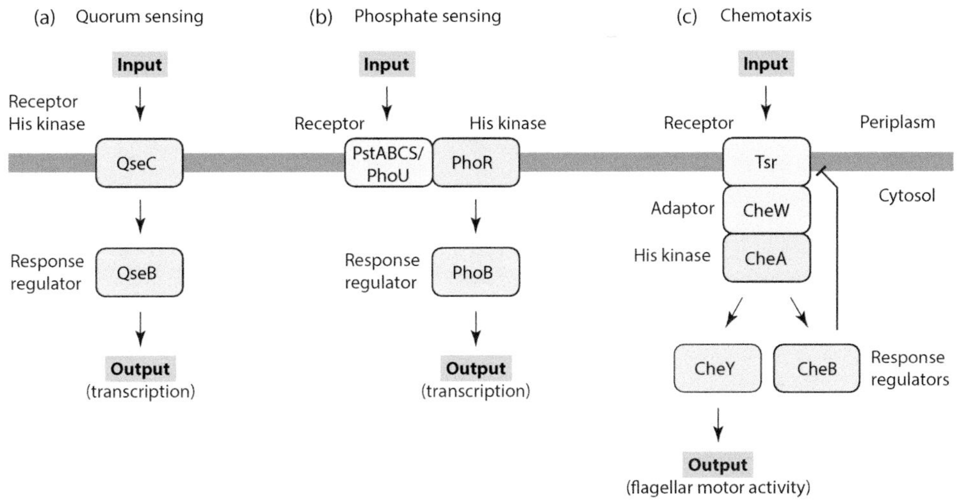

Figure 1.2 Three signal transduction pathways in *E. coli*. (a) The quorum-sensing pathway. (b) The phosphate-sensing pathway. (c) The chemotaxis pathway. The receptor protein shown here, Tsr, responds positively to serine and negatively to leucine, indole, and weak acids. Other receptors (Tsr, Trg, Tap, and Aer) that respond to other ligands couple into the same pathway. Note that all of these signaling pathways make use of a histidine-specific protein kinase and a response regulator—a so-called two-component regulatory system—highlighted in green.

pathway. The pathway is activated by one of several hormone-like small molecules, which include (1) a boron-containing compound called autoinducer-2 (AI-2), which is released by neighboring bacteria and used for both intraspecies and interspecies communication; (2) an as-yet unidentified bacterial factor termed autoinducer-3 (AI-3); and (3) the catecholamines epinephrine and norepinephrine, small molecules used as hormones and neurotransmitters by animals that can also mediate inter-kingdom communication between a host animal and the bacteria living within its gut. These stimulus molecules act by binding to a transmembrane receptor protein, QseC. QseC possesses a modular histidine kinase domain, and this domain is similar in sequence to the kinase domains of all of the 28 other *E. coli* histidine kinases (but not to the eukaryotic serine/threonine or tyrosine kinases). The activated receptor autophosphorylates at a specific conserved histidine residue and then transfers this phosphate to an aspartate residue on a response regulator, the DNA-binding protein QseB. When phosphorylated, QseB activates the transcription of specific flagellar genes. The protein phosphorylation is probably reversed by the kinase itself, in the case of the histidine autophosphorylation, and non-enzymatically, in the case of the aspartate phosphorylation of the response regulator. Thus, an input (the concentration of the small molecule receptor-binding ligand) gets converted into an output (changes in gene expression) via a simple linear signal transduction pathway with only two components (QseC and QseB).

The two other *E. coli* signaling systems shown in **Figure 1.2** are slightly more complicated. The phosphate-sensing system includes a transmembrane histidine kinase (PhoR), which regulates a single response regulator (PhoB), but in this case, the receptor is a complex of proteins (the PstA/B/C/S complex, which transports phosphate across the plasma membrane, plus the PhoU protein) that interact with PhoR, rather than PhoR itself. And the best-studied two-component signaling system, the chemotaxis system, includes not only separate receptor subunits (such as the serine-sensing receptor Tsr and the aspartate-sensing receptor Tar) but also two downstream targets rather than one: the response regulator CheY, which regulates flagellar motor activity, and the response regulator CheB, which feeds back on the receptor to regulate its sensitivity. But even with these elaborations, the basics of all of the *E. coli* histidine kinase signaling systems are the same: the pathways are short, linear or nearly linear, and mostly non-interconnected.

Two-component signaling is widespread (though not universal) in bacteria and also occurs in fungi, plants, and some basal eukaryotes. However, it appears to be absent from metazoans.

1.4 THE EGFR SYSTEM IS DEEP, INTERCONNECTED, AND COMPLICATED

Some pathways in eukaryotic signaling are probably almost as simple as those in bacteria, but many, including some of the best-understood pathways, are much more complicated. One good example is the mammalian EGF (epidermal growth factor) receptor (EGFR) system. EGF and its receptor have a storied place in the history of signal transduction. EGF was the second peptide growth factor to be discovered (after nerve growth factor, NGF), and the EGF receptor (alternatively called EGFR, ErbB-1 or, in humans, HER1) was one of the first growth factor receptors to be characterized and to be identified as a protein tyrosine kinase. The EGFR system is also implicated in human health and disease. The receptor is a proto-oncogene, meaning that it has given rise to a gene (v-erbB) that allows a tumor virus (the avian erythroblastosis virus from which the name v-erbB is derived) to cause cancer, and the receptor has been implicated in the pathogenesis of several types of human cancer, especially carcinomas and glioblastomas. Antibodies and small molecule inhibitors of the EGF receptor are used clinically as cancer chemotherapeutics. The signaling system downstream of the EGF receptor is riddled with proto-oncogenes (Ras, Raf, Akt, Myc, Fos, and Jun) as well. In addition to its role in regulating cell growth and replication, the EGF receptor plays critical roles in cell fate induction during development. These developmental roles were initially worked out through studies in *Drosophila* and *C. elegans*, but the system is involved in developmental decisions in mammals as well. All of these factors help explain why the EGF receptor and the signaling system downstream of it have been studied extensively for decades, and why the amount of information amassed on this system is so large.

The diagram in **Figure 1.3** starts with the mammalian EGFR and then depicts the direct downstream targets of the receptor, and the targets of those targets, and so on, until the first transcription factors are reached. The diagram omits the regulators of the downstream targets that are not themselves regulated by the EGFR; it is a little like a family tree that shows only the direct descendants of one ancestor. Some of the ramifying branches of the network have been lopped off in the interest of space, terminating in targets designated "others" before transcription factors have been reached (**Figure 1.3**). Thus the network has been simplified substantially. Simplification notwithstanding, it is clear that one cannot completely understand EGFR signaling by focusing on one or two proteins. Dozens of proteins contribute to EGFR signal transduction.

Like the bacterial signaling systems, EGF signaling makes use of reversible protein phosphorylation and brings about changes in transcription, motor protein activity, and other cellular processes. However, the kinases involved are not histidine-specific protein kinases. The EGF receptor is a tyrosine kinase, and it phosphorylates itself and other proteins at tyrosine residues. Several additional kinases function downstream of EGFR, and most of these (e.g. Raf and Akt) are serine and threonine specific. The protein tyrosine and

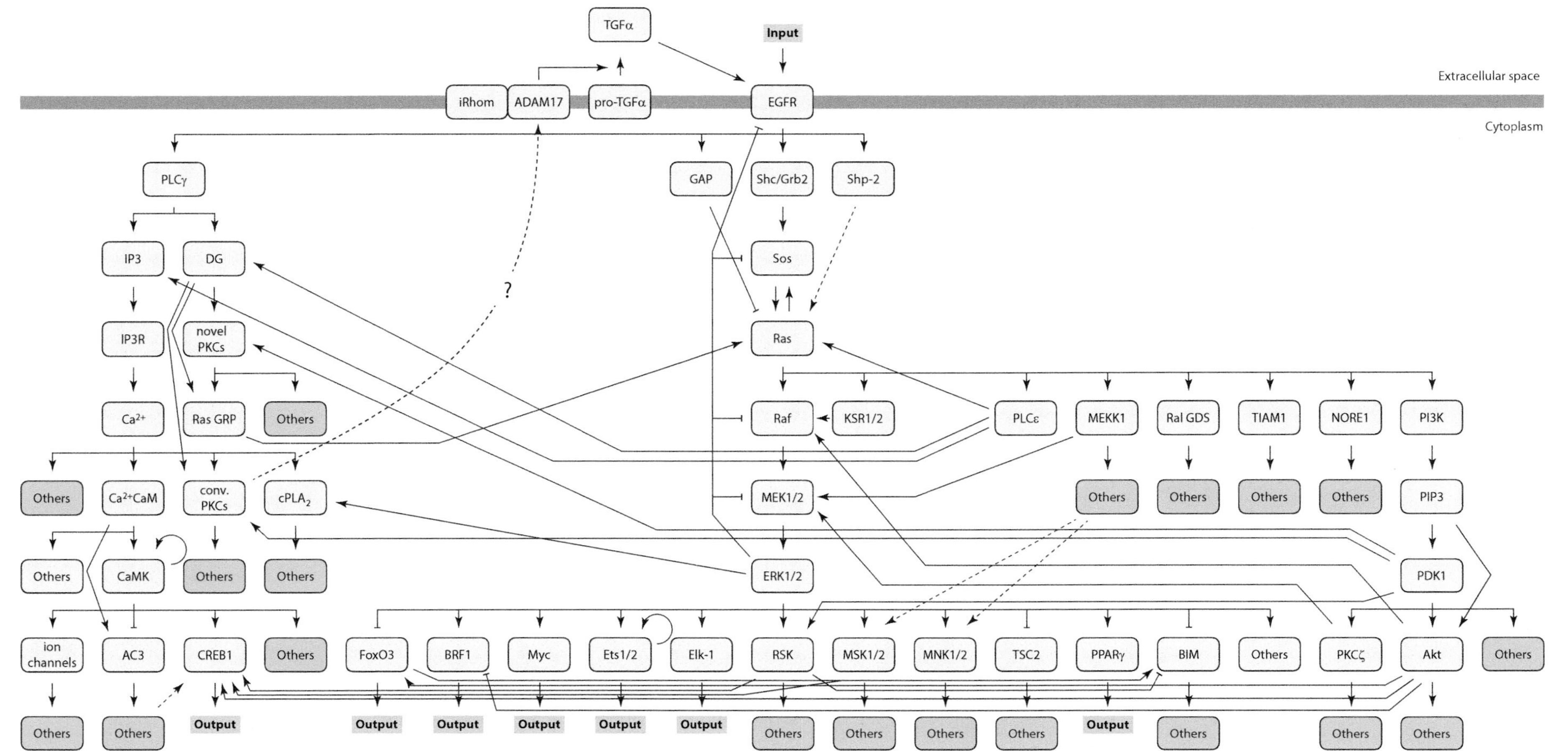

Figure 1.3 The EGFR signaling system in a typical mammalian cell. Only the proteins directly downstream of EGFR are shown. In many cases a group of related signaling proteins (e.g. the Raf proteins) are represented by a single species. The dashed lines represent regulatory processes where either the exact mechanism is not known or it is known but the intermediate proteins are not present on the diagram. More information about the signaling proteins and small molecules shown here can be found in **TABLE 1.2**.

TABLE 1.2 **Downstream Mediators of EGF Signaling**

Abbreviation	Full Name	Description and Comments
AC3	Adenylate cyclase 3	A membrane-bound adenylate cyclase present in many cell types. AC3 is inhibited by CaMKII and may also be directly activated by Ca^{2+}CaM.
ADAM17	A disintegrin and metalloproteinase 17	Also called TACE (tumor necrosis factor-α-converting enzyme). ADAM17 has been shown to be a sheddase, catalyzing the proteolytic shedding of various proteins, including TNF-α and TGFα, from the surface of cells.
Akt	Cellular homolog of the Ak-mouse strain thymoma virus onocogene	Also called PKB (protein kinase B). The Akt proteins (Akt1-3) are serine/threonine-specific protein kinases implicated in cell survival. Their activation depends on binding to PIP3 and phosphorylation by PDK1 (shown) and mTORC2 (not shown).
BIM	Bcl-2-interacting mediator of cell death	Also called BCL2-like protein 11 or BCL2L11. A pro-apoptotic BH3-domain-containing (BH3-only) protein thought to be negatively regulated by ERK1/2 and RSK1/2 phosphorylation in a coherent feedforward fashion. BIM is also positively regulated by JNKs and its transcription is induced by FoxO3.
BRF1	TFIIB-related factor 1	One of the three subunits of RNA polymerase III (together with TBP and BDP1). Pol III activity is thought to be stimulated through the phosphorylation of BRF1 by ERK. Akt also phosphorylates BRF1, and this is thought to inhibit both the activity and the degradation of BRF1.
Ca^{2+}	Calcium ion	A universal eukaryotic second messenger. At concentrations attained in cell signaling, most of the calcium is bound to calcium-binding proteins like calmodulin (CaM).
Ca^{2+}CaM	Calcium-calmodulin complex	An important mediator of the effects of calcium that stoichiometrically activates CaMKs and other downstream targets.
CaMK	Calmodulin-dependent protein kinase	A family of protein kinases (CaMKI through IV) activated by Ca^{2+}CaM.
conv. PKCs	Conventional protein kinase C proteins	A family of protein kinases (PKCα, two PKCβ isoforms, and PKCγ) that are activated by calcium plus diacylglycerol.
cPLA2	Cytosolic phospholipase A2	Cleaves arachidonic acid from the 2-position of phospholipids to yield this signaling metabolite and a lysolipid. cPLA2 localization is regulated by calcium and cPLA2 activity is regulated by phosphorylation by ERKs and other MAP kinases.
CREB1	Cyclic AMP-responsive element-binding protein 1	A DNA-binding transcription factor that dimerizes via a leucine zipper and binds to the cAMP response element, an enhancer sequence. CREB1 activity depends on Ser 133 phosphorylation, which can be carried out by PKA, Akt, CaMKIV, MSK, RSK, and number of other protein kinases.
DG	Diacylglycerol	A membrane-bound lipid second messenger formed by the cleavage of the inositol head group from PIP2.
EGFR	Epidermal growth factor receptor	Also called HER1 (for human EGF receptor 1) and ErbB1 (for avian erythroblastosis virus transforming gene B (v-erbB), which was transduced from the avian EGFR gene). The EGFR is a receptor tyrosine kinase and can be activated by EGF as well as TGFα, HB-EGF, amphiregulin, betacellulin, epigen, and epiregulin.
Elk-1	Ets-like protein 1	A member of the ternary complex factor subfamily of ETS domain transcription factors. Elk-1 can dimerize with serum response factor (SRF), bind to the serum response element (SRE), and activate transcription of various immediate-early response genes, including the c-fos proto-oncogene. Elk-1 can also regulate transcription independently of SRF.
ERK1/2	Extracellular signal-regulated kinases 1 and 2	Also called p44 MAPK and p42 MAPK (for 44 kDa and 42 kDa mitogen-activated protein kinases) or MAPK1 and MAPK3. A pair of proline-directed serine/threonine-specific protein kinases involved in a wide range of cellular responses.
Ets1/2	E twenty-six (cellular homolog of the E26 leukemia virus oncogene)	Two members of the ETS domain family of transcription factors that can activate their own transcription when overexpressed. Ets1 and 2 are orthologs of POINTED, a *Drosophila* gene implicated in the Sevenless/Ras/ERK pathway.
FoxO3	Forkhead box O3	A forkhead family transcription factor and tumor suppressor that can be inhibited and translocated from the nucleus to the cytoplasm through Akt-mediated phosphorylation.
GAP	p120 Ras GTPase-activating protein	One of a family of GTPase-activating proteins that serve as inactivators of small G-proteins. This founding member of the GAP family is recruited to the phosphorylated EGFR, promoting its interaction with, and inactivation of, Ras-GTP.
Ion channels		This is a general category. CaMKII has been implicated in the regulation of numerous ion channels, including L-type calcium channels, ryanodine receptors, sodium channels, and potassium channels in the heart, and AMPA receptors, L-type calcium channels, and potassium channels in neurons.
IP3	Inositol 1,4,5-trisphosphate	A calcium-mobilizing second messenger, produced by the cleavage of PIP2 by a phospholipase C protein.

(*Continued*)

TABLE 1.2 (*Continued*) **Downstream Mediators of EGF Signaling**

Abbreviation	Full Name	Description and Comments
IP3R	Inositol 1,4,5-trisphosphate receptor	This is a family of ligand-gated calcium channels (IP3R1-3 in humans) that release calcium from the endoplasmic reticulum when activated by the binding of IP3.
iRhom	Inactive Rhomboid-family protein	An enzymatically inactive relative of the Rhomboid proteases. iRhom proteins are involved in the transport and maturation of the ADAM17 protease, which cleaves pro-TGFα to produce soluble TGFα. In this way they promote the activation of the EGFR system. iRhom proteins are also involved in promoting the degradation of misfolded EGFR ligands.
KSR	Kinase suppressor of Ras	Two related proteins (KSR1 and KSR2) in humans with homologs in *C. elegans* and *Drosophila melanogaster*. The KSR proteins are close relatives of the Raf protein kinases and possess kinase domains; however, they have low catalytic activities and are thought to function as scaffolds, interacting with Ras, Raf, MEK, and ERK proteins.
MEK1/2	MAPK or ERK kinases 1 and 2	Also called MAP2K1 or MAPKK1 and MAP2K2 or MAPKK2. These components of the MAPK cascade are activated through phosphorylation by Raf and Mos proteins and in turn they activate the ERK1/2 MAP kinases. Activation involves the phosphorylation of two residues in the kinases' activation loops; either phosphorylation appears sufficient for activation.
MEKK1	MEK kinase 1	A MAPKKK protein unrelated to the Raf family kinases but related to the yeast MAPKKKs BCK1 and STE11. The downstream targets of MEKK1 probably include MEKs, which activate ERKs, as well as MAPKKs that regulate the JNKs and the p38 MAPKs. MEKK1 can interact with Ras, Rac, and Cdc42 in vitro and so may be a downstream target of these small G-proteins. MEKK1 can also be activated by the Ste20-related protein GCK and the adapter protein TRAF2.
MNK1/2	MAP kinase interacting kinases 1 and 2	Also called MKNK1 and MKNK2. A pair of related protein kinases that are activated by ERKs and other MAPKs. Downstream targets include the translation initiation protein eIF-4E.
MSK1/2	Mitogen and stress-activated protein kinases 1 and 2	Also called RPS6Kα5 and RPS6Kα4, respectively. A pair of protein kinases related to the RSK proteins. MSK1/2 can be activated by ERKs and other MAPKs. Downstream targets include numerous transcription factors.
Myc	Cellular homolog of the avian myelocytomatosis virus oncogene	A basic helix-loop-helix/leucine zipper (bHLH/LZ) transcription factor that is regulated by diverse upstream pathways, including Wnt, Hedgehog, and receptor tyrosine kinases.
NORE1	Novel Ras/Rap effector 1	Also called RASSF5 (for Ras association domain family member 5). NORE1 is a tumor suppressor gene and a mediator of the apoptotic effects of Fas. Its downstream targets include the MST1/2 protein kinases.
novel PKCs	Novel protein kinase C proteins	A family of four PKCs (PKCδ, PKCε, PKCη, and PKCθ) that require DG but not Ca^{2+} for activation.
PDK1	PIP3-dependent protein kinase 1	Also known as PDPK1. PDK1 binds to PIP3 molecules in the inner leaflet of the plasma membrane and thereby gains access to downstream targets, which include several ACG family protein kinases.
PI3K	Phosphatidylinositol (4,5)-bisphosphate 3-kinase	A family of lipid kinases, one group of which (the p85/p110 complexes) is activated by binding to specific phosphotyrosine residues on activated receptor tyrosine kinases.
PIP3	Phosphatidylinositol (3,4,5)-trisphosphate	A lipid second messenger that recruits PDK1 to the plasma membrane and participates in the activation of Akt through binding to these proteins' modular pleckstrin-homology (PH) domains. There are believed to be about 26 PIP3-binding PH domain proteins in humans.
PKCζ	Protein kinase C ζ	Also called PRKCZ. This is a so-called atypical PKC, which means it does not require either DG or Ca^{2+} for activation. It may be activated by protein–protein interaction plus phosphorylation by PDK1. It can bring about activation of MEK and ERK in the absence of Raf function.
PLCε	Phospholipase C ε	A bifunctional enzyme. Like other PLCs, PLCε can cleave the head group off PIP2 to yield IP3 plus DG. This activity is stimulated by Ras, Rho, and heterotrimeric G-proteins. In addition, PLCε can function as a guanine nucleotide exchange factor for Ras and Rap1.
PLCγ	Phospholipase C γ	A pair of phospholipase proteins (PLCγ 1 and 2) that are regulated through binding to specific phosphotyrosine residues on activated receptor tyrosine kinases and phosphorylation by the kinases. Like other PLCs, PLCγ can cleave the head group off PIP2 to yield IP3 plus DG.
PPARγ	Peroxisome proliferator-activated receptor γ	A nuclear receptor and transcription factor that can be regulated by fatty acids and through phosphorylation by a variety of protein kinases, including ERK1/2 and ERK5, AMPK, GSK3, PKA, and PKC.
Pro-TGFα	Pro-form of TGFα	See TGFα.

(*Continued*)

TABLE 1.2 (*Continued*) **Downstream Mediators of EGF Signaling**

Abbreviation	Full Name	Description and Comments
Raf	Rapidly accelerated fibrosarcoma oncogene	A family of three protein kinases, A-Raf, B-Raf, and Raf-1 or C-Raf. The activation of Raf proteins depends on binding of Ras-GTP. The best characterized targets of Raf are the MEK1 and MEK2 kinases. Thus Rafs are MEK kinases or MAPKKKs.
Ral GDS	Ral guanine nucleotide dissociation stimulator	Also called RalGEF for Ras-related (Ral) guanine nucleotide exchange factor. Ral GDS is activated by Ras, and in turn activates the Ras-related G-proteins Ral-A and Ral-B.
Ras	Cellular homologs of the rat sarcoma virus oncogenes	Three small G-proteins (H-Ras, K-Ras, and N-Ras) that are activated downstream of receptor tyrosine kinases and PKC-activating stimuli. The Ras proteins act as stoichiometric regulators of various mitogenic regulatory proteins, including the Raf proteins.
RasGRP	Ras guanine nucleotide releasing protein	A family of three Ras guanine nucleotide exchange factors (or dissociation stimulators) that are activated downstream of calcium and DG mobilization.
RSK	Ribosomal S6 kinase	A family of four protein kinases that are activated by ERK1/2 and PDK1.
Shc/Grb2	SH2 domain-containing protein/growth factor receptor-binding protein 2	A complex of two adaptor proteins, Shc and Grb2, that can bring Sos to tyrosine-phosphorylated EGFR. Grb2 itself, in the absence of Shc, can also link Sos to pY-EGFR.
Shp-2	SH2 domain-containing protein tyrosine phosphatase 2	Also called PTPN11, PTP-1D, and PTP-2C. One of two orthologs (with Shp-1) of the *Drosophila* Corkscrew protein. Shp2 is activated by binding to specific phosphotyrosine residues and positively regulates Ras through an incompletely understood mechanism.
Sos	Son of sevenless	A family of two guanine nucleotide exchange factors (Sos1 and Sos2) that are orthologs of *Drosophila* Sos, which was originally identified as a downstream mediator of sevenless activation. Human Sos proteins can activate Ras and can also be allosterically activated by the binding of active Ras.
TGFα	Transforming growth factor α	One of seven ligands for the mammalian EGF receptor. It is translated as a transmembrane pro-form and then proteolytically cleaved to yield active soluble TGFα.
TIAM1	T-lymphoma invasion and metastasis-inducing protein 1	A Ras-binding, Rac-specific guanine nucleotide exchange factor.
TSC2	Tuberous sclerosis complex 2	Also called tuberin. Forms a complex with the TSC1 protein (hamartin) that acts as a negative regulator of the small G-protein Rheb, which is a positive regulator of mTORC1. TSC2 is negatively regulated by Akt phosphorylation; TSC1 is negatively regulated by Rsk. As the name suggests, mutations in either TSC1 or TSC2 can result in tuberous sclerosis.

serine/threonine kinases are referred to as classical protein kinases, and the classical protein kinases are all evolutionarily related to each other but not to the bacterial histidine kinases.

Furthermore, whereas the bacterial system is shallow, the EGFR system is deep. In the three bacterial examples shown in **Figure 1.2**, the histidine kinase directly phosphorylates the terminal effector of the pathway, the response regulator protein. In some bacterial pathways (e.g. the RcsCBD system, not shown in **Figure 1.2**), there is a third protein (a phosphotransferase protein) interposed between the kinase and the terminal effector. But still-longer pathways have not been found, and in general bacterial signaling makes use of a small number of intermediaries.

This is decidedly not the case in the EGFR system. The EGFR does not directly phosphorylate terminal effectors; rather, it regulates signaling proteins that regulate other signaling proteins that regulate others... and on and on. One particularly striking example of this is the MAP kinase cascade. EGFR activation brings about the activation of the Ras GTPases through the intermediacy of the Shc and Grb2 adaptors and the Sos (shown in **Figure 1.3**) and Vav (not shown) guanine nucleotide exchange factors. Active Ras then stoichiometrically activates the Raf family protein kinases A-Raf, B-Raf and Raf-1 (Raf-1 is also termed C-Raf). The Raf proteins can be thought of as MAP kinase kinase kinases or MAPKKKs. Activated Raf probably does not directly phosphorylate any terminal effector proteins. Instead, it phosphorylates and activates the MEK1/2 protein kinases, which are MAPKKs, and these proteins then phosphorylate and activate the ERK1/2 MAPKs, with the Raf/MEK/ERK system constituting a **protein kinase cascade**. ERK1/2 does regulate some terminal effectors, such as the Ets-family transcription factors, but it also regulates a number of signaling intermediaries as well, such as the RSK, MSK, and MNK protein kinases. Thus, the EGF receptor signaling system is deep, with many signaling transduction proteins being interposed between the receptor and its terminal effectors.

The pathway also fans out substantially. The typical bacterial signaling system has one terminal effector, the response regulator, or maybe two. **Figure 1.3** shows seven transcription factors downstream of EGFR activation, and if we had traced some of the lopped off branches of the system further and included a more comprehensive list of ERK substrates, we could have included scores of EGF-regulated transcription factors. One can think of each of these transcriptional terminal effectors as an individual output, or, alternatively, one can take all of the changes together as a collective biochemical and phenotypic output state.

Finally, in the EGFR system there are numerous interconnections between the various signaling pathways that emerge from the receptor. The PLCγ/PKC pathway feeds into the Ras/ERK branch through activation of the guanine nucleotide exchange factor Ras GRP (**Figure 1.3**). The Ras side of the network feeds into the PLCγ/PKC side through PDK1, the MEKKs, and PLCε. At least five different protein kinases downstream of EGFR are thought to feed into the regulation of the transcription factor CREB1.

Instead of starting with the EGFR and mapping its downstream effectors, we could have started with one of the downstream proteins and then traced upstream all of the proteins that regulate it, that regulate its regulators, and so on. This would be like a family tree that shows all of the ancestors of one descendant, rather all of the descendants of one ancestor. The result is a different kind of pathway map, but, as it turns out, it is similar in depth, breadth, and complexity to the one shown in **Figure 1.3**. So it is not just that the system fans out; there are many inputs feeding into the system as well.

Some of the components listed in **Figure 1.3** may not be important for EGFR function in all contexts, and some may be important only when the receptor is overexpressed (as is the case in some cancer cells). But still, the EGFR signaling system is much more complicated than bacterial signaling systems are. This raises a serious question: how can one possibly understand such a complex system?

1.5 COMPLICATED SYSTEMS CAN BE SIMPLIFIED BY ASSUMING MODULARITY

One way to tackle the problem of complexity is to assume that one can learn something about the function of the whole system by examining key subcircuits. This is a decidedly reductionistic approach to systems biology, and there is no a priori guarantee that it will be successful, because of two potential complications. First, plugging a subcircuit into its upstream regulators and downstream targets can fundamentally change the way the subcircuit behaves. A reductionistic approach only makes sense if signal transduction is, at least to some degree, modular, with there being some plug-and-play character to the subcircuits that constitute the system. Second, it is not certain that the interesting behaviors of the system emerge at the level of

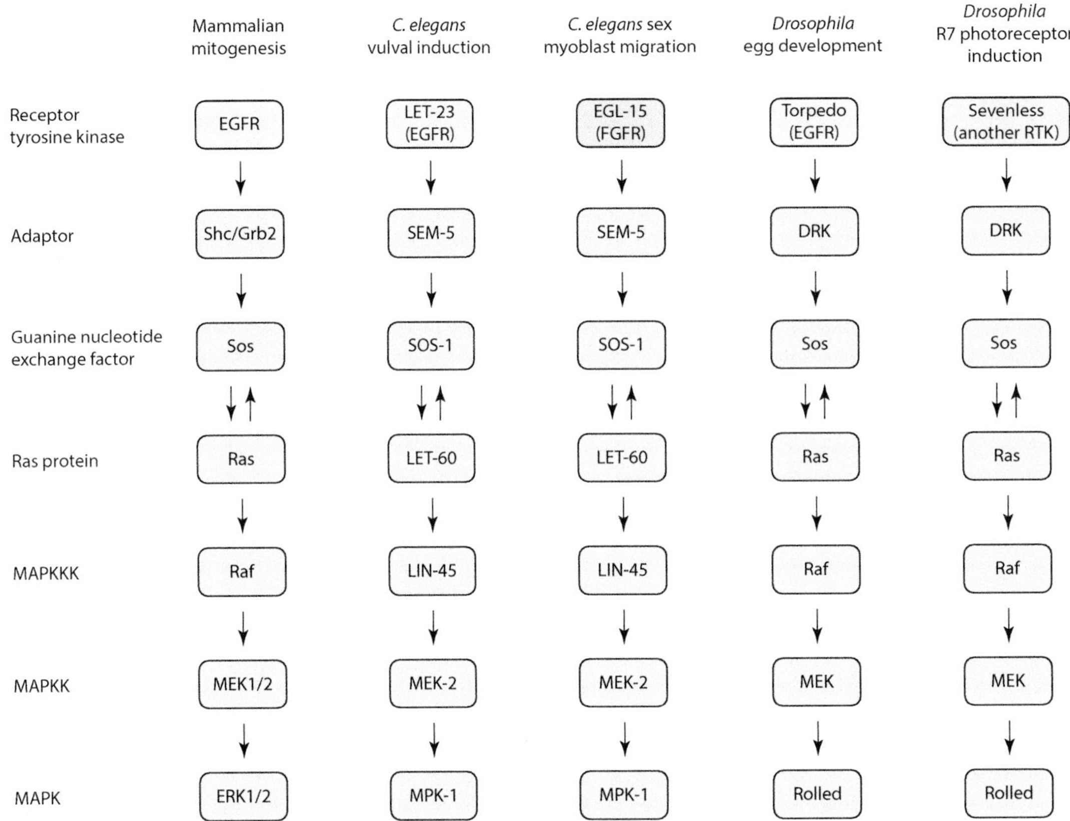

Figure 1.4 Modularity in the Ras/Erk pathway. The proteins from the adaptor through the MAPK form a linear pathway that can plug into different upstream receptor tyrosine kinases. In addition, these pathways can mediate a variety of biological processes, presumably by plugging into different ultimate effector proteins.

small subcircuits. Perhaps the whole system must be present in order for the system to perform even its most rudimentary functions.

So, first, is signal transduction modular, with some plug-and-play character? In the case of EGFR signaling, the answer appears to be yes. As shown in **Figure 1.4**, large chunks of the EGFR system plug into other receptors in other biological contexts. For example, during *C. elegans* development, orthologs of mammalian Grb2, Sos, Ras, Raf, MEK, and Erk mediate the effects of EGFR signaling during induction of vulval cell fates in the nematode hypodermis. But the same proteins are also critical for mediating the effects of a different receptor, the ortholog of the mammalian FGF receptors, during sex myoblast migration. Likewise, orthologs of the same proteins mediate the effects of the Torpedo EGF receptor protein during *Drosophila* oocyte development and the effects of the Sevenless tyrosine kinase during eye development. In addition, since Ras-MAPK activation has different downstream consequences in these different biological settings, it presumably plugs into different ultimate effectors. Thus the Ras-MAPK pathway does appear to be modular, plugging into different upstream inputs and downstream outputs.

In general then, if signaling tends to be modular, what generates and enforces the modularity? In part the answer is compartmentalization; signal transducers interact more often and more strongly with those components they are close to. Another part of the answer is time scales; if some signaling events take place in minutes and others require hours, one is usually safe in thinking about the processes separately. Perhaps in other cases, modularity is a consequence of the biochemistry of individual components. For example, if an upstream regulator directly activates a downstream target by, say, phosphorylating it, and a negligible fraction the regulator and target molecules are bound to each other at any given time, one is probably safe in considering the two proteins as being parts of separate modular subcircuits.

The next question is whether interesting behaviors emerge at the level of subcircuits, or alternatively, signal processing is so highly distributed that all parts of the system are roughly equally important. One way of addressing this question is through genetic analysis, and in several genetically tractable systems, the answer is that the most critical downstream components constitute a small fraction of the total network. For example, in *C. elegans* vulval induction and *Drosophila* R7 photoreceptor induction, the Ras-MAPK pathway seems to be a particularly important strand in the web of receptor tyrosine kinase signaling: loss of function mutations in these downstream components prevent the receptor from carrying out its normal developmental function, and gain of function mutations can abrogate the requirement for the upstream receptor. Thus, the web downstream of EGFR and other receptor tyrosine kinases may be exceedingly complicated, but one individual thread in the web is particularly important functionally.

There are dissenting opinions on the issue of how modular signal transduction is, both within and without the systems biology community. But for the purposes of this book, we will make the assumption that complex signaling systems can be understood by breaking them down into small subcircuits.

HOW SHOULD WE MODEL SIGNAL TRANSDUCTION SYSTEMS AND WHY?

1.6 ORDINARY DIFFERENTIAL EQUATIONS PROVIDE A POWERFUL FRAMEWORK FOR UNDERSTANDING MANY SIGNALING PROCESSES

Once we have broken a signaling system down into simpler subcircuits, the next question is to figure out how the outputs of these circuits change as the inputs are dialed up and down. We all have ways of trying to intuit such behaviors. Perhaps the most common strategy is to consciously or unconsciously consider the system to be a digital Boolean network, with the components either off or on, and then trace the effects of a change in one component's activity through the system—when this protein turns on it causes the next one to turn on, which causes the next one to turn off, and so on. Useful though it can be, this type of approach has its limitations, particularly if the proteins involved do not respond in a digital fashion or the circuit includes multiple feedback loops.

More detailed mathematical models can allow one to develop a richer picture of how a signaling circuit should behave. Both numerical simulation—the equivalent of computational experiments—and theoretical approaches can yield not-completely-intuitive, or even counterintuitive, predictions about the behavior of a biological system. And such predictions can be extremely valuable. They provide a way of challenging or testing the model, and they can also lead to the discovery of previously unsuspected or overlooked behaviors of importance to biology.

In this book we will concentrate on ordinary differential equation models (ODEs) for understanding biochemical reactions. Other types of modeling certainly have their place in biology, but ODEs are a particularly good place to start. They are well-suited to aspects of cell signaling where the numbers of molecules are not too small, and they allow us to draw on the beautiful and powerful techniques of nonlinear dynamics. Even when the numbers of molecules are very small—for example, in the transcription of a gene from two alleles in a G1-phase cell or two copies of two alleles in a G2-phase cell—and a different approach, non-deterministic stochastic modeling, is the proper framework for understanding the process, it is still usually worth starting with an ODE model to which the stochastic simulations can be compared. With ODE models, especially ODE models with small numbers of variables, it is comparatively easy to understand why the models behave the way they do. Understanding, rather than just reproducing the behaviors of biological circuits is one of the main goals of systems biology.

1.7 THEORY CAN HELP HIGHLIGHT THE COMMONALITIES OF DIVERSE BIOLOGICAL PHENOMENA

There is a famous quotation attributed to Ernest Rutherford that "all science is either physics or stamp collecting" (**Figure 1.5**). Perhaps a more accurate and less dismissive-sounding version of this provocative statement might be that in all science, physics included, progress is made through the discovery and description of phenomena and through theory. Theory attempts to organize, unify, and simplify the descriptions of the phenomena, and the discovery of new phenomena serves to test the theory. Many of us are rather fond of the stamp collecting aspect of biology—just watching cells divide, or neutrophils chase bacteria, or flowers bloom can be immensely satisfying. Biology is blessed with truly fascinating phenomena. But biology includes theory as well, most famously the theory of evolution, which unifies an astonishing diversity of biological phenomena and deepens our understanding of them. Thus, for example, we understand why the process of cell division is orchestrated by the same core regulatory components in all eukaryotic cells; the mechanism apparently

Stamp collecting

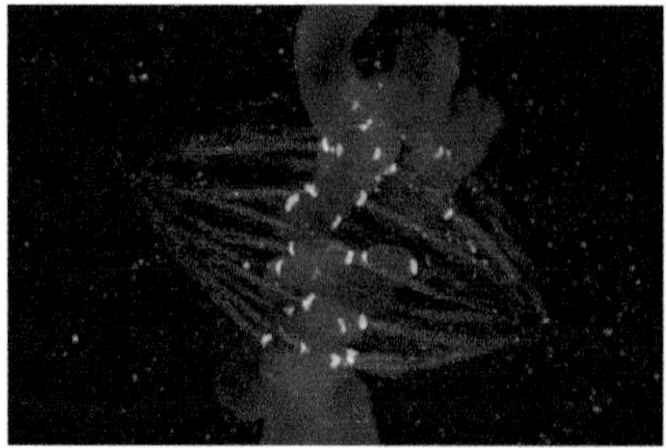

Biology

Physics

Figure 1.5 Stamp collecting, biology, and physics. The discovery and classification of phenomena (stamp collecting) plays a critical role in the advancement of all science. So does theory. Photo credits: 1. Stamp collection photo from http://www.bucketsandspadesblog.com/2012/12/american-stamp-collection.html, used with permission. 2. Metaphase epithelial cell in metaphase stained for microtubules (red), kinetochores (green) and DNA (blue), from Jane Stout and Claire Walczak, Indiana University, GE Healthcare 2012 Cell Imaging Competition, used with permission. 3. Einstein photo originally published in the Pittsburgh *Sun-Telegraph* in 1934. Taken from Topper D, Vincent D, Einstein's 1934 two-blackboard derivation of energy-mass equivalence. *American Journal of Physics* 75 (2007), 978.

evolved once, and all descendants of that original cell make use of some variation on that original mechanism.

Modeling can push this unification further. Even if the biological processes are very different and the proteins that regulate the processes are unrelated to each other, there can be a fundamental similarity between the processes that modeling can help to reveal. For example, the G2/M transition in eukaryotic cell cycle progression at first glance bears no resemblance to the transition of an *E. coli* bacterium between metabolic states when the availability of lactose changes, and the proteins and genes involved in regulating the processes are unrelated to each other. But both G2/M progression and *lac* induction involve the toggling between two discrete, alternative states of a control system that includes positive feedback loops. This toggling can, through modeling and theory, be attributed to the traversal of a **saddle-node bifurcation** in the phase space of the two processes' control systems. At first this might not seem like a very useful insight, particularly if one does not know what a saddle-node bifurcation is. But once one does know (and by the end of Chapter 8 we will!), it is indeed a very useful insight. All saddle-node bifurcations are fundamentally the same, at some level, and so these apparently different phenomena, involving different proteins, different biochemical reactions, and different time scales, are fundamentally the same as well.

1.8 SIX BASIC TYPES OF RESPONSE ARE SEEN OVER AND OVER AGAIN IN CELL SIGNALING

In the experimental cell signaling literature, one observes the same handful of basic types of response over and over again. One common type of response is that shown in **Figure 1.6**. If the input goes up, the output goes up, and if the input goes back down, the output goes back down. Different amounts of input typically lead to different amounts of output; the steady-state input/output relationship is monotonic and graded, with the system behaving like a rheostat. This is the simplest, most basic sort of signaling response: a graded, **monostable**, reversible response, where the term monostable means that for any given level of input, the system has one **steady-state** level of output, and that the steady state is stable, meaning that if one pushes the system a small ways away from the steady state, it will return toward it.

Alternatively, sometimes the response is monostable and reversible, but switch-like rather than graded, with the initial increments of input producing very little change in the output until a threshold is reached (**Figure 1.7**). These are often termed **ultrasensitive** responses. Both graded and ultrasensitive monostable responses are common in cell signaling.

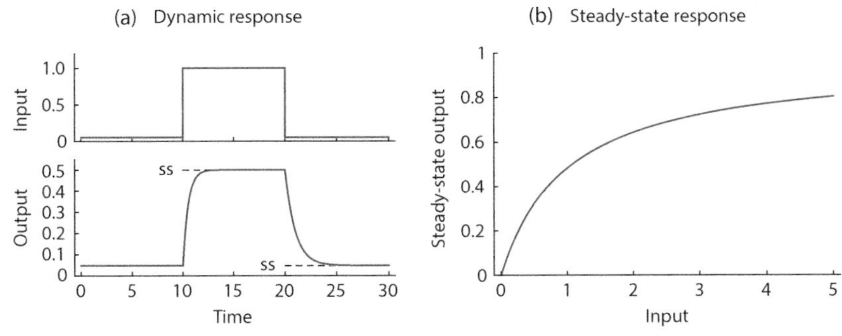

Figure 1.6 A graded monostable response. (a) Time course. A step up in the input level drives the system to approach a higher steady-state (ss) output; a step back down drives the system back to where it started. (b) The steady-state output of the system rises gradually as the input is increased.

But not all biological responses are reversible. Sometimes a response persists long after the stimulus is removed, or even indefinitely; the system actively "remembers" the input (**Figure 1.8**). Typically this type of response is all-or-none in character, and the system has a point-of-no-return, so that once an input of sufficient magnitude has been present for a sufficient duration, the response becomes self-sustaining. This type of behavior is the hallmark of **bistability**, and bistability is thought to be important for cell fate decisions, for transitions between phases of the cell cycle and for the laying down of memories in neural circuits.

Oftentimes in cell signaling the responses are pulses. A constant input causes the output to rise, but only transiently, with the output falling back to a low level even if the input persists (**Figure 1.9**), and the amplitude of the pulse is related to the strength of the input. This type of response is termed **adaptation**, and it is exhibited by many G-protein-coupled receptors; the system downstream of the input adapts, at least partially, to a given constant level of input. This is also the way receptor tyrosine kinases like the EGFR respond, and it is typical of chemotactic responses in bacteria.

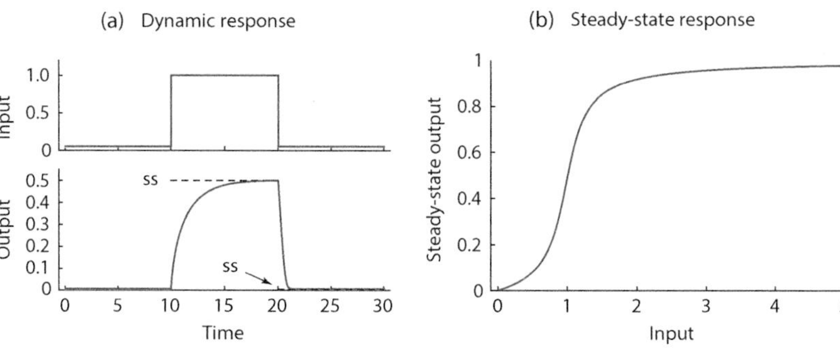

Figure 1.7 A switch-like monostable response. (a) The dynamical response looks qualitatively similar to that shown in Figure 1.5; once again, a step up in the input level drives the system to approach a higher steady-state (ss) output and a step back down drives the system back to where it started. (b) The steady-state response curve is sigmoidal. Compared to the curve shown in Figure 1.5b, there is less response at low input levels and more at high input levels, and the system switches from off to on over a narrower range of input values.

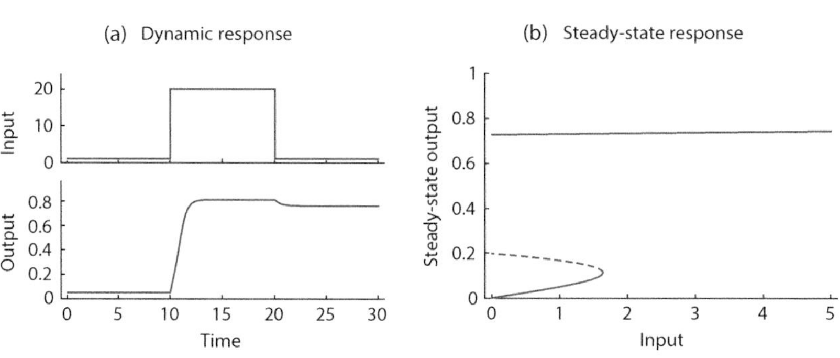

Figure 1.8 Bistability. (a) A sufficiently large change in input causes the system to switch between two discrete states. When the input is removed, the output stays high. (b) If the system starts out in the low output state, the steady-state output will rise slightly with the input until the input exceeds a threshold, which is at an input level of about 1.7. At that point, the steady-state output jumps to the upper branch of the response curve. The dashed part of the curve represents an unstable steady state; the smallest perturbation would drive the system up the upper branch or down to the lower branch.

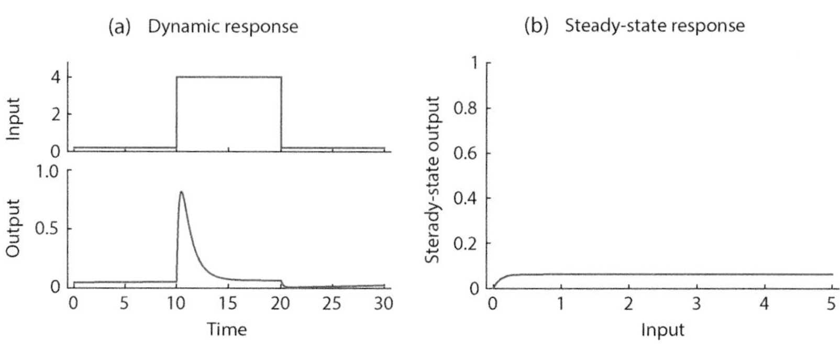

Figure 1.9 A pulsatile response. In this example, the transient response is large (a) but the steady-state response is small and nearly constant (b).

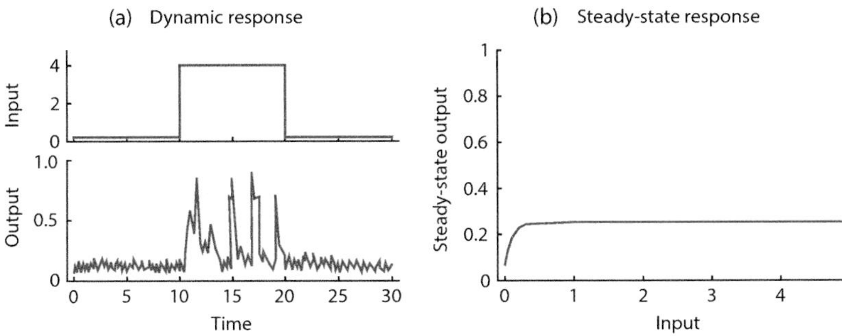

Figure 1.10 Irregular spikes. (a) In this example, increasing the input stimulus causes spikes of output to appear at irregular intervals. (b) The steady-state response, averaged over time or over many cells.

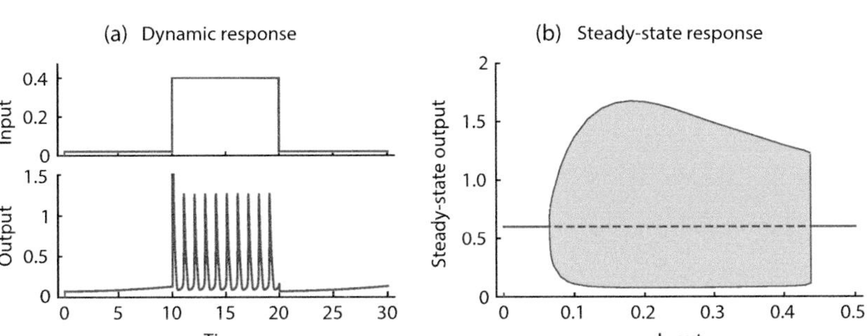

Figure 1.11 Oscillations. (a) In this example, the input is initially too low to permit oscillations. When the input is increased, oscillations begin; when it is lowered back down, the oscillations are extinguished. (b) Steady-state response. For a range of inputs, the system oscillates rather than settling into a steady state. Oscillations are "born" at an input level of just below 0.1, where the solid line changes to a dashed line. At this point, stable steady state (solid line) becomes unstable (dashed line) and so the system never settles down. The oscillations are extinguished once the input level gets a bit higher than 0.4. The boundaries of the shaded region show the peaks and troughs of the oscillations.

Sometimes the output is not just a single pulse but rather a succession of irregular, all-or-nothing spikes, and increasing the input increases the frequency or density of the spikes but has little effect on the spikes' amplitudes. This is the behavior generally seen when signaling outputs are examined at the level of individual receptors, as can be accomplished by patch clamping when the receptors are ion channels. But all-or-none, irregular spikes are sometimes even seen at the level of whole cells, where tens of thousands of signaling molecules burst together in a seemingly random but coordinated fashion (**Figure 1.10**). Systems that generate responses like this are called **excitable systems**. Excitability is typical in calcium signaling and is seen in the responses of the Raf/ERK pathway in some biological contexts.

And, finally, some systems function as biochemical oscillators, with the output varying periodically with time indefinitely. In these cases an input can trigger the oscillations, or it can adjust the frequency, amplitude, or phase of the oscillations (**Figure 1.11**). This is the way that calcium signaling works in some cell types and it is the way the sinoatrial node pacemaker cells, which drive the heartbeat, behave.

So that comes to six types of response—graded reversible monostable responses, switch-like reversible monostable responses, bistability, pulses, irregular repeated spikes, and oscillations. Of course there are others, but these six constitute much of what is seen in signal transduction.

1.9 FIVE OR SIX BASIC CIRCUIT MOTIFS ARE SEEN OVER AND OVER AGAIN IN SIGNALING SYSTEMS

Just as we have classified the basic types of signaling response that are seen over and over again in cell signaling, we can identify and classify the common signaling subcircuits or motifs (**Figure 1.12**). As it happens, all of these can be found in the unusually well-characterized

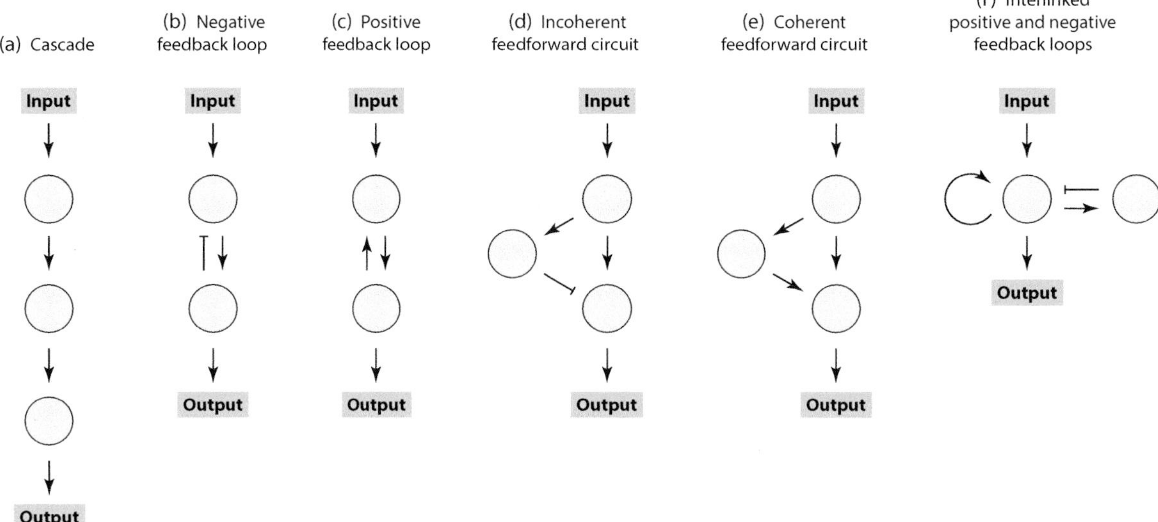

Figure 1.12 Recurring motifs in signal transduction systems. (a) A signaling cascade. (b) A negative feedback loop. (c) A positive feedback loop. (d) An incoherent feedforward circuit. (e) A coherent feedforward circuit. (f) A composite system with interlinked positive and negative feedback loops.

EGFR system (**Figure 1.3**), and it seems possible that most or all of these motifs will be found in many different signaling systems.

The first motif is the **signaling cascade** (**Figure 1.12a**)—a chain of signaling proteins reminiscent of a chain of little waterfalls in a cascading stream. The best-studied of such cascades is the MAP kinase cascade, the Raf, MEK, and ERK proteins, mentioned in Sections 1.5 and 1.6 above and shown in Figures 1.3 and 1.4. In principle a cascade can be of any length, but the best-studied ones consist of few signal transducers.

Feedback loops are also common in cell signaling. These are subcircuits where a nominally downstream protein regulates something that is nominally upstream of it. For example, the ERK1/2 MAP kinases not only pass signals on to downstream targets but also phosphorylate and negatively regulate several of their upstream regulators, including the Raf protein kinases, the Sos guanine nucleotide exchange factors, and the EGF receptor itself (Figure 1.3). This constitutes **negative feedback**.

Positive feedback (**Figure 1.12c**) is also common. For example, in the EGFR system, Sos activates the Ras GTPases. In turn, active Ras can bind to and allosterically activate Sos (Figure 1.3). This constitutes a short, simple positive feedback loop. There are longer positive feedback loops built into the system as well. For example, EGFR activation leads to the rapid and reversible activation of ADAM proteases, including ADAM17. ADAM17 functions as a "sheddase," and it can cleave the membrane-bound precursor forms of several EGFR ligands, including TGFα, to release active, diffusible growth factors that can reinforce EGFR activation and spread it to neighboring cells (Figure 1.3). The details of ADAM17 activation are not yet worked out; PKC is a likely intermediary, although how it regulates ADAM17 is not clear. Nevertheless, it is clear that the EGFR and ADAM17 are part of a long positive feedback loop, with each protein activating the other, and it is clear that this loop is important in the overall functioning of the EGFR signal transduction system.

Although in the bacterial signaling systems there is generally a single pathway from the input to the output (Figure 1.2), in the EGFR system there are several instances where the signal splits and then reconverges. For example, the EGFR affects the Ras protein both positively and negatively, through the guanine nucleotide exchange factors Sos and Ras GRP, the GTPase-activating protein GAP, and Shp-1, a phosphotyrosine phosphatase that positively regulates Ras through an incompletely understood mechanism (Figure 1.3). This constitutes **incoherent feedforward regulation**—incoherent because some of the regulators have opposite effects on their targets, and feedforward, a term from control theory, because an upstream protein (EGFR) affects a downstream protein (Ras) through more than one pathway, shown schematically in **Figure 1.12d**. There are also examples of **coherent feedforward regulation** (**Figure 1.12e**). For example, two pathways from PIP3 activates the protein kinase Akt through two interactions: a direct one and an indirect one through PDK1 (Figure 1.3). Both coherent and incoherent feedforward regulation are common in cell signaling.

These elementary motifs can, of course, be combined to produce composite motifs. One particularly interesting and important one is shown in **Figure 1.12f**: a system with interlinked positive and negative feedback loops. This is the basic architecture of the circuit that produces action potentials in neurons, cell cycle oscillations in embryos, and the Turing patterns thought to give rise to a leopard's spots and a tiger's stripes.

SUMMARY

Signal transduction is an important and fascinating area of biology. However, given the large number of signaling proteins in eukaryotic cells (thousands) and the complexity of signaling systems (see Figure 1.3), the question of how best to try to understand how these signaling systems function is far from trivial. One reasonable way forward is to assume that these complex systems are at least somewhat modular (Figure 1.4) and so can be understood by examining the behaviors of subcircuits of the whole system. And by examining the behaviors, we mean a combination of experimental studies and theory (or, if you prefer, stamp collecting and physics; Figure 1.5). Modeling and theory help us to understand how the phenomena of cell signaling arises.

We surveyed six common types of signaling responses: graded monostable responses (Figure 1.6), switch-like monostable responses (Figure 1.7), bistable responses (Figure 1.8), pulses (Figure 1.9), stochastic spikes (Figure 1.10), and oscillations (Figure 1.11). We also looked at five elementary signaling motifs, plus one important composite motif (Figure 1.12).

MOVING FORWARD

This sets the stage for the next 15 chapters, which explore how the various types of biological responses shown in Figures 1.6–1.11 can and do arise, and what types of signal processing can be accomplished by the various simple circuits shown in **Figure 1.12**. We will begin Chapter 2 with the proteins that initiate signaling responses: the receptors.

FURTHER READING

SIGNAL TRANSDUCTION COMPONENTS AND SYSTEMS

Cantley LC, Hunter T, Sever R, Thorner J. *Signal Transduction: Principles, Pathways, and Processes*. Cold Spring Harbor Laboratory Press, Cold Spring Harbor, NY, 2014.

Capra EJ, Laub MT. Evolution of two-component signal transduction systems. *Annu Rev Microbiol*. 2012;66:325–47.

Fredriksson R, Lagerström MC, Lundin LG, Schiöth HB. The G-protein-coupled receptors in the human genome form five main families. Phylogenetic analysis, paralogon groups, and fingerprints. *Mol Pharmacol*. 2003 Jun;63(6):1256–72.

Hartwell LH, Hopfield JJ, Leibler S, Murray AW. From molecular to modular cell biology. *Nature*. 1999 Dec 2;402(6761 Suppl):C47–52.

Lemmon MA, Schlessinger J. Cell signaling by receptor tyrosine kinases. Cell. 2010 Jun 25;141(7):1117–34.

Lim W, Mayer B, Pawson T. *Cell Signaling: Principles and Mechanisms*. Garland Science, Taylor and Francis Group LLC, New York, 2015.

Marks KMDF, Klingmüller U. *Cellular Signal Processing: An Introduction to the Molecular Mechanisms of Signal Transduction*. 2nd Edition. Garland Science, Taylor & Francis Group LLC, New York, 2017.

Schlessinger J. Receptor tyrosine kinases: legacy of the first two decades. *Cold Spring Harb Perspect Biol*. 2014 Mar 1;6(3): a008912.

HOW CAN WE MODEL SIGNAL TRANSDUCTION SYSTEMS, AND WHY?

Alon U. *An Introduction to Systems Biology: Design Principles of Biological Circuits*. Chapman and Hall/CRC, Taylor & Francis Group, London, 2007.

Covert MW. *Fundamentals of Systems Biology: From Synthetic Circuits to Whole-Cell Models*. CRC Press, Taylor & Francis Group, London, 2014.

Endres RG. *Physical Principles in Sensing and Signaling: With an Introduction to Modeling in Biology*. Oxford University Press, Oxford, 2013.

Goldbeter A. *Biochemical Oscillations and Cellular Rhythms: The Molecular Bases of Periodic and Chaotic Behaviour*. Cambridge University Press, Cambridge, 1996.

Heinrich R, Schuster S. *The Regulation of Cellular Systems*. Chapman and Hall, International Thomson Publishing, New York, 1996.

Klipp E, Liebermeister W, Wierling C, Kowald A. *Systems Biology: A Textbook*. Second Edition. Wiley-Blackwell, New York, 2016.

Milo R, Shen-Orr S, Itzkovitz S, Kashtan N, Chklovskii D, Alon U. Network motifs: simple building blocks of complex networks. *Science*. 2002 Oct 25;298(5594):824–7.

Murray JD. *Mathematical Biology: I. An Introduction*. 3rd Edition. Springer, New York, 2002.

Murray JD. *Mathematical Biology: II. Spatial Models and Biomedical Applications*. 3rd Edition. Springer, New York, 2003.

Nelson P. *Physical Models of Living Systems*. W. H. Freeman & Company, New York, 2015.

Sneppen K. *Models of Life*. Cambridge University Press, Cambridge, 2014.

Strogatz SH. *Nonlinear Dynamics and Chaos*. Westview Press, Perseus Books Group, Cambridge, MA, 1994.

RECEPTORS 1

Monomeric Receptors and Ligands

2

IN THIS CHAPTER . . .

Signal transduction typically begins with the binding of a ligand to a receptor. In the simplest cases, a monomeric ligand binds to a monomeric receptor, and this produces a change in the receptor's conformation and activity, as shown schematically in **Figure 2.1a.** In this chapter, we will focus on one well-studied example of this type of process, the

DOI: 10.1201/9781003124269-2

activation of the β_2-adrenergic receptor by a hormone such as epinephrine, a neurotransmitter like norepinephrine, or a structurally related drug like albuterol (Table 1.1). We will begin with some background on what these receptors do, then examine the basic characteristics of their steady-state and dynamical responses as seen in experiments, and finally model the processes of ligand binding and receptor activation to gain some insight into why the responses look the way they do.

2.1 THE β_2-ADRENERGIC RECEPTOR CAN FUNCTION AS A MONOMERIC RECEPTOR THAT BINDS MONOMERIC LIGANDS

The largest class of receptors in humans is the **G-protein-coupled receptors** (GPCRs), a diverse family of proteins that are evolutionarily related to bacterial rhodopsin proteins and that share a common topology—they span the plasma membrane seven times, with the N-terminus of the protein outside the cell and the C-terminus inside. Probably the best-studied of the GPCRs are the adrenergic receptors, so-named because one of the hormones that activates them—epinephrine or adrenaline—is synthesized and released by cells in the medulla of the adrenal glands. Adrenergic receptors function in the central nervous system, the peripheral nervous system, and in organs such as the heart and the lung. In humans there are nine types of adrenergic receptors, which are divided into two groups (α and β) based on their pharmacology. Adrenergic receptors regulate blood pressure, cardiac contractility, pupil size, the smooth muscles in the bronchial tree, and intermediary metabolism. Studies of adrenergic signaling, from the late 19th century through the present day, have yielded and continue to yield enormous insights into physiology, disease, and the general principles of cellular regulation.

One of the best-studied adrenergic receptors is the β_2-adrenergic receptor, which plays a particularly important role in the lungs, and which is the target of the bronchodilator drug albuterol (Table 1.1). In the absence of an activating ligand, or **agonist**, the β_2-adrenergic receptor mainly adopts an inactive conformation that is unable to productively interact with G-proteins (**Figure 2.1b**, blue). Agonists like the circulating adrenal hormone epinephrine and the neurotransmitter norepinephrine bind to a site in the middle of the seven transmembrane (TM) helices and cause the receptor to adopt an active conformation (**Figure 2.1b**, pink). The main difference between the inactive and active conformations is the position of one of the helices, TM6. The outward displacement of TM6 in the active conformation allows the receptor to bind to and activate a trimeric G-protein, which then can activate downstream proteins. A second effector protein, β-arrestin, probably binds to a different active conformation of the receptor and brings about a different set of downstream responses.

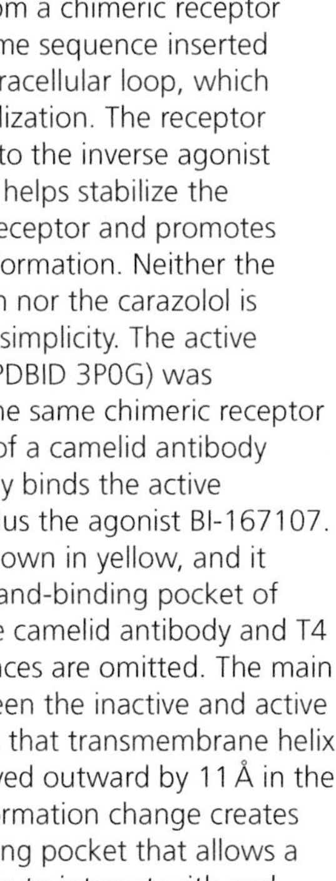

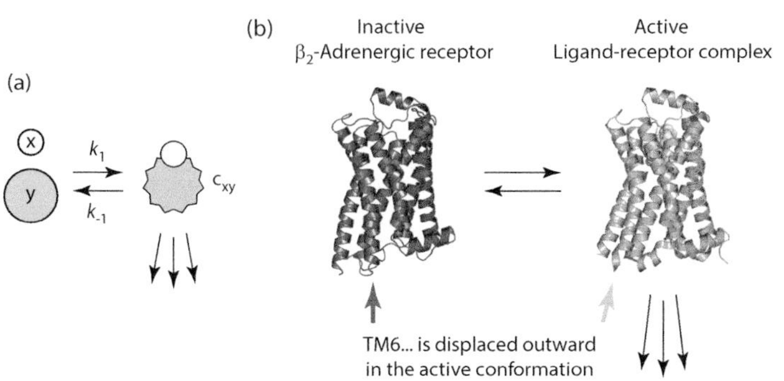

Figure 2.1 Activation of a monomeric receptor by a monomeric ligand. (a) Schematic view of a generic ligand–receptor interaction. (b) Crystal structures of the human β_2-adrenergic receptor in inactive and active conformations. The receptor possesses a bundle of seven membrane-spanning α-helices, and the ligand-binding pocket sits in the middle of this helical bundle. The inactive structure (blue; PDBID 2RH1) was obtained from a chimeric receptor with a T4 lysozyme sequence inserted into the third intracellular loop, which promotes crystallization. The receptor was also bound to the inverse agonist carazolol, which helps stabilize the top part of the receptor and promotes the inactive conformation. Neither the lysozyme domain nor the carazolol is shown here, for simplicity. The active structure (pink; PDBID 3P0G) was obtained from the same chimeric receptor in the presence of a camelid antibody that preferentially binds the active conformation, plus the agonist BI-167107. The agonist is shown in yellow, and it resides in the ligand-binding pocket of the receptor. The camelid antibody and T4 lysozyme sequences are omitted. The main difference between the inactive and active conformations is that transmembrane helix 6 (TM6) has moved outward by 11 Å in the latter. This conformation change creates an effector-binding pocket that allows a trimeric G-protein to interact with and become activated by the receptor.

Adrenergic drugs and hormones are monomeric, and reconstitution studies have established that the β_2-adrenergic receptor can signal as a monomer. Some other members of the GPCR family must be dimeric to function, and there is evidence that even the β_2-adrenergic receptor may oligomerize in vivo as well. Oligomerization is indeed common in receptor signaling, and in signaling in general, and will be discussed further in Chapter 3. But for present purposes, we will analyze the signaling that results from the interaction of a monomeric ligand with a monomeric β_2-adrenergic receptor.

The activation of a receptor by a ligand constitutes **stoichiometric regulation**, meaning that signaling involves a fixed integral ratio of the stimulus (the ligand) to its target (the receptor). In this case, one molecule of an agonist ligand brings about the activation of one molecule of receptor. Once the receptor is bound and activated, it acts as an enzyme; one molecule of active receptor can yield any number of molecules of activated G-proteins, depending on how long you wait.

2.2 EXPERIMENTS SHOW THE RECEPTOR'S EQUILIBRIUM AND DYNAMICAL BEHAVIORS

What happens when some concentration of ligand is incubated with a dish of cells that possess, say, a few thousand β_2 adrenergic receptors each, or a dish containing millions of recombinant receptor molecules? As shown in **Figure 2.2a**, the number of receptors bound by the ligand increases steadily until an equilibrium level of binding is approached. The time scale for binding is typically seconds or tens of seconds. If the ligand is washed away, the fraction of receptors bound decreases steadily and they return to their unbound state, although, curiously, the unbinding is always slower than the binding (**Figure 2.2a**). The equilibrium level of binding has a law-of-diminishing-returns quality: the first increment of ligand produces a good amount of binding, the next increment less, and the next even less (**Figure 2.2b**).

The consequences of this binding depend on the particular ligand in question. Some ligands—agonists, like the physiological ligands epinephrine and norepinephrine—increase the receptor's activity above its measurable basal level. Other ligands (such as the drug carazolol, which was used to stabilize the inactive conformation shown in **Figure 2.1b**) cause a decrease in the receptor's activity, and these are termed **inverse agonists**, and some do something in between. If they increase activity a bit, but not as much as the most efficacious agonists, they are termed **partial agonists**. If they bind but leave the basal activity of the receptor unaffected, they are called

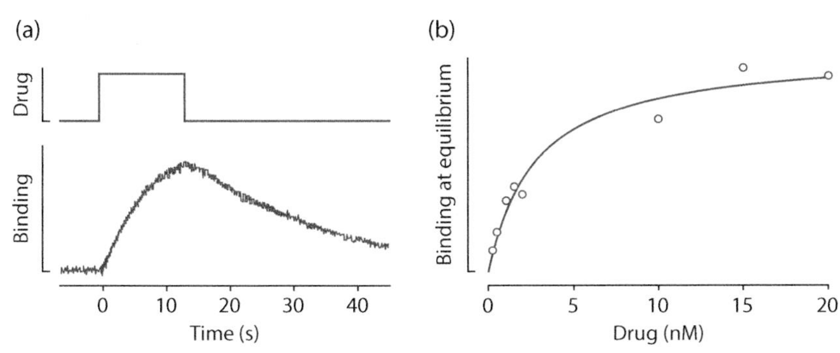

Figure 2.2 Typical behavior of a monomeric ligand interacting with receptors in vitro. (a) The kinetics of the binding and dissociation of a high-affinity agonist (fenoterol) to the β_2-adrenergic receptor in vitro. The binding and drug concentration data are expressed in arbitrary units. (b) Equilibrium binding of fenoterol to a β_2-adrenergic receptor-$G\beta_s$ chimera in vitro, again with binding expressed in arbitrary units. (Redrawn from Aristotelous et al. *ACS Med Chem Lett.* 2013 and Reinartz et al. *Naunyn Schmiedebergs Arch Pharmacol.* 2015).

neutral antagonists or just **antagonists**. In principle, there could be drugs that decrease the basal activity, but less effectively than the best inverse agonists; these would be called partial inverse agonists.

These are the basic characteristics of the interactions of β_2-adrenergic receptors with their ligands—fast binding, slower unbinding, a law-of-diminishing-returns equilibrium response, basal receptor activity, and a variety of relationships between binding and activity. The goal in our modeling is to understand how these behaviors arise.

2.3 A SIMPLE BINDING-DISSOCIATION MODEL EXPLAINS THE HYPERBOLIC EQUILIBRIUM RESPONSE

We could think of binding and activation as occurring either sequentially or simultaneously. The latter is simpler to model, so that is how we will start. Let us call the free ligand concentration x, the free receptor concentration y, and the concentration of the (active) complex c_{xy}. We begin by writing down the rate equation for the net production of c_{xy}:

$$\underbrace{\frac{dc_{xy}}{dt}}_{\text{Net rate}} = \underbrace{k_1 x \cdot y}_{\text{Forward rate}} - \underbrace{k_{-1} c_{xy}}_{\text{Back rate}} \tag{2.1}$$

This equation includes three time-dependent variables, although for simplicity we write them as x, y, and c_{xy} rather than $x[t]$, $y[t]$, and $c_{xy}[t]$, and two parameters, k_1 and k_{-1}. The right-hand side consists of two terms that contribute to the net production rate: a positive term for the forward reaction, in which x and y associate to form the c_{xy} complex, and a negative term for the back reaction, the dissociation of the complex into its constituents. The forward rate depends on the concentration of x, the concentration of y, and a proportionality constant, which is the rate constant for the association reaction and which we have denoted as k_1 (it is customary to use lower case k's for rate constants). The back rate depends on the concentration of the complex c_{xy} and another proportionality constant, the rate constant for the dissociation reaction, denoted k_{-1}. Both the forward and back reactions are examples of **mass action processes**, where the rate of the process is directly proportional to the concentration of the reactant or reactants.

To solve Eq. 2.1, we need to decrease the number of time-dependent variables from three to one. If we assume that the total concentration of the receptor y is unchanging with time, at least on the time scale of the binding and dissociation reactions, we can write a conservation equation:

$$y_{tot} = y + c_{xy} \tag{2.2}$$

This equation can be used to express y as a function of the constant y_{tot} and the time-dependent variable c_{xy}, and so eliminate one variable from the rate equation:

$$\frac{dc_{xy}}{dt} = k_1 x\left(y_{tot} - c_{xy}\right) - k_{-1} c_{xy}. \tag{2.3}$$

We could similarly eliminate x by invoking a second conservation equation ($x = x_{tot} - c_{xy}$), and in fact we will do this in Chapter 3 when we consider the stoichiometric regulation of downstream signaling proteins. But for now, we will assume that the ligand x is present in vast excess over the receptor y, so that the binding of x to y changes the concentration of x by a negligible amount. This assumption is a good approximation of the truth in many experimental situations. For example, to measure the response of mammalian cells to epinephrine in vitro, one might have a 10-cm dish containing 10^7 cells, with each cell expressing, say, 5,000 β_2-adrenergic receptors, which means that there is a total of 5×10^{10} receptors in the dish. If the cells are incubated with 10 mL of medium containing 1 μM epinephrine (a concentration that yields about half-maximal binding), there will be 10^{-8} mol of epinephrine present, or 6×10^{15} molecules. Since the total number of receptor molecules is 5×10^{10}, whereas the total number of epinephrine molecules is 6×10^{15}, the binding of epinephrine to half (or even all) of the receptor molecules will change the free concentration of epinephrine by less than 0.001%, a negligible amount for our purposes. Thus, for now, we will assume that x is unchanging with respect to time and that $x \approx x_{tot}$.

The rate equation (Eq. 2.3) is an **ordinary differential equation** (ODE), with the term "ordinary" meaning that there is only one independent variable (time t). It can be solved, yielding an expression for c_{xy} as a function of time, and we will do this a little later, in Section 2.3. But even without solving the rate equation, we can extract some interesting information from Eq. 2.1: we can derive an expression for the equilibrium concentration of c_{xy} as a function of the input stimulus (the ligand concentration (x)) and the system's parameters (the total concentration of the receptor (y_{tot}) and the rate constants (k_1 and k_{-1})). From the resulting expression, we can learn something about both the qualitative and quantitative character of the equilibrium response.

For the system to be in equilibrium, the concentration of c_{xy} has to be constant with respect to time, which means that:

$$\frac{dc_{xy}}{dt} = 0. \tag{2.4}$$

Combining Eqs. 2.3 and 2.4 yields:

$$k_1 x\left(y_{tot} - c_{xy}\right) - k_{-1} c_{xy} = 0. \tag{2.5}$$

Putting the terms containing c_{xy} on one side gives us:

$$k_1 x \cdot y_{tot} = k_{-1} c_{xy} + k_1 x \cdot c_{xy}. \tag{2.6}$$

It follows that the fraction of the receptor molecules bound by ligand at equilibrium is:

$$\left(\frac{c_{xy}}{y_{tot}}\right)_{eq} = \frac{k_1 x}{k_{-1} + k_1 x}, \tag{2.7}$$

where we have included the subscript "*eq*" here to emphasize that we are talking about the fraction of c_{xy} in the complex *at equilibrium*—we are no longer considering the complex as a time-dependent, dynamical quantity. Equivalently, we can write this equation as:

$$\left(\frac{c_{xy}}{y_{tot}}\right)_{eq} = \frac{x}{K_1 + x}, \tag{2.8}$$

where $K_1 = \frac{k_{-1}}{k_1}$, which by definition is the **equilibrium constant** for the binding-dissociation reaction (equilibrium constants are traditionally denoted by capital K's). This form emphasizes that it is only the ratio of the rate constants, rather than their individual values, that bears on the equilibrium level of complex formation. The individual values do affect how quickly the system approaches equilibrium, as we will show below, but not what the equilibrium concentration of c_{xy} is.

Equation 2.8 is the **Langmuir equation** (or the Langmuir binding isotherm), named in honor of Irving Langmuir, who in 1916 derived an equation of the same form for the adsorption of non-interacting gas molecules to a checkerboard surface of independent gas-binding sites—a process that is conceptually equivalent to our stoichiometric regulation process.

It is not obvious from just looking at it, but Eq. 2.8 describes a rectangular hyperbola, so this type of response is sometimes referred to as a **hyperbolic response**. In addition, Eq. 2.8 is similar in form to the famous **Michaelis–Menten equation**:

$$V = V_{max} \frac{S}{K_M + S}, \tag{2.9}$$

which relates the initial velocity or rate of an enzyme-catalyzed reaction (V) to the concentration of substrate present (S). Because of this similarity, equilibrium or steady-state responses that are described by equations of the form of Eq. 2.8 are commonly called **Michaelian responses**. Note, however, that we have not assumed that Michaelis–Menten kinetics govern the binding of a ligand to a receptor; we assumed mass action kinetics. For this reason, the term "Michaelian response" is a bit of a misnomer, but it is in common usage, and so we will use it throughout this book.

Note that there is a law-of-diminishing-returns quality to this hyperbolic or Michaelian response. The first increment of ligand yields some binding; the second increment of ligand yields a bit less incremental binding (Figure 2.3a); the third less still; until, once the binding is near-maximal, changes in ligand concentration have almost no effect on receptor binding. When the concentration of the ligand x equals K_1, the fraction of y in the complex is equal to 0.5, so the ***EC50***—the concentration of x required for 50% binding—is equal to K_1.

So, does Eq. 2.8 actually account for the equilibrium binding of ligands to β_2-adrenergic receptors? The answer is usually yes; in fact the curve in **Figure 2.2b** is a Michaelian curve, fitted to experimental data for the binding of a high-affinity β_2-adrenergic agonist (fenoterol) to a uniform population of receptors (chimeras of the β_2-adrenergic receptor and a G-protein β-subunit) in vitro. The fitted value for K_{eq} is 2.8 nM.

Michaelian responses turn up in other cell signaling processes—for example, in phosphorylation-dephosphorylation cycles, as long as the kinase and phosphatase are operating far from saturation (Chapter 3). For this reason, the Michaelian response can be used as a sort of benchmark to which other types of response can be compared.

If the ligand inhibits the receptor rather than activating it, so that y rather than c_{xy} is the active species, then the fraction of the receptors that are active at equilibrium is given by:

$$\left(\frac{y}{y_{tot}}\right)_{eq} = 1 - \left(\frac{c_{xy}}{y_{tot}}\right)_{eq} \tag{2.10}$$

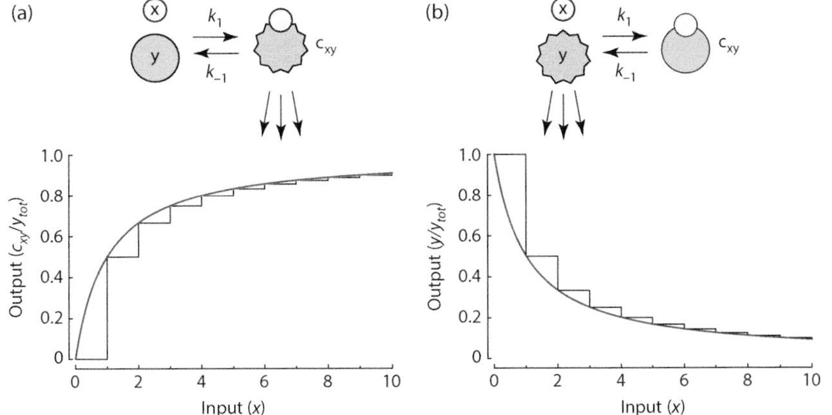

Figure 2.3 Equilibrium response curves for stoichiometric activation (a) and inhibition (b) reactions. The red curves show the responses; responses shaped like this are sometimes termed Michaelian. The black stair steps illustrate how much change in the response is obtained from each unit increment of stimulus. We have assumed that the equilibrium constant K_1 equals 1, so one can think of the input concentration as being measured in multiples of K_1.

$$\left(\frac{y}{y_{tot}}\right)_{eq} = \frac{K_1 + x}{K_1 + x} - \frac{x}{K_1 + x} \tag{2.11}$$

$$\left(\frac{y}{y_{tot}}\right)_{eq} = \frac{K_1}{K_1 + x} \text{ or } \frac{1}{1 + \frac{x}{K_1}}. \tag{2.12}$$

This type of response is sometimes called **hyperbolic** or **Michaelian inhibition**.

Probably the best way to get a feel for the character of the responses defined by Eqs. 2.8 and 2.12 is to plot them. There is only one adjustable parameter—the equilibrium constant K_1, which is in concentration units. If we choose units such that $K_1 = 1$, we obtain the red curve shown in **Figure 2.3a** for the fraction of the receptor molecules that are bound by the ligand at equilibrium. Each incremental step upward in x gives a smaller incremental change in receptor binding—the law-of-diminishing-returns character—and the maximal response $(c_{xy} = y_{tot})$ is approached very gradually. This is one of the qualities of ligand–receptor interaction that we were hoping to account for in our modeling (see Section 2.1) and, lo and behold, we have.

Michaelian inhibition, as defined by Eq. 2.12, exhibits the same sort of diminishing returns, except that the curve is flipped upside down, and each successive increment of stimulus yields a diminishing increment of inhibition (**Figure 2.3b**).

2.4 A SEMILOG PLOT EXPANDS THE RANGE BUT DISTORTS THE GRADED CHARACTER OF THE RESPONSE

When stoichiometric binding curves are plotted in the experimental literature, very often the ligand concentration is plotted on a logarithmic scale rather than a linear scale. This is sometimes because the binding curves for more than one ligand are being plotted, and a semilog plot allows ligands with dissimilar binding affinities to be plotted together and compared (**Figure 2.4b**). When a Michaelian response is plotted on a semilog plot, it is a sigmoidal curve, and changing the assumed value of the equilibrium constant shifts the curve to the left or right without stretching it or otherwise changing its shape (**Figure 2.4b**).

Note that there are some potentially misleading features in the picture presented by the semilog plot. It looks like there is a threshold,

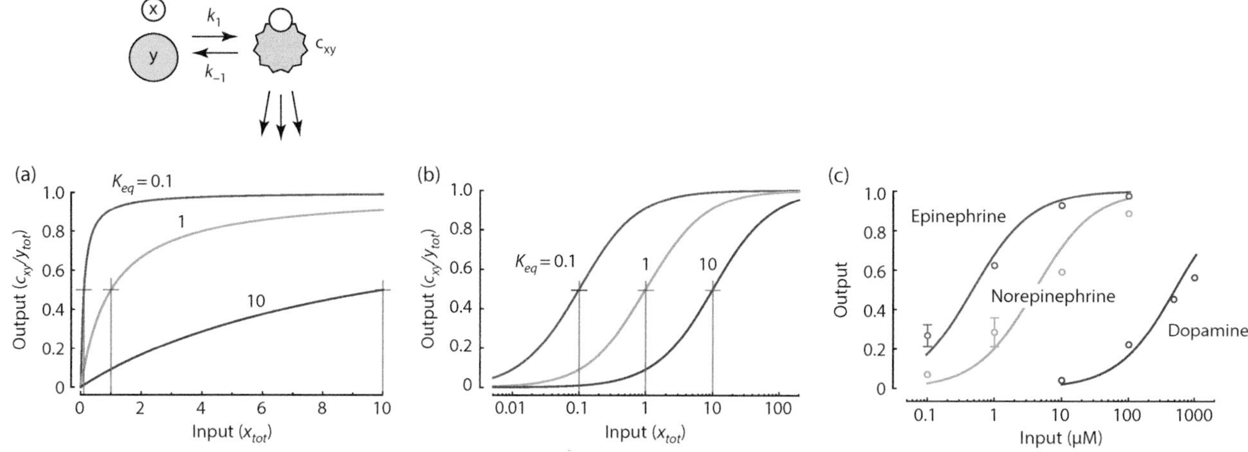

Figure 2.4 Michaelian responses depicted on a linear plot (a) or a semilog plot (b, c). Panels A and B show the modeled equilibrium responses for K_{eq} values of 0.1, 1, and 10. Note that when the input (x) equals K_{eq}, the output is half-maximal. Panel C is an example of experimental data plotted on a semilog plot together with fitted Michaelian curves. (Experimental data are adapted from Yao et al. *Nat Chem Biol.* 2006 and used with permission).

with the response being nearly flat for small increments of input and then steepest when the input equals the equilibrium constant. But of course this is not correct; the real slope (**Figures 2.2b**, **2.3a**, and **2.4a**) is maximal when x is zero, and it is the initial increments of input that produce the largest increments of output.

Thus, while semilog plots are commonplace and useful, it is important to not let them fool you. Some responses do exhibit thresholds, and thresholds are important for generating many of the complex responses we will discuss later in the book, but they do not arise out of this stoichiometric activation mechanism, despite what the semilog plot seems to show. We will examine several mechanisms that actually do produce thresholds in Chapter 3, when we discuss cooperativity, and in Chapters 4 and 5, when we examine some other mechanisms that yield ultrasensitivity.

2.5 THE SYSTEM APPROACHES EQUILIBRIUM EXPONENTIALLY

So far we have concerned ourselves only with the equilibrium of the receptor–ligand system. We can go further by solving the rate equation (Eq. 2.3) to deduce an equation for the time course for approaching equilibrium from some out-of-equilibrium initial condition. In the general case, where the input x might be a function of time, we may have to do this numerically with a numerical ODE-solver like the ones in *Mathematica* or MATLAB®. However, for the special case where we imagine that the input (x) changes instantaneously from one constant value to another, we can solve Eq. 2.3 analytically, yielding a relatively simple formula for the time course of the net formation of c_{xy} (if the input steps up) or the net dissociation of c_{xy} (if the input steps down).

Let us start again with Eq. 2.3, the rate equation for the stoichiometric regulation process:

$$\frac{dc_{xy}}{dt} = k_1 x\left(y_{tot} - c_{xy}\right) - k_{-1}c_{xy}. \tag{2.3}$$

Rearrange the right-hand side:

$$\frac{dc_{xy}}{dt} = -\left(k_1 x + k_{-1}\right)c_{xy} + k_1 x \cdot y_{tot}, \tag{2.13}$$

$$\frac{dc_{xy}}{dt} = -(k_1x + k_{-1})\left(c_{xy} - \frac{k_1x \cdot y_{tot}}{k_1x + k_{-1}}\right), \tag{2.14}$$

$$\frac{dc_{xy}}{dt} = -(k_1x + k_{-1})\left(c_{xy} - \left(c_{xy}\right)_{eq}\right). \tag{2.15}$$

Next, we replace the variable c_{xy} with a new variable z that represents the difference between the c_{xy} and the equilibrium value of c_{xy}, $(c_{xy})_{eq}$:

$$z = c_{xy} - \left(c_{xy}\right)_{eq} \text{ and } \frac{dz}{dt} = \frac{dc_{xy}}{dt}. \tag{2.16}$$

Plugging this into Eq. 2.15 yields

$$\frac{dz}{dt} = -(k_1x + k_{-1})z. \tag{2.17}$$

We can solve this. First we write

$$\frac{dz}{z} = -(k_1x + k_{-1})dt. \tag{2.18}$$

We then integrate both sides of Eq. 2.18:

$$\ln z = -(k_1x + k_{-1})t + C, \tag{2.19}$$

where C is a constant of integration whose value can be determined from the initial conditions. Exponentiating both sides yields:

$$z = e^C e^{-(k_1x + k_{-1})t}. \tag{2.20}$$

We can assign a value to the constant of integration C in the term e^C by noting that at $t = 0$, the value of z is:

$$z[0] = e^C \cdot e^0 = e^C. \tag{2.21}$$

Combining Eqs. 2.20 and 2.21, we obtain an equation for the time evolution of z in terms of the input stimulus x, the rate constants k_1 and k_{-1}, and the initial condition $z[0]$:

$$z[t] = z[0]e^{-(k_1x + k_{-1})t}. \tag{2.22}$$

Note that to emphasize that the difference between $z[0]$, a number, and the z on the left-hand side of Eq. 2.20, a time-dependent function, we have written the time dependence explicitly in Eq. 2.22.

Eq. 2.22 shows that $z[t]$, which represents how far c_{xy} is from its equilibrium value, decreases exponentially with time. We can say the system approaches equilibrium exponentially. But keep in mind that this is not an explosive, positive exponential but rather an exponential with a negative exponent, which therefore decreases with time.

We can convert back from z to c_{xy} by substituting the definition for z (Eq. 2.16) into Eq. 2.23:

$$c_{xy}[t] = \left(c_{xy}\right)_{eq} + \left(c_{xy}[0] - \left(c_{xy}\right)_{eq}\right)e^{-(k_1x + k_{-1})t}. \tag{2.23}$$

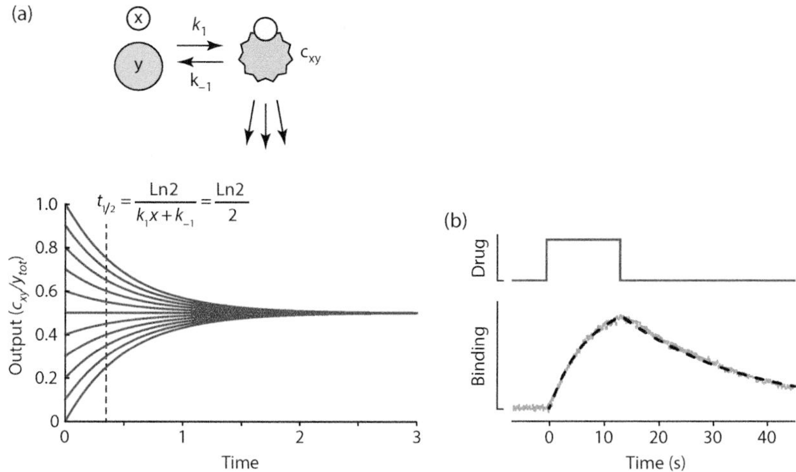

Figure 2.5 Exponential approach to equilibrium. (a) Calculated time courses from Eq. 2.22, assuming the following parameter values: $k_1 = k_{-1} = x = 1$, and taking $c_{xy}[0]/y_{tot}$ to be various values between 0 and 1. (b) Comparison of modeled time courses for binding and dissociation (dashed curve) to the experimental binding/dissociation data from Figure 2.2a (here shown in pink). The model accounts for the data quite well.

We can get a feel for this equation by plotting it. To make things simple, let us first assume that the rate constants are both equal to 1 (which means the equilibrium constant K_1 is also equal to 1), that $x = 1$ (so we should expect a half-maximal response), and that $y_{tot} = 1$, and let us start the system at a range of initial values of c_{xy}. As shown in **Figure 2.5**, when c_{xy} is initially above its equilibrium concentration of $c_{xy} = 0.5$, the time courses are monotonic decreases; when c_{xy} is initially below its equilibrium concentration, the time courses are monotonic increases; and when it starts at 0.5, it remains unchanged. The trajectories never overshoot the equilibrium. For each of these curves (except for the flat $c_{xy} = 0.5$ curve), the halftime for approaching equilibrium is:

$$t_{1/2} = \frac{\text{Ln2}}{k_1x + k_{-1}} = \frac{\text{Ln2}}{1+1} = 0.347. \tag{2.24}$$

The quantity $k_1x + k_{-1}$ is sometimes called $k_{apparent}$, the apparent rate constant for the exponential approach to equilibrium.

Note that whereas the expression for the equilibrium level of c_{xy} depended on the ratio of the two rate constants, not the absolute values of the individual rate constants (Eq. 2.8), the dynamics of the system (from Eqs. 2.22 and 2.23) does depend on the individual rate constants. Thus doubling both k_1 and k_{-1} has no effect on the equilibrium level of c_{xy} but does increase the speed at which the system approaches equilibrium.

So do experimental binding data agree with the exponential approach to equilibrium that the model predicts? The answer is often yes. In **Figure 2.5b** we have overlaid the experimental binding-dissociation time course from **Figure 2.2a** with best-fit exponential curves, and the agreement is quite good.

2.6 INCREASING THE ASSOCIATION RATE DECREASES $t_{1/2}$; SO DOES INCREASING THE DISSOCIATION RATE

From Eq. 2.24 we can see that the speeds of both the association reaction (which yields the k_1x term) and the dissociation reaction (k_{-1}) contribute to $k_{apparent}$ and the halftime of the binding reaction. The faster the association, the smaller the halftime, and the faster the dissociation, the smaller the halftime. Intuitively, it seems plausible

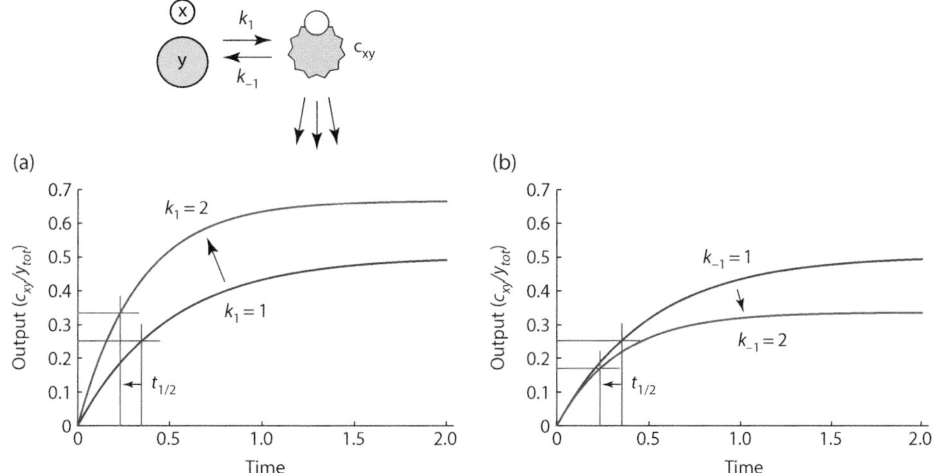

Figure 2.6 The speeds of both the association and dissociation reactions contribute to the halftime for stoichiometric regulation. (a) Increasing the association rate constant (k_1) decreases the halftime by increasing the initial speed of association. (b) Increasing the dissociation rate constant (k_{-1}) does not affect the initial speed, but makes the system level off earlier.

that speeding up the association should quicken the process, but why should speeding up the dissociation do the same?

To answer this question, we first look at what happens to the time course when we increase the association rate. For simplicity, we will start with the parameters we used for **Figure 2.5** and then double the assumed value of k_1 (or, equivalently, of x). We will also take the initial concentration of c_{xy} to be zero. As shown in **Figure 2.6a**, when k_1 is doubled (from 1 to 2), the initial slope of the time course is doubled. This allows the system to reach its half-maximal response faster, even though it has a little farther to go to get to the half-maximum than it did when k_1 was 1.

Next consider when happens when $k_1 = 1$ and the dissociation rate constant k_{-1} is increased from 1 to 2 (**Figure 2.6b**). The initial rate of complex formation is unchanged; it depends only on k_1, so the system does not start off toward equilibrium binding any faster. But the equilibrium level of response is lower (1/3 rather than 1/2), and so for this reason the system gets to the half-maximal level of response sooner than it did when k_1 was 1.

Thus, increasing the forward reaction rate constant decreases the halftime because you go faster, and increasing the back reaction rate constant decreases the halftime because you have less far to go.

2.7 GOING UP IS FASTER THAN COMING DOWN

Suppose that we start with zero ligand and an unoccupied receptor, increase the total ligand concentration to some value x, let the system equilibrate, and then decrease the ligand concentration back to zero. How will the time courses going up and coming down compare?

We can answer this by looking at the expression for $t_{1/2}$ or $k_{apparent}$; here we pick $k_{apparent}$:

$$k_{apparent} = k_1 x + k_{-1}. \tag{2.25}$$

Going up, both terms on the right-hand side of Eq. 2.25 contribute to $k_{apparent}$. Coming down, x is zero, so only the k_{-1} term contributes.

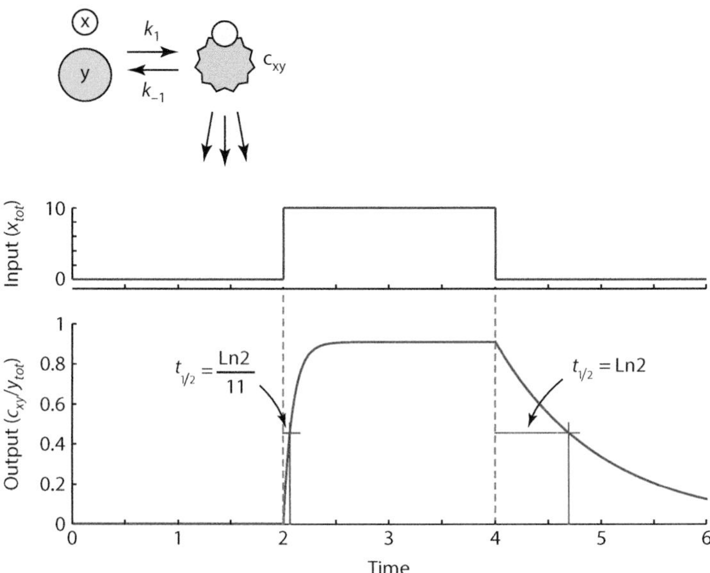

Figure 2.7 The binding of a ligand *x* to a receptor *y* equilibrates more quickly in response to a step up in ligand concentration than it does to a step back down. The input (x_{tot}) is shown in blue and the output (c_{xy}, in fractions of y_{tot}) is shown in red. We assumed that $k_1 = k_{-1} = 1$ and that x_{tot} goes from 0 to 10 and back. The amount by which the two $t_{1/2}$ values differ depends on the assumed values of the rate constants and input, but going up is always faster than coming down.

Thus the new equilibrium will be approached faster ($k_{apparent}$ is larger) on the way up than on the way down. This is shown in **Figure 2.7**. We have taken $k_1 = k_{-1} = y_{tot} = 1$ for the model's parameters and have let x switch from 0 to 10 and back. The halftime on the way up is and on the way down is $\frac{\text{Ln2}}{11}$, and on the way down is Ln2 eleven times longer. In general, for this sort of reaction, going up is faster than coming down.

2.8 THE DISSOCIATION RATE CONSTANT k_{-1} DETERMINES THE HALF-LIFE AND MEAN LIFETIME OF A LIGAND–RECEPTOR COMPLEX

Note that when c_{xy} is some nonzero concentration and $x = 0$, $k_{apparent}$ is equal to k_{-1}, and the length of time it takes for half of the receptor molecules to lose their ligand molecules to the $x = 0$ void is $\frac{\text{Ln2}}{2k_{-1}} \approx \frac{0.693}{k_{-1}}$. This quantity has a special meaning; it represents the half-life of the receptor–ligand complex.

We can also derive an expression for the average lifetime of the complex, a quantity often just called the lifetime of the complex or the dwell time of the ligand. Equation 2.23 reduces to:

$$c_{xy}[t] = c_{xy}[0]e^{-k_{-1}t}. \tag{2.26}$$

The lifetime can be calculated as follows. The number of the c_{xy} complexes that decay within some tiny time interval dt around t is given by:

$$c_{xy}[0]e^{-k_{-1}t}dt. \tag{2.27}$$

The fraction of the complexes within this time interval is:

$$\frac{c_{xy}[0]e^{-k_{-1}t}}{\int_0^\infty c_{xy}[0]e^{-k_{-1}t}\,dt}. \tag{2.28}$$

Note that the $c_{xy}[0]$'s cancel. The decay time for these complexes is t, so the contribution of this interval's complexes to the overall average decay time is:

$$\frac{te^{-k_{-1}t}}{\int_0^\infty e^{-k_{-1}t}\,dt}. \tag{2.29}$$

And now to get an expression for the average lifetime, which we call τ, we integrate over all t. The numerator can be evaluated using integration by parts (or the Integrate command in a program like *Mathematica*); the denominator is easier:

$$\tau = \frac{\int_0^\infty te^{-k_{-1}t}\,dt}{\int_0^\infty e^{-k_{-1}t}\,dt} = \frac{\left.\frac{e^{-k_{-1}t}(1+k_{-1}t)}{k_{-1}^2}\right|_0^\infty}{\left.\frac{e^{-k_{-1}t}}{k_{-1}}\right|_0^\infty} = \frac{0-\frac{1}{k_{-1}^2}}{0-\frac{1}{k_{-1}}} = \frac{1}{k_{-1}} \tag{2.30}$$

Thus, the lifetime of a complex is simply the reciprocal of the dissociation rate constant k_{-1}.

2.9 PARTIAL AGONISTS, ANTAGONISTS, AND INVERSE AGONISTS CAN BE EXPLAINED BY ASSUMING THAT BINDING AND ACTIVATION OCCUR IN DISTINCT STEPS

So far we have assumed that ligand binding and receptor activation occur simultaneously; there are only two species of receptor: inactive, unbound receptors and active, ligand-bound receptors. What if we instead assume that ligand binding and receptor activation take place in separate, distinct steps?

We could imagine that the binding of the ligand promotes a conformation change in the receptor, in the spirit of Daniel Koshland's induced fit model for allosteric activation, which we will discuss in Chapter 3. In this scheme, ligand binding occurs first and is followed by receptor activation (**Figure 2.8a**). Alternatively, we could assume that the empty receptor flips between an inactive and an active conformation, and the ligand essentially selects and stabilizes the activated conformation of the receptor (**Figure 2.8b**). This conformational selection mechanism is at the heart of the Monod–Wyman–Changeux (MWC) models of allostery and cooperativity, which again we will see more of in Chapter 3. Of course both routes to activation could be occurring in parallel (**Figure 2.8c**). Here we will look at the first scheme (**Figure 2.8a**) first, because it is the easiest to analyze mathematically.

There are three forms of the receptor y in this model: unliganded y, the ligand-bound but as yet inactive species c_{xy}, and the active species $c_{xy}*$ (**Figure 2.8a**). We can write rate equations for the net production of each of these species in the presence of a constant concentration of the ligand x:

$$\frac{dy}{dt} = -k_1x\cdot y + k_{-1}c_{xy} \tag{2.31}$$

$$\frac{dc_{xy}}{dt} = k_1x\cdot y - k_{-1}c_{xy} - k_3c_{xy} + k_{-3}c_{xy}^{*} \tag{2.32}$$

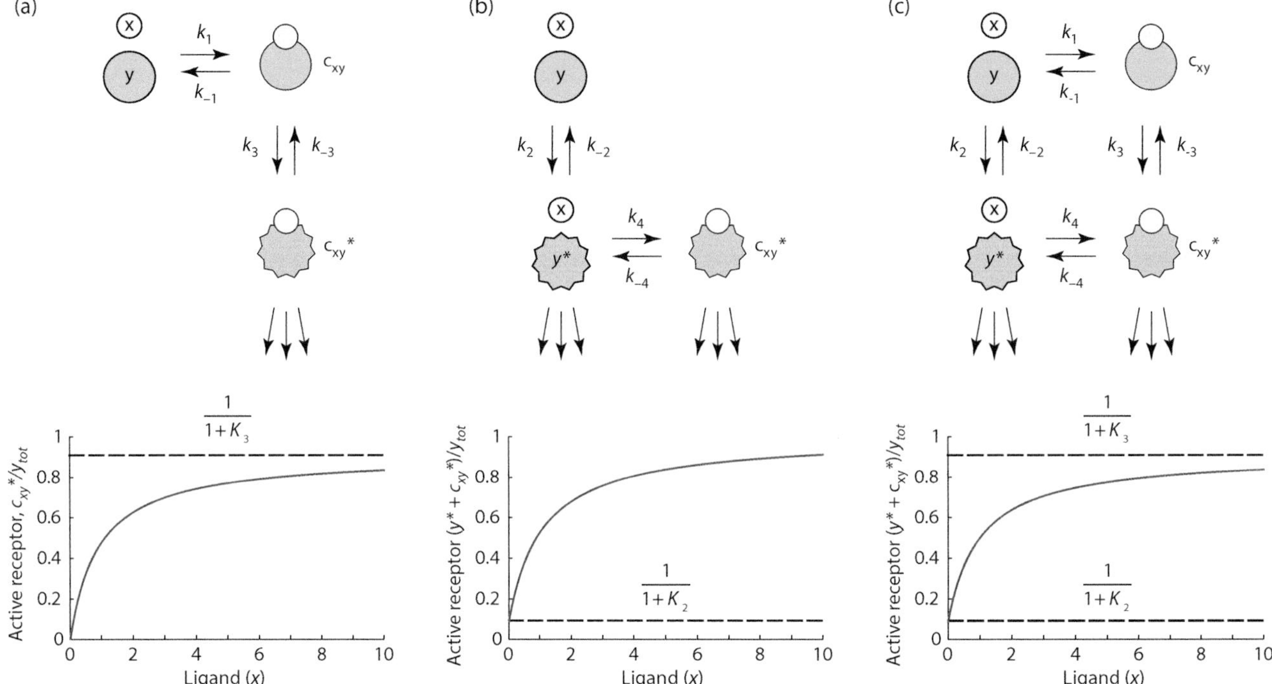

Figure 2.8 Sequential binding and activation. (a) This model assumes that x binds y first and then induces a conformation change that activates y. If a drug can bring about a near maximum level of receptor activation (i.e., $K_3 = k_{-3}/k_3$ is very small), the drug is called a full agonist. If the maximum level of activation is zero (i.e., K_3 is large), the drug is called an antagonist. And if the maximum level of activation is intermediate between these extremes, the drug is called a partial agonist. (b) Here we assume that the free receptor equilibrates between the active and inactive conformation, and the ligand x captures the active receptor. (c) The full model assuming both pathways contribute. This is often termed the "two-state model" of receptor activation. Note that if, for some drug, the level of activation when the drug is bound to the receptor is actually lower than the basal level of activity ($K_2 < K_3$), the drug is called an inverse agonist. In all cases we have taken $K_1 = K_2 = 10$, $K_3 = K_4 = 0.1$, and $y_{tot} = 1$.

$$\frac{dc_{xy}^*}{dt} = k_3 c_{xy} - k_{-3} c_{xy}^*. \tag{2.33}$$

At equilibrium, each of these net rates must equal zero, yielding three algebraic equations:

$$0 = -k_1 x \cdot y + k_{-1} c_{xy}, \tag{2.34}$$

$$0 = k_1 x \cdot y - k_{-1} c_{xy} - k_3 c_{xy} + k_{-3} c_{xy}^*, \tag{2.35}$$

$$0 = k_3 c_{xy} - k_{-3} c_{xy}^*. \tag{2.36}$$

Note that Eq. 2.35 = – Eq. 2.34 – Eq. 2.36, so we only have two independent equations. We can obtain a third independent equation from the conservation relationship:

$$y_{tot} = y + c_{xy} + c_{xy}^*. \tag{2.37}$$

We can now solve for each of the three species of receptor as a function of the rate constants and y_{tot}:

$$\left(\frac{y}{y_{tot}}\right)_{eq} = \frac{K_1 K_3}{K_1 K_3 + K_3 x + x} \tag{2.38}$$

$$\left(\frac{c_{xy}}{y_{tot}}\right)_{eq} = \frac{K_3 x}{K_1 K_3 + K_3 x + x} \tag{2.39}$$

$$\left(\frac{c_{xy}^*}{y_{tot}}\right)_{eq} = \frac{x}{K_1 K_3 + K_3 x + x}. \tag{2.40}$$

The parameters K_1 and K_3 are the equilibrium constants for the binding and conformation-change steps, with $K_1 = \frac{k_{-1}}{k_1}$ and $K_3 = \frac{k_{-3}}{k_3}$. The equilibrium concentrations of y, c_{xy}, and c_{xy}* depend on the equilibrium constants, and not on the individual values of the rate constants, as usual.

To see how this response function (Eq. 2.40) compares to a Michaelian response, we can start by plotting it for some choice of parameters. Let us take $K_1 = 10$, and let us suppose the equilibrium between c_{xy} and c_{xy}* favors c_{xy}*, with $K_3 = 0.1$. The resulting curve (**Figure 2.8a**) looks a lot like a Michaelian response curve, except that it levels off below an output of 1 (10/11 or 0.909... for this choice of rate constants). In fact we can rearrange Eq. 2.35 to show that this is exactly the case:

$$\left(\frac{c_{xy}^*}{y_{tot}}\right)_{eq} = \frac{1}{K_3+1}\frac{x}{\frac{K_1 K_3}{K_3+1}+x}. \tag{2.41}$$

The right-hand term in Eq. 2.41 describes a Michaelian response with an $EC50$ of $\frac{K_1 K_3}{K_3+1}$ and a maximal response of $\frac{1}{K_3+1}$. Note that since K_3 is a positive number, the maximal response will be less than 1.

This model predicts that activation maxes out at less than 100%, with the propensity of a ligand to induce the receptor to adopt its activated conformation being determined by K_3, which could vary from ligand to ligand. This explains why some ligands act as full agonists—presumably the equilibrium in the K_3 step very much favors the activated conformation—while other ligands act as partial agonists (promoting the conformation change more weakly) and still others act as antagonists, binding but not promoting activation at all. The μ-opioid receptor-binding drugs morphine, heroin, and fentanyl are all regarded as full agonists, with maximal binding leading to maximal receptor activation, whereas buprenorphine is a high-affinity partial agonist, with maximal binding causing less-than-maximal activation. For this reason, buprenorphine is sometimes used to treat opioid addiction; it can maintain an addict in a less-than-maximally intoxicated state that can allow the addict to function more normally. The high-affinity μ-opioid receptor-binding drugs naltrexone and naloxone function as antagonists, binding to receptors without promoting receptor activation. For this reason they can save the life of someone overdosing on an opioid agonist, competing with a death-inducing agonist for access to the receptor and thus decreasing receptor activity to levels compatible with life.

Next let us examine the second scheme (**Figure 2.8b**), with the receptor equilibrating between an inactive and an active conformation, and an agonist ligand x essentially selecting and stabilizing the active conformation. The rate equations for this model are:

$$\frac{dy}{dt} = -k_2 y + k_{-2} y^*, \tag{2.42}$$

$$\frac{dy^*}{dt} = k_2 y - k_{-2} y^* - k_4 x \cdot y^* + k_{-4} c_{xy}^*, \tag{2.43}$$

$$\frac{dc_{xy}^*}{dt} = k_4 x \cdot y^* - k_{-4} c_{xy}^*. \tag{2.44}$$

Setting all of these equations equal to zero and solving for y, y^*, and c_{xy}^* yields:

$$\left(\frac{y}{y_{tot}}\right)_{eq} = \frac{K_2 K_4}{K_2 K_4 + K_4 + x}, \tag{2.45}$$

$$\left(\frac{y^*}{y_{tot}}\right)_{eq} = \frac{K_4}{K_2 K_4 + K_4 + x}, \tag{2.46}$$

$$\left(\frac{c_{xy}^*}{y_{tot}}\right)_{eq} = \frac{x}{K_2 K_4 + K_4 + x}. \tag{2.47}$$

In this case, the total active receptor is $y^* + c_{xy}^*$, which yields:

$$\left(\frac{y^* + c_{xy}^*}{y_{tot}}\right)_{eq} = \frac{K_4 + x}{K_2 K_4 + K_4 + x}. \tag{2.48}$$

Now the minimal response—the response when the concentration of the ligand x is zero—is nonzero, equal to $\frac{1}{1+K_2}$. The response approaches a maximum value of 1 as x approaches infinity, and the shape of the response curve is again Michaelian.

Note that this model can explain why many receptors, including the β_2-adrenergic receptor, exhibit basal activity, and it explains the existence of **inverse agonists**; in the context of this model, an inverse agonist is a drug that binds better to the inactive conformation than to the active conformation. In fact the structure of the inactive conformation of the receptor shown in **Figure 2.1b** was obtained in the presence of an inverse agonist, the drug carazolol. The ligand evidently firms up the positions of the transmembrane helices, and without such a ligand, only structures of the lower portions of the helices are obtained. Carazolol not only blocks the effects of agonists like epinephrine on receptors and cells, it also decreases the basal levels of β_2-adrenergic receptor signaling seen in cells by binding more strongly to the inactive receptor than to the active receptor. In this way, inverse agonists select the receptor's inactive conformation, whereas agonists select the active conformation.

Perhaps not too surprisingly, if one solves the full system with all four species and all four interconversion processes (**Figure 2.8c**), the result is a Michaelian response with a minimum greater than zero and a maximum less than one. The equations for the four species are pretty complicated:

$$\left(\frac{y}{y_{tot}}\right)_{eq} = \frac{K_1 K_2 K_4}{K_1 K_4 + K_1 K_2 K_4 + (K_1 + K_2 K_4)x}, \tag{2.49}$$

$$\left(\frac{c_{xy}}{y_{tot}}\right)_{eq} = \frac{K_2K_4x}{K_1K_4 + K_1K_2K_4 + (K_1 + K_2K_4)x}, \quad (2.50)$$

$$\left(\frac{y^*}{y_{tot}}\right)_{eq} = \frac{K_1K_4}{K_1K_4 + K_1K_2K_4 + (K_1 + K_2K_4)x}, \quad (2.51)$$

$$\left(\frac{c^*_{xy}}{y_{tot}}\right)_{eq} = \frac{K_1x}{K_1K_4 + K_1K_2K_4 + (K_1 + K_2K_4)x}. \quad (2.52)$$

But the graph of the response (which is given by $\frac{y^* + c_{xy}}{y_{tot}}$) is simple (**Figure 2.8c**), and it is just as expected—there is a nonzero minimum response and a maximum response of less than 1—and in between the response curve is Michaelian in shape. This is often termed the **two-state model** of receptor–ligand interaction, and it has become something of a standard model in the receptor field.

Thus, adding a separate activation step to a binding reaction accounts for the existence of partial agonists, antagonists, and inverse agonists. Although the equilibrium binding equations are a bit more complicated than we obtained for concerted binding and activation, they can be viewed as variations on the standard law-of-diminishing-returns Michaelian response, albeit a Michaelian response with a minimum greater than zero and a maximum less than 1.

SUMMARY

The simplest receptors are monomeric proteins that bind monomeric ligands. Here we have shown that a model of receptor–ligand interaction, where ligand binding and receptor activation happen simultaneously, can account for many of the basic qualities of ligand binding: a hyperbolic or Michaelian equilibrium response with a law-of-diminishing returns character, exponential approach to equilibrium, and a faster response after adding ligand than after washing it away. A more realistic model where ligand binding and receptor activation occur in separate steps tweaks the equilibrium response a bit: the standard Michaelian response acquires a nonzero basal activity and a less-than-100% maximal activity. But, more importantly, it provides an explanation for the existence of partial agonists, antagonists, and inverse agonists, classes of drugs with varying abilities to activate receptors that may not activate receptors fully, but nevertheless can be of great utility in clinical medicine.

FURTHER READING

RECEPTORS

Katritch V, Cherezov V, Stevens RC. Structure-function of the G protein-coupled receptor superfamily. *Annu Rev Pharmacol Toxicol.* 2013;53:531–56.

Katzung BG, Trevor AJ. *Basic and Clinical Pharmacology.* 13th Edition. McGraw-Hill, New York, 2015.

Kobilka BK. Structural insights into adrenergic receptor function and pharmacology. *Trends Pharmacol Sci.* 2011 Apr;32(4):213–8.

Parmar VK, Grinde E, Mazurkiewicz JE, Herrick-Davis K. Beta(2)-adrenergic receptor homodimers: Role of transmembrane domain 1 and helix 8 in dimerization and cell surface expression. *Biochim Biophys Acta Biomembr.* 2017 Sep;1859(9 Pt A):1445–55.

Scarselli M, Annibale P, McCormick PJ, Kolachalam S, Aringhieri S, Radenovic A, Corsini GU, Maggio R. Revealing G-protein-coupled receptor oligomerization at the single-molecule level through a nanoscopic lens: methods, dynamics and biological function. *FEBS J.* 2016 Apr;283(7):1197–217.

EXAMPLES OF LIGAND-BINDING STUDIES

Aristotelous T, Ahn S, Shukla AK, Gawron S, Sassano MF, Kahsai AW, Wingler LM, Zhu X, Tripathi-Shukla P, Huang XP, Riley J, Besnard J, Read KD, Roth BL, Gilbert IH, Hopkins AL, Lefkowitz RJ, Navratilova I. Discovery of β_2 adrenergic receptor ligands using biosensor fragment screening of tagged wild-type receptor. *ACS Med Chem Lett.* 2013 Oct 10;4(10):1005–1010.

Reinartz MT, Kälble S, Wainer IW, Seifert R. Interaction of fenoterol stereoisomers with β_2-adrenoceptor-Gsα fusion proteins: antagonist and agonist competition binding. *Naunyn Schmiedebergs Arch Pharmacol.* 2015 May;388(5):517–24.

Yao X, Parnot C, Deupi X, Ratnala VR, Swaminath G, Farrens D, Kobilka B. Coupling ligand structure to specific conformational switches in the beta2-adrenoceptor. *Nat Chem Biol.* 2006 Aug;2(8):417–22.

THE TWO-STATE MODEL

Leff P. The two-state model of receptor activation. Trends Pharmacol Sci. 1995 Mar;16(3):89–97.

RECEPTORS 2

Multimeric Receptors and Cooperativity

3

IN THIS CHAPTER . . .

DOI: 10.1201/9781003124269-3

INTRODUCTION

In Chapter 2 we examined the binding of a monomeric ligand to a monomeric receptor, a type of interaction that is commonplace in cell signaling. But multimeric receptors and ligands are common as well. Some G-protein-coupled receptors, such as the metabotropic glutamate receptor, function as dimers rather than monomers. Many receptor tyrosine kinases are multimeric and some of their ligands are as well. For example, the well-studied epidermal growth factor receptor (EGFR) is thought to function as a dimer in which two epidermal growth factor (EGF) molecules interact with two receptor molecules to yield an active EGF_2–$EGFR_2$ complex. The platelet-derived growth factor (PDGF) receptor is dimeric too, and it binds a PDGF dimer. The downstream Raf proteins also function as dimers, as do many of the further downstream transcription factors. Dimerization is not peculiar to receptor tyrosine kinase signaling—it has been estimated that, all told, perhaps 35% of proteins form homodimers, trimers, or larger polymeric complexes (**Figure 3.1**).

Whenever multiple ligand molecules (e.g., EGF, Ras, or transcription factors) bind a multimeric target (EGFR, Raf, or the transcription factors' target sequences), the equilibrium response may differ in important ways from the Michaelian response described in Chapter 2. In some cases, the shape of the binding curve is sigmoidal, with the first increments of ligand binding relatively poorly and subsequent increments binding better (**Figure 3.2**, red curve). This mechanism is termed **cooperativity**, or, more precisely, **positive cooperativity**. The sigmoidal binding curves are sometimes termed **ultrasensitive** curves, in part because mechanisms other than cooperativity, which we will encounter in Chapters 4 and 5, can yield very similar response curves.

Alternatively, a multimeric receptor may bind ligands more gradually than a monomeric receptor does, but over a larger range of input stimuli (**Figure 3.2**, blue curve). This mechanism is termed **negative cooperativity**, and the resulting binding curves are termed **subsensitive** response curves. Both ultrasensitivity and subsensitivity can contribute in important ways to the overall functioning of a signaling system.

Here we will begin by examining how ultrasensitive and subsensitive responses can arise in multimeric systems through positive and

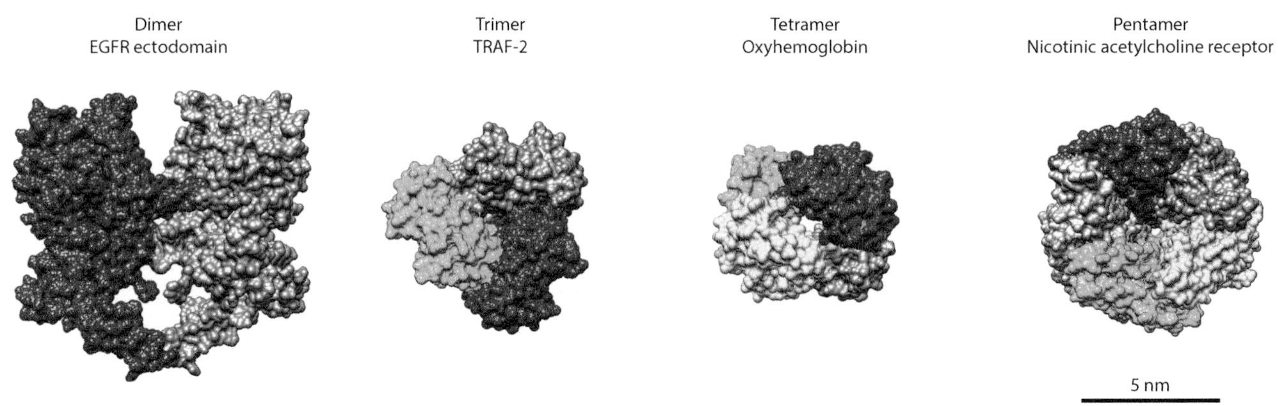

Figure 3.1 Examples of oligomeric proteins. (a) The dimeric extracellular portion of the *Drosophila* EGF receptor (PDB 3I2T). (b) The trimeric protein TRAF-2 (human), which stands for TNF (tissue necrosis factor) receptor-associated factor-2 (PDB 1CA4). (c) The tetrameric protein hemoglobin (human), bound to four oxygen molecules (PDB 1GZX). (d) The pentameric nicotinic acetylcholine receptor, from *Torpedo marmorata* (the marbled electric ray) (PDB 2BG9). All structures are shown to the same scale.

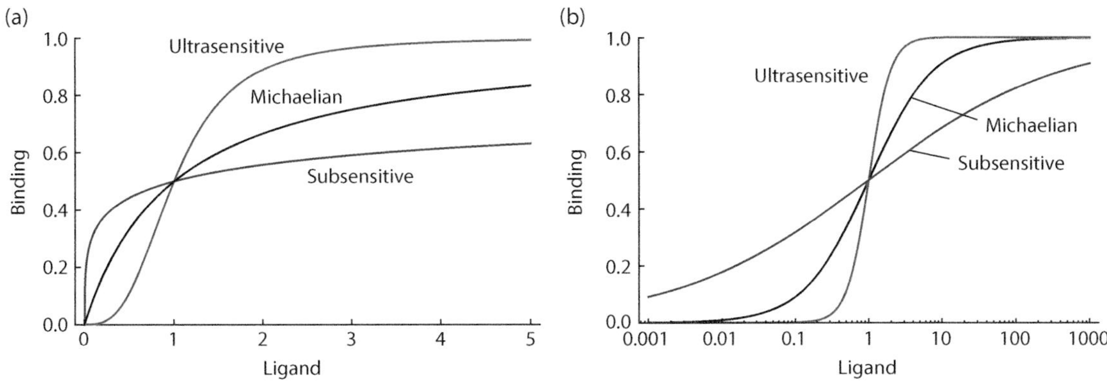

Figure 3.2 Three types of equilibrium binding curves: ultrasensitive, Michaelian, and subsensitive. Panel A shows a linear plot; panel B a semilog plot.

negative cooperativity. The theory behind cooperativity goes back to the early 20th century and was contributed to by many of that century's most famous biochemists. The original motivation for the theory was not cell signaling; it was the binding of oxygen to hemoglobin, and the puzzle was to understand why hemoglobin exhibited a sigmoidal oxygen-binding curve.

3.1 THE HILL EQUATION IS A SIMPLE EXPRESSION FOR THE EQUILIBRIUM BINDING OF LIGAND MOLECULES TO AN OLIGOMERIC RECEPTOR

Data for the equilibrium binding of oxygen to hemoglobin in vitro are shown in **Figure 3.3**. At low oxygen tensions, hemoglobin binds less oxygen than predicted by the Langmuir equation (**Figure 3.3a**), and at high oxygen tensions it binds more. Moreover, the deviation from theory is physiologically significant. It means that hemoglobin can pick up more oxygen in the high-pO_2 environment of the lungs, and then unload more of this oxygen in the low-pO_2 environment of the tissues, than it otherwise could. The question then was how this sigmoidal binding curve might arise.

In 1910, A.V. Hill proposed a mechanism that can account for the sigmoidal shape. He hypothesized that at least some of the hemoglobin in solution was present as a multimeric complex—there were indications from experiments that this was probably true, although the famous tetrameric structure of hemoglobin had not yet been elucidated. Next, he assumed that an *n*-mer of hemoglobin could bind *n* molecules of oxygen and that the *n* molecules of oxygen bound and dissociated from the protein simultaneously, so that no appreciable concentration of partially saturated hemoglobin molecules ever accumulated. With these assumptions, the binding reaction is *n*th order in O_2. The net rate of oxygen binding is therefore given by:

$$\frac{dy_n}{dt} = k_1 x^n . y_0 - k_{-1} y_n, \tag{3.1}$$

where x represents the concentration of the ligand oxygen, y_0 is the concentration of hemoglobin with no bound oxygens, and y_n is the concentration of hemoglobin with n bound oxygens. At equilibrium:

$$0 = k_1 x^n . y_0 - k_{-1} y_n. \tag{3.2}$$

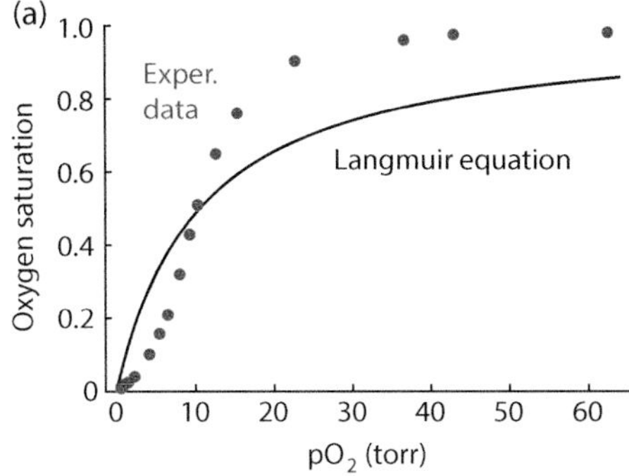

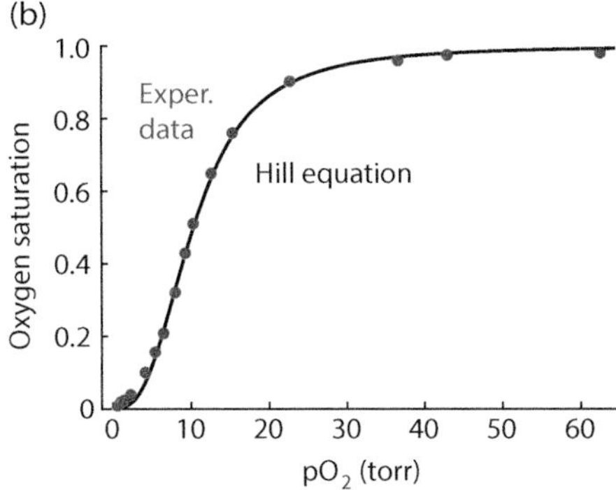

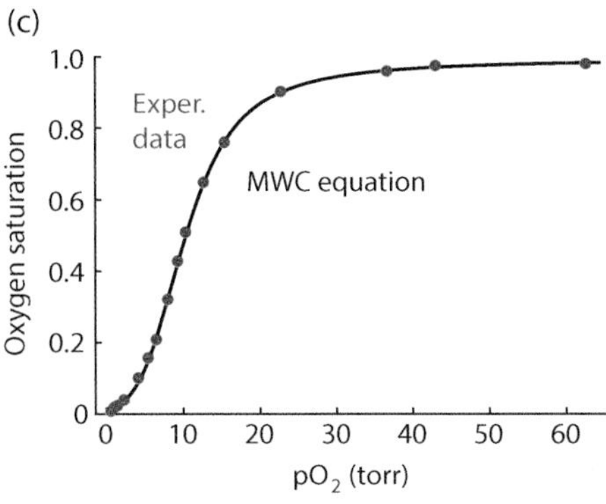

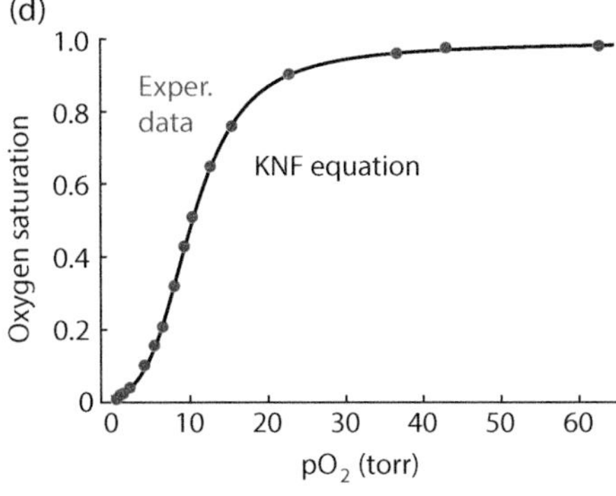

Figure 3.3 The binding of oxygen to hemoglobin in vitro. The experimental data are for the binding of oxygen to horse hemoglobin and are adapted from Monod et al., *J Mol Biol.* 1965. The curves are fits of the following equations to the data: (a) the Langmuir equation (Eq. 2.8); (b) the Hill equation (Eq. 3.7); (c) the Monod–Wyman–Changeux saturation equation (Eq. 3.39); and (d) the Koshland–Némethy–Filmer saturation equation (the four-subunit analog of Eq. 3.51). In all cases, the parameters for the curve fitting were obtained by nonlinear regression.

Since we have assumed that there are no intermediate hemoglobin–oxygen complexes, the conservation law for this system is $y_{tot} = y_0 + y_n$. Using this to eliminate the variable y_0 from Eq. 3.2 yields:

$$0 = k_1 x^n y_{tot} - k_1 x^n y_n - k_{-1} y_n \tag{3.3}$$

$$y_n\left(k_{-1} + k_1 x^n\right) = k_1 x^n y_{tot} \tag{3.4}$$

$$\frac{y_n}{y_{tot}} = \frac{k_1 x^n}{k_{-1} + k_1 x^n} \tag{3.5}$$

$$\frac{y_n}{y_{tot}} = \frac{x^n}{\frac{k_{-1}}{k_1} + x^n}. \tag{3.6}$$

If we introduce a new parameter K, where $K^n = k_{-1}/k_1$, we can rewrite the equation for the equilibrium fraction of y in the oxygen-bound state as:

$$\left(\frac{y_n}{y_{tot}}\right)_{eq} = \frac{x^n}{K^n + x^n}. \tag{3.7}$$

The advantage of this form is that K is equal to the *EC*50 (effective concentration-50), the concentration of x for which y achieves a 50%-maximal value. Equation 3.7 is the **Hill equation**, and the quantity on the left side—the fraction of the total hemoglobin molecules bound to n oxygen molecules—can be regarded as the oxygen saturation of the hemoglobin.

Although Hill derived the equation because of an interest in oxygen transport, not signal transduction, it is easy to see that a completely analogous equation could describe the simultaneous binding of n hormone molecules to a multimeric receptor or the simultaneous binding of n transcription factors to an enhancer sequence. A generic Hill equation for signal transduction would be:

$$\left(\frac{y}{y_{tot}}\right)_{eq} = \frac{x^n}{K^n + x^n}. \tag{3.8}$$

The exponent n is either called the **Hill coefficient** (the more common name) or the **Hill exponent** (the more correct name). For the binding of two EGF molecules (x) to the dimeric EGFR (y), the resulting Hill equation would be:

$$\left(\frac{c_{x_2y}}{y_{tot}}\right)_{eq} = \frac{x^2}{K^2 + x^2}, \tag{3.9}$$

where c_{x_2y} represents the concentration of dimeric EGFRs bound to two EGF molecules.

3.2 THE HILL EXPONENT IS A MEASURE OF HOW SWITCH-LIKE A SIGMOIDAL RESPONSE IS

One way to get a feel for the Hill equation is to plot it for various choices of the Hill exponent n. When $n = 1$, Eq. 3.9 is identical to the Langmuir equation (Eq. 2.8), which means that the response is hyperbolic and

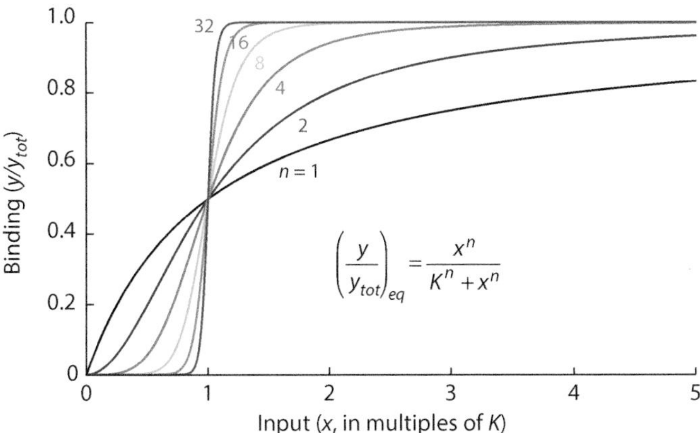

Figure 3.4 The Hill equation. Six equilibrium binding curves described by the equation $\left(\frac{y}{y_{tot}}\right)_{eq} = \frac{x^n}{K^n + x^n}$ are shown. As the Hill exponent increases, the binding becomes more all-or-none in character.

TABLE 3.1 **Fold-Change in Input Needed to Drive Output from 10%- to 90%-Maximal**

Hill Exponent (n)	Required Fold-Change in Input
1	81
2	9
3	4.33
4	3
8	1.73
16	1.32
32	1.15
64	1.07

Michaelian in character. Ligand binding increases linearly with x for small values of the stimulus x, and it gradually approaches a maximum of 1 as x becomes very large. Over the entire response range, the response curve is concave down (**Figure 3.4**). To drive the system from a 10%- to a 90%-maximal binding, one needs to increase the stimulus from $\frac{1}{9}K$ to $9K$, or 81-fold (**TABLE 3.1**).

When n is greater than one, the output is initially nth order in x and the binding curve is initially concave up. Overall, the curve is sigmoidal; it has an inflection point where it transitions from being concave up to concave down. The higher the value of n, the more steeply sigmoidal the curve becomes (**Figure 3.4**). With a Hill exponent of 2, it takes only a 9-fold-change (from $\frac{1}{3}K$ to $3K$) to drive the output from 10%- to 90%-maximal, and with a Hill exponent of 4, it takes only a 3-fold-change (**TABLE 3.1**). Once the Hill exponent reaches 46 (admittedly an astronomical number as Hill exponents go), a 10% change in input is enough to drive the output from 10%- to 90%-maximal. In general terms, the fold-change in input required to drive the output from 10%- to 90%-maximal is equal to $81^{1/n}$. Thus, in the context of signal transduction, high Hill exponents can make a response be highly switch-like. In the limit where n approaches infinity, the Hill curve approaches a step function.

3.3 THE HILL EQUATION ACCOUNTS FOR HEMOGLOBIN'S OXYGEN BINDING PRETTY WELL, BUT THE ASSUMPTIONS UNDERPINNING THE MODEL ARE DUBIOUS

For hemoglobin, with its four oxygen-binding subunits, the simultaneous binding of four oxygens should yield a binding curve with a Hill exponent of 4. In actuality, the Hill equation, with its two adjustable parameters n and K, can be fitted quite well to oxygen-binding data, but the Hill exponent obtained is around 2.7 (and the fitted value of K is 10.2 torr; **Figure 3.3b**). Certainly the fit is much better than that obtained with the one-parameter Langmuir equation. The best-fit Hill curve does deviate a bit from the experimental data at low oxygen tensions (**Figure 3.3b**), but overall the equation captures the character of the equilibrium binding.

The problem with the Hill model, of course, is its assumption that n molecules of ligand simultaneously interact with the multimeric protein. In reality, four oxygens (or even 2.7) will never collide exactly simultaneously with a hemoglobin molecule. It seems like a satisfactory theory would need to acknowledge that ligands may associate with and dissociate from their binding partner one at a time.

So why does the Hill equation, an equation that comes from an incorrect physical model, fit the data so well? The answer is that the more complicated sigmoidal binding equations that arise out of more physically reasonable models can usually be approximated by a Hill function—the Hill function serves as a pretty reasonable generic sigmoidal curve.

3.4 THE MORE-PLAUSIBLE MONOD–WYMAN–CHANGEUX (MWC) MODEL YIELDS SIGMOIDAL BINDING CURVES

What happens if we relax the Hill model's assumption of simultaneous ligand binding and assume that ligand molecules bind one at a time? Can we still get a sigmoidal binding curve like that seen experimentally for hemoglobin's oxygen binding? The answer is yes, and in fact there are a number of models that assume oxygen molecules (or, more generally, input ligands) bind one at a time to the tetrameric hemoglobin molecule (or, more generally, a multimeric receptor), the two best known of which are the **MWC model**, proposed in 1965, and the **Koshland–Némethy–Filmer (KNF) model**, proposed in 1966. Here we will start with the MWC model.

The basic idea behind the MWC model is shown in **Figure 3.5**. For the motivating example of the binding of oxygen to hemoglobin, it was assumed that four oxygen molecules bind one at a time to the hemoglobin tetramer. Hemoglobin is actually composed of two different subunit proteins, but for simplicity the two were considered equivalent. Next, it was assumed that the hemoglobin complex can exist in two distinct conformations, irrespective of whether oxygen is bound or not. In one conformation, shown in blue in **Figure 3.5**, the subunits bind oxygen with low affinity (this is customarily called the tense or T state); in the other (pink), they bind it with high affinity (the relaxed or R state). Monod and coworkers had already invoked this sort of two-conformation scheme in their work on the **allosteric** regulation (meaning regulation due to the binding of something to a site other than an enzyme's active site), and so it was natural to extend the idea

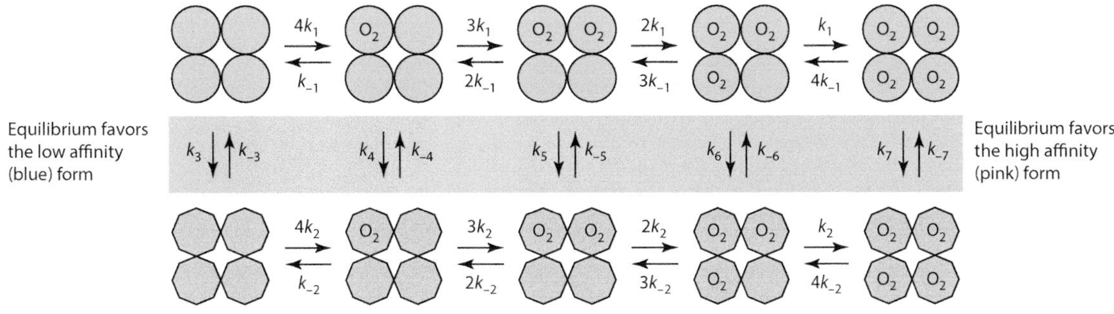

Figure 3.5 The Monod–Wyman–Changeux model for the binding of oxygen to hemoglobin. The blue hemoglobin molecules are in the low-affinity conformation (T state); the pink molecules are in the high-affinity conformation (R state). The rate constants k_1, k_{-1}, k_2, and k_{-2} are sometimes referred to as microscopic rate constants, whereas the expressions that include the number-of-ways factors, like $3k_1$ and $2k_{-1}$ for the second blue step, are termed macroscopic rate constants.

to hemoglobin. These assumptions—one-at-a-time oxygen binding and two preexisting, equilibrating conformations—are the conceptual core of the MWC model.

Note that there are four ways that the first oxygen molecule can bind to hemoglobin—it can bind to any one of the four hemoglobin subunits—but since the subunits are assumed to be identical, we can consider the four Hb-$(O_2)_1$ complexes equivalent. To account for the four ways of making the complex, we include a factor of 4 before the relevant rate constant (k_1) (**Figure 3.5**).

Next, to make the derivations a bit simpler, it was assumed that the four hemoglobin subunits flip in concert between the two conformations. Why assume the hemoglobin subunits flip together (**Figure 3.5**)? At the time the assumption may have seemed sort of magical, but we now know that oligomeric proteins are often, though not always, symmetrical, and it simplifies the algebra to consider only the symmetrical complexes. Since the subunits are assumed to flip in concert, the MWC model is sometimes referred to as the **concerted model**. A word of caution though—although the subunit flipping is assumed to be concerted, the oxygen-binding step is assumed to be separate from the subunit flipping, so in that sense this is a **sequential model**, a term that is usually reserved for the KNF model that we will discuss in the next section.

It was also assumed that the low-affinity conformation predominates when there are no oxygens bound to hemoglobin, but because the binding of oxygen to the high-affinity conformation is more energetically favorable, the equilibrium between the two conformations shifts more and more in favor of the high-affinity conformation as the number of oxygens bound to the hemoglobin increases. This is shown schematically in **Figure 3.5**.

To make the binding curve turn upward, one might think of assuming that the binding of one oxygen to the low-affinity (blue) conformation of hemoglobin would facilitate the binding of the next one to the same blue conformation. However, Monod, Wyman, and Changeux realized that such an assumption was not required. Instead they supposed that the equilibrium constants for the binding of oxygen to any of the low-affinity species (the blue species in **Figure 3.5**) were identical, and that the equilibrium constants for the binding of oxygen to any of the high-affinity species (pink) were identical. Even with this assumption, the first oxygen to bind would still promote the binding of the next oxygen, though indirectly. This is because once the first oxygen binds, the equilibrium between the low-affinity (blue) and high-affinity (pink) form shifts in favor of the high-affinity form, which in turn makes the

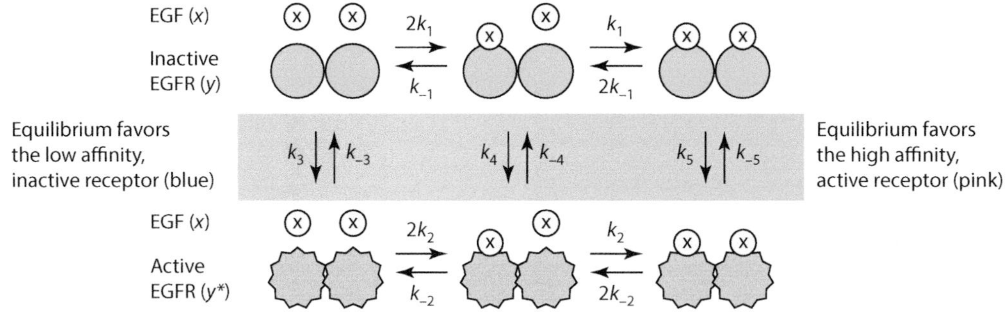

Figure 3.6 A Monod–Wyman–Changeux model for the activation of the EGF receptor by EGF.

binding of the next oxygen more energetically favorable (**Figure 3.5**). Likewise, each subsequent binding event shifts the equilibrium further in favor of the high-affinity form. This shift from one conformation to the other makes the binding of one oxygen molecule promote binding of the next, resulting in an upswing in the binding curve.

With these assumptions specified, we can now derive the MWC equation. But the algebra is simpler if one considers the binding of a ligand like EGF to a dimeric receptor like the EGF receptor (**Figure 3.6**), rather than the binding of oxygen to the tetrameric hemoglobin protein, so that is how we will begin.

First, we write rate equations for each of the six species shown in **Figure 3.6** and set each of the derivatives equal to zero, as they must be for the system to be in equilibrium. We call the low-affinity form of the dimeric receptor y and the high-affinity form y^* and we denote the number of EGF molecules (0, 1, or 2) bound to the receptor by a subscript. Finally, we simplify the system by noting that, as far as the equilibrium behavior is concerned, it does not matter whether we consider all three routes between the blue forms and the pink forms, or just one. By considering just one route—say the route between y_0 and y_0^*—we make the algebra a lot simpler. We then have the following algebraic equations:

$$0 = k_{-3}y_0^* - k_3y_0 - 2k_1x \cdot y_0 + k_{-1}y_1, \tag{3.10}$$

$$0 = 2k_1x \cdot y_0 - k_{-1}y_1 - k_1x \cdot y_1 + 2k_{-1}y_2, \tag{3.11}$$

$$0 = k_1x \cdot y_1 - 2k_{-1}y_2, \tag{3.12}$$

$$0 = k_3y_0 - k_{-3}y_0^* - 2k_2x \cdot y_0^* + k_{-2}y_1^*, \tag{3.13}$$

$$0 = 2k_2x \cdot y_0^* - k_{-2}y_1^* - k_2x \cdot y_1^* + 2k_{-2}y_2^*, \tag{3.14}$$

$$0 = k_2x \cdot y_1^* - 2k_{-2}y_2^*. \tag{3.15}$$

These equations can be simplified further by noting that Eq. 3.12 can be used to eliminate the last two terms from the Eq. 3.11; the resulting simplified version of Eq. 3.11 can then be used to eliminate the last two terms from Eq. 3.10. Likewise, Eq. 3.15 can be used to eliminate the last two terms from Eq. 3.14, and so on. Thus:

$$0 = k_{-3}y_0^* - k_3y_0, \tag{3.16}$$

$$0 = 2k_1x \cdot y_0 - k_{-1}y_1, \tag{3.17}$$

$$0 = k_1 x \cdot y_1 - 2k_{-1} y_2, \tag{3.18}$$

$$0 = k_3 y_0 - k_{-3} y_0^*, \tag{3.19}$$

$$0 = 2k_2 x \cdot y_0^* - k_{-2} y_1^*, \tag{3.20}$$

$$0 = k_2 x \cdot y_1^* - 2k_{-2} y_2^*. \tag{3.21}$$

This simplification is equivalent to saying that the system can be in equilibrium only if each pair of opposing arrows is balanced.

Note that Eqs. 3.16 and 3.19 are equivalent, so we really have only five independent linear equations that constrain the six variables (the y's and y^*'s). There is one more algebraic equation, the conservation equation:

$$y_{tot} = y_0 + y_1 + y_2 + y_0^* + y_1^* + y_2^*. \tag{3.22}$$

This provides us with enough constraints to solve for all of the six variables in terms of the rate constants and y_{tot}. We start by using Eq. 3.16 and solving for y_0:

$$y_0 = \frac{k_{-3}}{k_3} y_0^* = K_3 y_0^*, \tag{3.23}$$

where $K_3 = \frac{k_{-3}}{k_3}$, the equilibrium constant for the transition between the low- and high-affinity conformations of the free receptor. We can plug this result into Eq. 3.17 to derive an expression for y_1 in terms of y_0^*:

$$y_1 = 2\frac{x}{K_1} y_0 = 2k_3 \frac{x}{K_1} y_0^*. \tag{3.24}$$

Likewise for y_2:

$$y_2 = \frac{x}{2K_1} y_1 = K_3 \frac{x^2}{K_1^2} y_0^*. \tag{3.25}$$

For the active receptor species, we begin with Eq. 3.21 and rearrange it to solve for y_1^*:

$$y_1^* = 2\frac{x}{K_2} y_0^*. \tag{3.26}$$

Likewise for y_2^*:

$$y_2^* = \frac{x}{2K_2} y_1^* = \frac{x^2}{K_2^2} y_0^*. \tag{3.27}$$

We can now combine Eqs. 3.23–3.27 with the conservation equation (3.22) to produce expressions for each of the six forms of the EGFR dimer as a function of the EGF concentration. For example, the fraction of the receptor in the inactive, unbound form (y_0) is:

$$\left(\frac{y_0}{y_{tot}}\right)_{eq} = \frac{y_0}{y_0 + y_1 + y_2 + y_0^* + y_1^* + y_2^*}$$
$$= \frac{K_3}{K_3 + 2K_3 \frac{x}{K_1} + K_3 \frac{x^2}{K_1^2} + 1 + 2\frac{x}{K_2} + \frac{x^2}{K_2^2}}. \tag{3.28}$$

Note that there is a factor of y_0^* in both the numerator (from Eq. 3.23) and each term in the dominator (from Eqs. 3.23–3.27), which all cancel.

We can make this equation look a little simpler by factoring the first three and the last three terms in the denominator:

$$\left(\frac{y_0}{y_{tot}}\right)_{eq} = \frac{K_3}{K_3\left(1+\frac{x}{K_1}\right)^2+\left(1+\frac{x}{K_2}\right)^2}. \tag{3.29}$$

And we can quickly write down analogous terms for the rest of the species, each with a different numerator but the same denominator:

$$\left(\frac{y_1}{y_{tot}}\right)_{eq} = \frac{2K_3\frac{x}{K_1}}{K_3\left(1+\frac{x}{K_1}\right)^2+\left(1+\frac{x}{K_2}\right)^2}, \tag{3.30}$$

$$\left(\frac{y_2}{y_{tot}}\right)_{eq} = \frac{K_3\frac{x^2}{K_1^2}}{K_3\left(1+\frac{x}{K_1}\right)^2+\left(1+\frac{x}{K_2}\right)^2}, \tag{3.31}$$

$$\left(\frac{y_0^*}{y_{tot}}\right)_{eq} = \frac{1}{K_3\left(1+\frac{x}{K_1}\right)^2+\left(1+\frac{x}{K_2}\right)^2}, \tag{3.32}$$

$$\left(\frac{y_0^*}{y_{tot}}\right)_{eq} = \frac{2\frac{x}{K_2}}{K_3\left(1+\frac{x}{K_1}\right)^2+\left(1+\frac{x}{K_2}\right)^2}, \tag{3.33}$$

$$\left(\frac{y_2^*}{y_{tot}}\right)_{eq} = \frac{\frac{x^2}{K_2^2}}{K_3\left(1+\frac{x}{K_1}\right)^2+\left(1+\frac{x}{K_2}\right)^2}. \tag{3.34}$$

We can combine these equations to obtain theoretical expressions for the binding of EGF to the EGFR as a function of the EGF concentration (x), the equilibrium constants (K's), and the total receptor concentration (y_{tot}). We will express the binding as the fraction of the maximum of 2 moles of EGF per mole of EGFR—the fractional saturation of the receptor:

$$Saturation = \frac{1}{2}\frac{y_1+y_1^*}{y_{tot}} + \frac{y_2+y_2^*}{y_{tot}}, \tag{3.35}$$

$$Saturation = \frac{K_3\frac{x}{K_1}\left(1+\frac{x}{K_1}\right)+\frac{x}{K_2}\left(1+\frac{x}{K_2}\right)}{K_3\left(1+\frac{x}{K_1}\right)^2+\left(1+\frac{x}{K_2}\right)^2}. \tag{3.36}$$

If we assume that all of the y^* species are active and all of the y species are inactive, we can obtain a formula for the fractional receptor activity as a function of the EGF concentration x:

$$Activity = \frac{y_0^* + y_1^* + y_2^*}{y_{tot}} \tag{3.37}$$

$$Activity = \frac{\left(1+\frac{x}{K_2}\right)^2}{K_3\left(1+\frac{x}{K_1}\right)^2 + \left(1+\frac{x}{K_2}\right)^2}. \tag{3.38}$$

In general, the binding of a substrate x to an n subunit, two-state protein that obeys the MWC assumptions, is given by:

$$Saturation = \frac{K_3\frac{x}{K_1}\left(1+\frac{x}{K_1}\right)^{n-1} + \frac{x}{K_2}\left(1+\frac{x}{K_2}\right)^{n-1}}{K_3\left(1+\frac{x}{K_1}\right)^n + \left(1+\frac{x}{K_2}\right)^n}, \tag{3.39}$$

and the fractional activity is:

$$Activity = \frac{\left(1+\frac{x}{K_2}\right)^n}{K_3\left(1+\frac{x}{K_1}\right)^n + \left(1+\frac{x}{K_2}\right)^n}. \tag{3.40}$$

No matter how many subunits are present, the resulting equations have four adjustable parameters: the three equilibrium constants (K_1, K_2, and K_3), and the number of subunits n.

To get an idea of the behavior of the MWC model, let us look at how the concentrations of the six different EGFR species shown in **Figure 3.6** change as the concentration of EGF increases. We will assume that K_3, which describes the equilibrium between the unliganded EGFR in its loose-binding or inactive (blue) and tight-binding or active (pink) conformations, is a fairly large number (100), which means that ~99% of the EGFR will be in the blue conformation in the absence of ligand. Furthermore, we will assume that the pink form binds ligand 100× as tightly as the blue form, with $K_1=10$ and $K_2=0.1$.

Under these assumptions, as the concentration of EGF is increased, the amount of unliganded, inactive EGFR falls, and the amount of doubly bound, active EGFR increases (**Figure 3.7a**). At intermediate EGF concentrations, there is some singly bound EGFR as well, and for this particular choice of the equilibrium constants, half of the singly bound EGFR is active and half is inactive (**Figure 3.7a**). The concentrations of the other two EGFR species—the unliganded active receptor and the doubly bound inactive receptor—are negligible at all EGF concentrations for this choice of parameters and are not shown in the plot. Thus, EGF pushes the EGFR dimer from an unbound and inactive state to a doubly bound active state.

The next question concerns the shape of the binding curve and of the input/output relationship—do these curves look like hyperbolic, Michaelian curves, or like Hill curves, or like something else? As shown

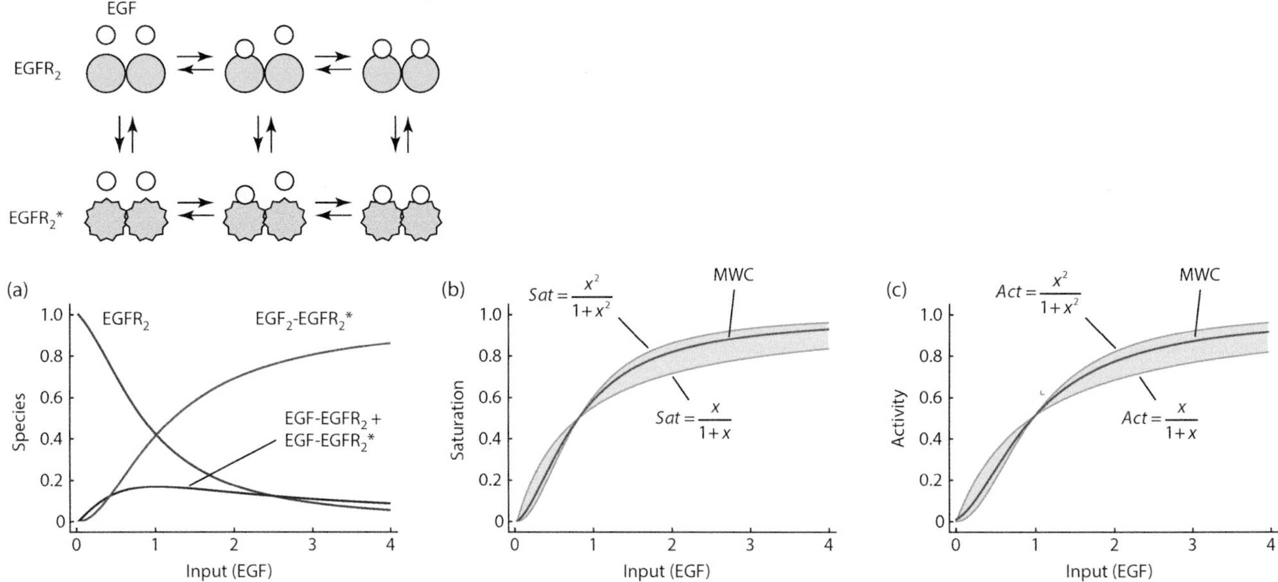

Figure 3.7 The binding of EGF to the EGFR and the activation of the EGFR according to the Monod–Wyman–Changeux model. (a) As the concentration of EGF increases, the dimeric receptor is converted from the inactive, unliganded form (blue curve), to the singly bound form, which for this choice of parameters includes equal concentrations of the active and inactive forms of the singly bound receptor (purple curve), and then to the doubly bound, active receptor (red curve). For the parameters chosen here, the concentrations of the active, unliganded form of the receptor and of the inactive, doubly bound form are negligible at all EGF concentrations. (b) Receptor saturation as a function of the EGF concentration. Hill curves with $n=1$ and 2 are shown for comparison. (c) Activity of the EGFR as a function of the EGF concentration. Again, Hill curves with $n=1$ and 2 are shown for comparison. For panel a and for the solid red MWC curves in panels b and c, it is assumed that $K_1=10$, $K_2=0.1$, and $K_3=100$. Regardless of the parameter choice, the saturation and activity curves will always lie between the $n=1$ and 2 Hill curves, the orange regions shown in panels b and c.

in **Figure 3.7b,c**, both curves are slightly sigmoidal, with half-maximal binding and half-maximal activation obtained when the concentration of EGF is equal to 1. In fact, for any choice of parameters, if one adopts concentration units such that 1 unit of EGF yields 50%-maximal binding and activation, then the binding curve will be guaranteed to fall somewhere in the shaded region between a Michaelian curve and an $n=2$ Hill curve (**Figure 3.7b,c**). The activity curve can be made to fall between the Michaelian and $n=2$ curves as well, provided one expresses the activity relative to its EGF=0 value (there will be some receptor in the active conformation even when no ligand is present, especially if one assumes K_3 is not too large) and its EGF $\rightarrow \infty$ value. Making both K_3 and the ratio between K_1 and K_2 large yields curves that approach the $n=2$ limit; making K_3 small and the ratio between K_1 and K_2 closer to 1 yields curves closer to the Michaelian limit. In between, the response curves are usually well approximated by Hill functions with a fractional Hill exponent between 1 and 2.

For the binding of ligands to a receptor with some other number of subunits, the saturation and activity curves will lie somewhere between a Michaelian curve and a Hill curve with a Hill exponent equal to the number of subunits.

3.5 THE MWC MODEL ACCOUNTS FOR THE BINDING OF OXYGEN TO HEMOGLOBIN, BUT NOT THE BINDING OF EGF TO THE EGFR

The problem that originally motivated the MWC model was the binding of oxygen to hemoglobin, and, gratifyingly, the MWC saturation equation can be fitted beautifully to the experimental data for the binding of oxygen to hemoglobin (**Figure 3.3c**). However, note that this agreement by itself does not really validate the model, because there are alternative models and equations that can be fitted just about as well to the data, including the Hill equation we already discussed and the KNF model that we will examine in the next section. Really the best way to test the model is to see how well the equilibrium constants inferred from the model agree with more direct measurements, for example, by freezing hemoglobin into one state or the other and measuring oxygen binding. The parameters for the fit shown in **Figure 3.3c** are: $K_1=67\pm5$ torr, $K_2=0.81\pm0.11$ torr (so that the high-affinity pink species shown in **Figure 3.5** bind oxygen ~83 times as tightly as the low-affinity blue species), and $K_3=22065\pm11177$ (so that in the unbound state, the equilibrium very strongly favors the low-affinity (blue) T state). And, in general, the more direct measurements of these parameters yield similar values. Thus, the original MWC model holds up quite well.

MWC-type models have been successfully applied to a variety of signaling processes as well as to hemoglobin's oxygen binding. For example, from patch-clamping experiments, we know that individual pentameric, ligand-gated ion channels generally flip between two states—one where the channel is fully closed and one where it is fully open—which likely correspond to the T- and the R-states in the MWC model, and it is easy to find parameters for the MWC equations (Eqs. 3.41 and 3.42) to account for the slightly sigmoidal response curves seen experimentally for these receptors.

So then does the MWC model—specifically, the MWC saturation equation—also account for the binding of EGF to the EGF receptor? The answer is no, or at least not always. In a variety of cell types, the binding of EGF to EGFR rises to half-maximal levels over ~nanomolar EGF concentrations, but then rises much more gradually, with binding not approaching saturation until micromolar EGF concentrations are used. This behavior is not always seen with purified recombinant EGFR preparations, but in one case—the interaction of the *Drosophila* EGF homolog Spitz with the extracellular domain of the *Drosophila* EGFR—it is (**Figure 3.8**). The MWC model cannot account for this behavior no matter what the equilibrium constants are assumed to be; the model always yields a binding curve that is somewhere between a Michaelian response (the most graded MWC behavior) and an $n=2$ Hill equation response (the most switch-like MWC behavior) (**Figure 3.8a**).

In contrast, the simple Hill equation does account reasonably well for the observed binding (**Figure 3.8b**), and the fitted Hill exponent turns out to be approximately 0.24. This is a bit perplexing. Recall that in Hill's model, the Hill exponent nominally represents the number of ligand molecules that simultaneously interact with the receptor; what it would physically mean to have a quarter of an EGF molecule interact with the dimeric receptor is unclear.

One simple way to account for the observed extremely graded binding curve is to assume that there is not just one homogenous receptor but two: one with a high-affinity for ligand and one with a low affinity. However, in the case of the *Drosophila* EGFR, the unbound extracellular domain crystallizes as a symmetrical dimer. This observation suggests that the complexes are homogeneous and that within the complexes the two binding sites are equivalent, at least initially.

Instead, it appears that the binding of the first EGF molecule to the *Drosophila* EGFR makes it harder for the second EGF molecule to bind. The unbound EGFR is symmetrical, but the ligand-bound receptor is

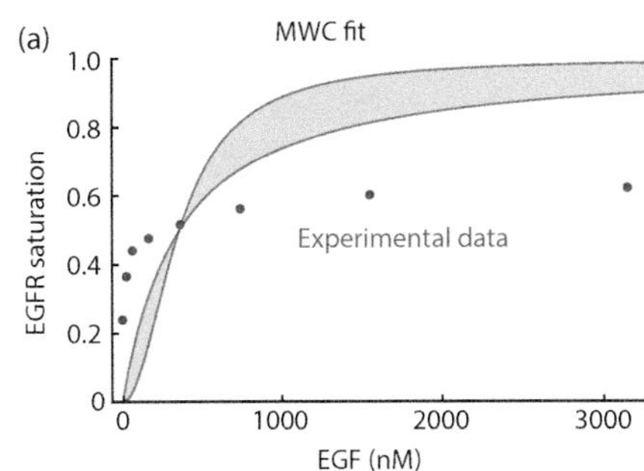

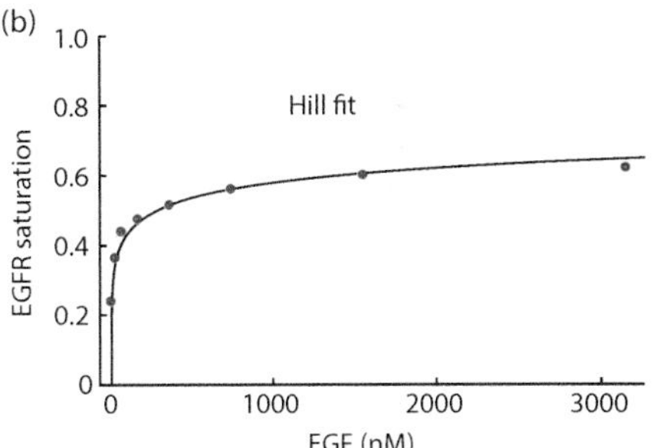

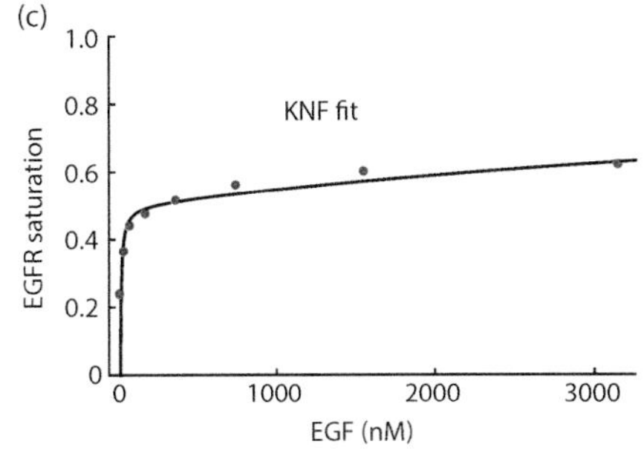

Figure 3.8 The binding of a *Drosophila* EGF protein to the EGFR extracellular domain in vitro. (a) The experimental data are not satisfactorily accounted for by a Monod–Wyman–Changeux model. (b, c) The experimental data can be accounted for by the Hill equation (b) or the Koshland–Némethy–Filmer model (c). For the Hill equation, the best fit is obtained when $n=0.24$ and the $EC50=3.8$ nM. For the KNF equation, the best fit is obtained when the first equilibrium constant (K_1) is 12 ± 2 nM and the second (K_2) is 4454 ± 1259 nM, which means that the binding exhibits high negative cooperativity ($K_1/K_2=0.0027$). The experimental data are taken from lvarado D, Klein DE, Lemmon MA. Structural basis for negative cooperativity in growth factor binding to an EGF receptor. *Cell*. 2010 Aug 20;142(4):568–79, with permission.

asymmetrical, suggesting that the binding of one EGF molecule has induced a conformation change and that the conformation change made the second site worse at binding ligand. One simple model that allows for either a positive or a negative interaction between two binding sites in a multimeric receptor protein was proposed by Koshland, Némethy, and Filmer shortly after the MWC model was proposed.

3.6 THE KNF MODEL CAN ACCOUNT FOR EITHER ULTRASENSITIVE OR SUBSENSITIVE BINDING

The KNF model harkens back to models proposed decades earlier by Adair and Pauling. Again, the motivating example was originally the binding of oxygen to hemoglobin (**Figure 3.9a**), and again we will first consider the binding of EGF to the EGFR because the algebra is simpler and because that is the example where we ran into trouble with the MWC model.

We assume that the interaction of EGFR with the first EGF molecule is characterized by an association rate constant k_1, a dissociation rate constant k_{-1}, and the usual number-of-ways factor (since there are two equivalent ways the first EGF molecule can be added). Next we assume that the binding immediately results in a conformation change in the bound receptor subunit (**Figure 3.9b**). Furthermore, we assume that this conformation change impacts on the neighboring subunits. This can make the second binding become more or less favorable than the first was, and hence we use a different pair of association and dissociation constants, k_2 and k_{-2} (again with a number-of-ways factor). For the EGFR, that is all there is to the KNF model (**Figure 3.9b**). For hemoglobin, each additional oxygen binds with its own set of rate constants (**Figure 3.9a**).

For the EGFR, there are three receptor species, with 0, 1, or 2 EGFs bound (denoted y_0, y_1, and y_2), and we can write three rate equations:

$$\frac{dy_0}{dt} = -2k_1x \cdot y_0 + k_{-1}y_1 \tag{3.41}$$

$$\frac{dy_1}{dt} = 2k_1x \cdot y_0 - k_{-1}y_1 - k_2x \cdot y_1 + 2k_{-2}y_2 \tag{3.42}$$

$$\frac{dy_2}{dt} = k_2x \cdot y_1 - 2k_{-2}y_2. \tag{3.43}$$

At equilibrium, each time derivative must equal zero:

$$0 = -2k_1x \cdot y_0 + k_{-1}y_1 \tag{3.44}$$

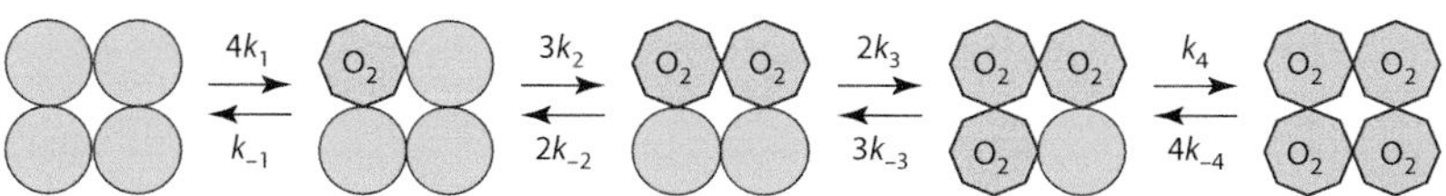

Figure 3.9 Koshland–Némethy–Filmer models for the binding of oxygen to hemoglobin (a) and the binding of EGF to the EGF receptor (b). Each ligand molecule induces a conformation change in the receptor subunit to which it binds. In turn this conformation change influences, either positively or negatively, the affinity of the other subunits for the ligand.

$$0 = 2k_1x \cdot y_0 - k_{-1}y_1 - k_2x \cdot y_1 + 2k_{-2}y_2 \tag{3.45}$$

$$0 = k_2x.y_1 - 2k_{-2}y_2. \tag{3.46}$$

From these three equations and the conservation relationship:

$$y_{tot} = y_0 + y_1 + y_2, \tag{3.47}$$

we can obtain solutions for each receptor species in terms of the concentration of free ligand (x), the total receptor concentration y_{tot}, and the rate constants:

$$\left(\frac{y_0}{y_{tot}}\right)_{eq} = \frac{K_1K_2}{K_1K_2 + 2K_2x + x^2} \tag{3.48}$$

$$\left(\frac{y_1}{y_{tot}}\right)_{eq} = \frac{2K_2x}{K_1K_2 + 2K_2x + x^2} \tag{3.49}$$

$$\left(\frac{y_2}{y_{tot}}\right)_{eq} = \frac{x^2}{K_1K_2 + 2K_2x + x^2}. \tag{3.50}$$

As usual, $K_1 = k_{-1} / k_1$ and $K_2 = k_{-2} / k_2$. The fractional receptor saturation is therefore:

$$Saturation = \frac{K_2x + x^2}{K_1K_2 + 2K_2x + x^2}. \tag{3.51}$$

The ratio of K_1 to K_2 is sometimes called c, the cooperativity of the binding. When $c>1$, the binding is positively cooperative, meaning that the first binding event makes the second event become more favorable. When $c<1$, the binding is negatively cooperative, meaning that the first event makes the second event become less favorable, and when $c=1$, the binding is noncooperative. Note that there are only two adjustable parameters for this model of binding of x to a dimeric protein, but for each additional subunit in a higher oligomer, there is one new parameter.

Figure 3.10 shows the saturation curves for various assumed values of c. If $c=1$ (the two equilibrium constants are equal), the result is a Michaelian binding curve (**Figure 3.10**, dashed black curve). If $c>1$, the result is a sigmoidal curve (**Figure 3.10**, blue curves), and the larger the value of c, the more steeply sigmoidal the curve is. In the

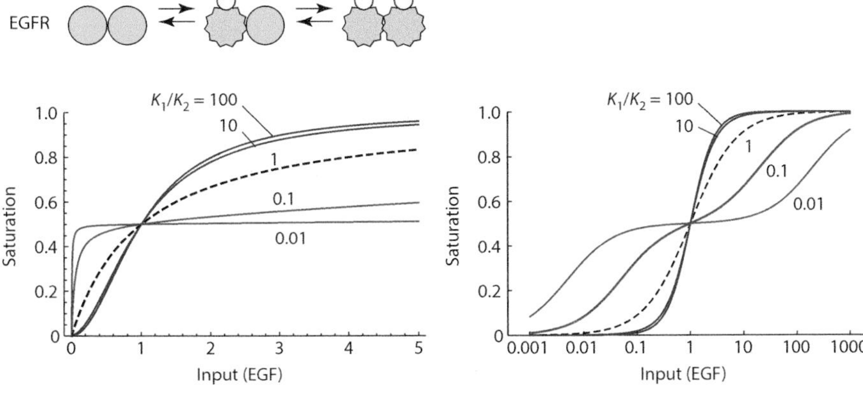

Figure 3.10 The binding of EGF to the EGFR according to the Koshland–Némethy–Filmer model. The saturation curves are shown on a regular linear plot (a) and on a semilog plot (b). If the two equilibrium constants are equal, the binding is Michaelian (dashed black curves). If $K_1>K_2$ (i.e., the second EGF molecule binds more strongly than the first), the saturation curve is ultrasensitive (blue curves). If $K_2>K_1$ (i.e., the first EGF molecule binds more strongly than the second), the saturation curve is subsensitive (red curves).

limit as $K_1 / K_2 \to \infty$, the saturation curve approaches a Hill function with $n=2$ and an $EC50$ of $\sqrt{K_1K_2}$.

If there is negative cooperativity and $c<1$—the first ligand-binding event makes the next event less favorable—the result is subsensitive binding, with a binding curve that is more graded than a Michaelian curve (**Figure 3.10**, red curves). In the limit where $K_1 / K_2 \to 0$, half of the receptor's subunits get filled as soon as the free EGF concentration reaches a nonzero value, and the remaining half never gets filled. Changing the affinities while keeping the value of c constant changes the $EC50$ but leaves the basic shape of the curve unaltered. On a linear plot this results in stretching or contracting the x-axis; on a semilog plot, the curve is shifted left or right.

Either an ultrasensitive or a subsensitive binding relationship can provide a signaling system with interesting properties. Ultrasensitivity makes the system more switch-like, with the first increments of stimulus being essentially ignored by the receptor, and then, once a threshold is exceeded, producing a decisive response. Subsensitivity trades some of this decisiveness for the ability to continue to respond differently to different stimulus concentrations at concentrations that would saturate an ultrasensitive system.

3.7 RESPONSE SENSITIVITY IS CUSTOMARILY DEFINED IN FOLD-CHANGE TERMS

At the beginning of this chapter we introduced the adjectives ultrasensitive and subsensitive to describe the shapes of the binding curves we have encountered in these cooperative systems. This terminology probably merits a little further discussion.

In common parlance, the term **sensitivity** usually refers to the minimum level of input that is required to produce a reliable output. Thus a highly sensitive (or ultrasensitive) assay is one that can detect small (or ultra-small) quantities or concentrations of a substance. In this sense of the word, the $EC50$ value of a system is an appropriate measure of sensitivity. The lower the $EC50$ value, the higher the sensitivity.

However, this is not what systems biologists usually mean by sensitivity. Instead of describing how much input is required to produce some level of output, they (we) are referring to how much of a change in input, measured in fold-change terms, is required to produce some given fold-change in output. This can be assessed either locally or globally. The most commonly used global measure of this type of sensitivity is the fold-change in an input that is required to drive a system from a 10%-maximal response to a 90%-maximal response, i.e., the $EC90/EC10$. For a Michaelian response given by:

$$y = \frac{x}{K+x}, \tag{3.52}$$

the $EC10$ is $\frac{1}{9}K$ (since $\frac{\frac{1}{9}K}{K+\frac{1}{9}K} = \frac{1}{10}$) and the $EC90$ is $9K$ (since $\frac{9K}{K+9K} = \frac{9}{10}$).

Therefore:

$$EC90 / EC10 = 81. \tag{3.53}$$

This number (81) can be thought of as a benchmark $EC90/EC10$ ratio. For an ultrasensitive response, the ratio of $EC90$ to $EC10$ will be less than 81. For example, for the $K_1/K_2=100$ binding curve in **Figure 3.10**, it turns out that the $EC10$ is about 0.29 and the $EC90$ is about 3.0, so that the ratio is about 9.2—less than 81, and not too far from the ratio for a Hill function with $n=2$ (which is 9; see **TABLE 3.1**). And for a subsensitive response, the ratio of $EC90$ to $EC10$ will be greater than 81. For example, for the $K_1/K_2=0.01$ binding curve in **Figure 3.10**, the ratio is an enormous number, about 640,000. The larger the $EC90/EC10$ ratio is, the smaller the sensitivity of the system.

Another common way of expressing global response sensitivity is by calculating the Hill exponent that would yield a Hill curve with the same $EC90/EC10$ ratio as the given response function. With a little algebra, one can show that this **effective Hill exponent** is given by:

$$n = \frac{\mathrm{Log}_{10}[81]}{\mathrm{Log}_{10}[EC90 / EC10]}. \tag{3.54}$$

For the ultrasensitive case where $K_1/K_2=100$, the effective Hill exponent n is 1.98; for the subsensitive case where $K_1/K_2=0.01$, $n=0.33$. The effective Hill exponent is a particularly reasonable gauge of the switch-like character of a response when the response curves look at least qualitatively like Hill curves, as MWC and KNF curves invariably do. When this is not the case, this metric may not be so good. For example, for a linear response the $EC90/EC10$ is 9 and the effective Hill exponent is approximately 2, but an $n=2$ Hill curve does not look much like a linear response. In general it is probably best to confine the use of the terms ultrasensitivity, subsensitivity, and effective Hill exponents to situations where the response looks like a Hill response.

Of course, a response is not equally switch-like or graded throughout its entire range. For a Michaelian response or a Hill function, the response becomes progressively more graded as the input increases. It therefore makes sense to define a local measure of the fold-change of output per fold-change of input; for example:

$$S = \frac{\Delta Output / Output}{\Delta Input / Input}. \tag{3.55}$$

S is sometimes referred to as the local fold-sensitivity, but it is more often just called sensitivity. Note that S is a function rather than a single number; its value varies with *Input*. Its value also depends on how large of a fold-change in *Input* one is talking about. Typically the fold-change is chosen to be infinitesimal, and the sensitivity function therefore becomes:

$$S = \frac{dOutput / Input}{dInput / Output}. \tag{3.56}$$

Note that since $\frac{d\ln Output}{dOutput} = \frac{1}{Output}$ and $\frac{d\ln Input}{dInput} = \frac{1}{Input}$, it follows that:

$$S = \frac{d\ln Output}{d\ln Input}, \tag{3.57}$$

and one sees Eq. 3.57 fairly frequently in the systems biology literature. The higher the value of S, the more switch-like or decisive the response is at that value of *Input*; the lower the value of S, the more graded the response.

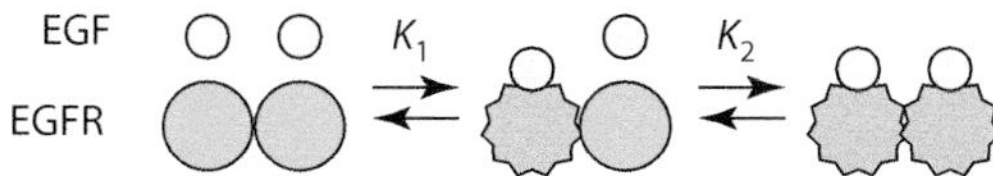

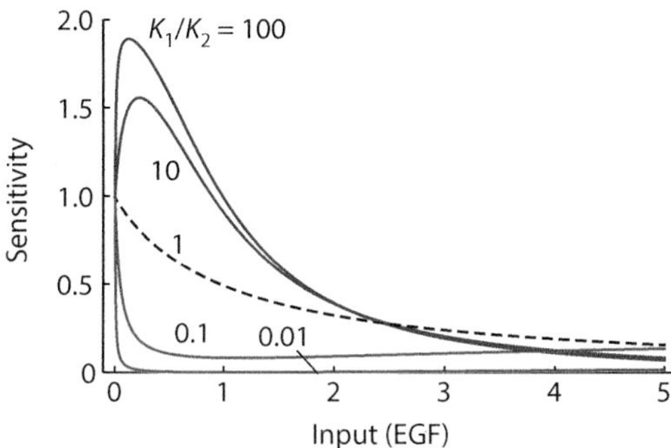

Figure 3.11 The sensitivity of a KNF-binding response assuming positive cooperativity (blue curves), negative cooperativity (red curves), or noncooperative binding (dashed black curves).

Equation 3.57 shows that the sensitivity *S* is the slope of a log–log plot of the response function, which is also the same as the polynomial order of the response. A Michaelian response is first order when the input is infinitesimal, is 0.5 order when the input equals the *EC*50, and approaches zero order when the input is very large. For a Hill function, the sensitivity is *n*th order when the input is small, is *n*/2 order when the input equals the *EC*50, and again approaches zero order when the input is large. The sensitivities for the KNF saturation curves in **Figure 3.10** are shown in **Figure 3.11**. The sensitivities are always less than *n*, and the highest sensitivities are achieved when the inputs and outputs are relatively small. Thus, there is a trade-off between how large and how decisive a KNF response can be.

3.8 THE RELATIONSHIP BETWEEN BINDING AND ACTIVATION YIELDS A VARIETY OF POSSIBLE RESPONSES

In applying the MWC model to a signaling protein-like a receptor, it is natural to assume that one conformation or the other—say, the tight-binding conformation—is the active form of the receptor (**Figure 3.7**). In the KNF model, one plausible hypothesis would be that the induced conformation change produced by the binding of the first ligand results in the activation of one receptor subunit, and the binding of the second ligand results in the activation of the second subunit (and so on if the receptor is bigger than a dimer). If this independent activation mechanism is applicable, the activity curves would be identical to the saturation curves and would vary from subsensitive to ultrasensitive depending on the value of K_1/K_2 (**Figure 3.12a**):

$$Activity = \frac{K_2x + x^2}{K_1K_2 + 2K_2x + x^2}. \tag{3.58}$$

However, independent activation is not the only way that ligand binding might be coupled to receptor activation, and it may not be the way the EGFR works. From structural studies, it is known that the binding of EGF to the extracellular ligand-binding domain results in a symmetry-breaking change in the conformation of the intracellular part of the receptor, with one kinase domain allosterically activating the other. This results in the autophosphorylation of one domain by the other, and the autophosphorylated domain then transmits the activation signal downstream by recruiting phosphoepitope-binding

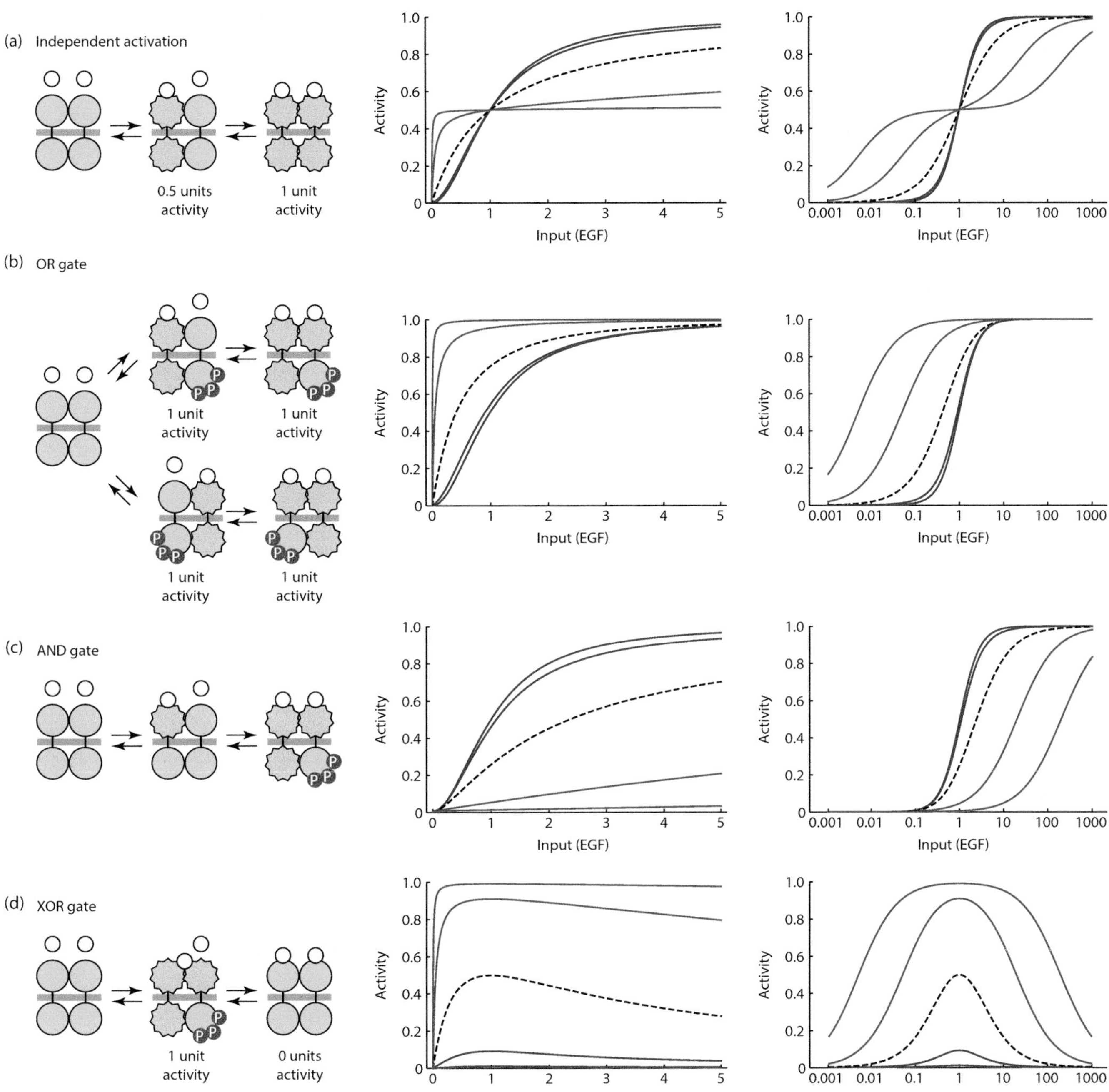

Figure 3.12 The relationship between ligand binding and receptor activation in four variations on the KNF model. (a) Independent activation. Each binding event activates one receptor molecule. (b) An OR gate response. Maximal activation (which could be either two receptors activated, as shown, or one) is assumed to result from the binding of the first ligand molecule. (c) An AND gate response. The singly bound species is assumed to be inactive and the doubly bound species active. (d) An XOR (exclusive OR) gate response. The first ligand molecule is assumed to activate the receptor maximally (activating either both receptors, as shown, or just one), and the second ligand molecule inactivates it. In all cases, the activities are plotted for 5 choices of K_1 and K_2: 100 and 0.01 (red); 10 and 0.1 (red); 1 and 1 (dashed black); 0.1 and 10 (blue); and 0.01 and 100 (blue). The activation curves are shown on regular linear plots (left) and on semilog plots (right).

proteins. One could imagine then that the binding of a single molecule of EGF results in a maximal response, the autophosphorylation of one intracellular domain, with the binding of a second EGF molecule having no further effect. This would be OR gate logic (**TABLE 3.2**), in the lexicon of electronic circuits and Boolean algebra; a binary switch from off to on is achieved by the binding of one molecule of EGF to either receptor subunit (**Figure 3.12b**).

TABLE 3.2 **Some Logic Gates Relevant to Dimeric Signaling Proteins**

Term	Input/Output Relationship
OR gate	No ligand → OFF Ligand bound to subunit 1 → ON Ligand bound to subunit 2 → ON Ligand bound to both → ON
AND gate	No ligand → OFF Ligand bound to subunit 1 → OFF Ligand bound to subunit 2 → OFF Ligand bound to both → ON
XOR gate	No ligand → OFF Ligand bound to subunit 1 → ON Ligand bound to subunit 2 → ON Ligand bound to both → OFF

$$Activity = \frac{2K_2x + x^2}{K_1K_2 + 2K_2x + x^2}. \tag{3.59}$$

One could also imagine that a single molecule of EGF results in the autophosphorylation of both intracellular domains; perhaps the two kinase domains would take turns allosterically activating each other. This would still be OR gate logic, and the shape of the activity curve would be the same.

Figure 3.12b shows Eq. 3.59 plotted for various ratios of K_1 to K_2. For the red curves there is negative cooperativity in the binding, and for the blue curves, positive cooperativity. The *EC*50 of the response changes as the ratio of K_1 to K_2 is varied, which makes it perhaps easier to appreciate the intrinsic shapes of the response curves on a semilog plot (**Figure 3.12b**, right). The curves range from Michaelian, when there is strong negative cooperativity in the binding, to ultrasensitive (when there is strong positive cooperativity). Surprisingly, perhaps, subsensitive responses are not possible.

One could also imagine that two EGFs must bind before the intracellular part of the receptor becomes activated and autophosphorylated. Such a system would behave like a binary AND gate (**Figure 3.12c**; **TABLE 3.2**), and the activity would be:

$$Activity = \frac{x^2}{K_1K_2 + 2K_2x + x^2}. \tag{3.60}$$

Again the result is a sigmoidal curve intermediate in shape between an $n=1$ and an $n=2$ curve. In general, for a KNF activation process where a protein with n subunits must bind n ligand molecules to be activated, the equilibrium activity is given by:

$$\begin{aligned} Activity &= \left(\frac{y_n}{y_{tot}}\right)_{eq} \\ &= \frac{x^n}{\binom{n}{0}K_1K_2K_3\cdots K_n + \binom{n}{1}K_2K_3\cdots K_nx + \binom{n}{2}K_2K_3\cdots K_nx + \cdots \binom{n}{n}x^n}, \end{aligned} \tag{3.61}$$

where the coefficients in the denominator of the form $\binom{n}{i}$ represent the combinations of n things taken i at a time, as given by $\binom{n}{i} = \frac{n!}{i!(n-i)!}$, and are the binomial coefficients from the nth row of Pascal's triangle.

There are of course other possible relationships between binding and activation, and we will consider one more here—not for its relevance to EGFR signaling but for its possible significance in other situations. For example, growth hormone exerts its effects through a dimeric transmembrane receptor, the growth hormone (GH) receptor. The GH receptor, which does not possess an enzymatic domain, then activates an associated tyrosine kinase, a Jak protein. At low concentrations, one molecule of GH binds and dimerizes two molecules of the receptor, with each receptor subunit providing half of the ligand binding site, resulting in a binding stoichiometry of 1:2 (ligand to receptor). In principle, it should be possible for very high concentrations of GH to bind 2:2 to the receptor, with each molecule engaging what would normally be half of a binding site (**Figure 3.12d**). It is not clear that this occurs, but for the sake of argument let us assume that it does.

Next, suppose that the 1:2 complex is active but the 2:2 complex is inactive. The result would be a biphasic binding curve (**Figure 3.12d**). The receptor acts like an exclusive OR gate (XOR gate) (TABLE 3.2), turning on when bound to one ligand molecule and turning back off when bound to two. The ratio of the two binding constants determines how high and broad the activity peak is. If there is strong positive cooperativity in the binding ($K_1 >> K_2$), hardly any mono-ligated, active receptor will be produced at any concentration of GH, because the receptor tends to skip from unligated to doubly ligated. On the other hand, if there is strong negative cooperativity—which seems plausible here—then the peak equilibrium receptor activation will be expected to show a high, broad concentration peak (**Figure 3.12d**, red curves).

This type of mechanism may or may not be important in GH signaling, but something very similar does pop up in the regulation of lysis vs. lysogeny in *Escherichia coli* infected with bacteriophage λ. The relevant ligand is the lambda repressor protein λcI, which actually can either activate or repress transcription, and the relevant targets are DNA sequences upstream and downstream of the λcI coding sequence. The binding of the first four λcI dimers promotes the recruitment of RNA polymerase and so activates transcription—in this concentration regime, the lambda repressor is actually a transcriptional activator. But adding two more λcI dimers turns the gene back off by blocking RNA polymerase binding.

This sort of biphasic response is important in diverse biological contexts. It constrains the behavior of adaptor and scaffold proteins; it underlies the so-called prozone effect in antibody–antigen interactions; it is thought to account for the phenomenon of transcriptional squelching; and it is critical to the formation of some microdomains and separated phases in the plasma membrane and cytoplasm.

SUMMARY

Many receptors, including receptor tyrosine kinases, are multimeric, and this can generate a variety of possible behaviors. The ligand binding curves can become sigmoidal rather than hyperbolic, through positively cooperative binding, where the first molecule of ligand to bind makes the next binding event more favorable. The MWC model, which is built on the idea of two interconverting conformation states in the receptor and the preferential binding of the ligand to one of the two, can account for such sigmoidal binding curves and so can the KNF model, which assumes that the first binding event causes the second binding site to have a higher (for positive cooperativity) or lower (for negative cooperativity) affinity for ligand. Both the MWC and KNF models can account for the equilibrium binding of oxygen to hemoglobin, but only the KNF model can account for the binding of EGF to the *Drosophila* EGFR, which bears all the hallmarks of negative cooperativity. Positive cooperativity can make receptor activation more switch-like and decisive; negative cooperativity can make receptor activation less decisive but allows the receptor to operate over a greater range of ligand concentrations without becoming saturated.

There is one important caveat to all of this: we have assumed that the ligand molecules are present in huge excess over the receptor molecules, so that the production of a ligand–receptor complex depletes the available pool of receptors but has no significant effect on the free ligand. This is almost always true in ligand-binding studies carried out in vitro. But it is not always true in vivo, and it is almost never true when one examines the interactions between two intracellular signaling proteins (as opposed to an extracellular ligand and a cell surface receptor). In the next three chapters, we will turn to the elementary responses of intracellular signaling proteins, beginning with stoichiometric regulation processes where the upstream protein is not in infinite supply.

FURTHER READING

COOPERATIVITY

Adair GS. The hemoglobin system. VI. The oxygen dissociation curve of hemoglobin. J Biol Chem. 1925;63:529–45.

Alvarado D. Klein DE, Lemmon MA. Structural basis for negative cooperativity in growth factor binding to an EGF receptor. Cell. 2010;142:568–79.

Eaton WA, Henry ER, Hofrichter J, Mozzarelli A. Is cooperative oxygen binding by hemoglobin really understood? Nat Struct Biol. 1999;6:351–8.

Hill AV. The possible effects of the aggregation of the molecules of haemoglobin on its dissociation curves. J Physiol. 1910;40: Proceedings, iv–vii.

Koshland DE Jr, Némethy G, Filmer D. Comparison of experimental binding data and theoretical models in proteins containing subunits. Biochemistry. 1966;5:365–85.

Levitzki A, Stallcup WB, Koshland DE Jr. Half-of-the-sites reactivity and the conformational states of cytidine triphosphate synthetase. Biochemistry. 1971;10:3371–8.

Monod J, Wyman J, Changeux JP. On the nature of allosteric transitions: a plausible model. J Mol Biol. 1965;12:88–118.

Pauling L. The oxygen equilibrium of hemoglobin and its structural interpretation. Proc Natl Acad Sci USA. 1935;21:186–91.

DOWNSTREAM SIGNALING 1

Stoichiometric Regulation

4

IN THIS CHAPTER . . .

The signaling pathways downstream of receptors may be dauntingly complicated, but even the most complicated signaling pathways, such as the epidermal growth factor receptor (EGFR) pathway discussed in Chapter 1, are built out of a handful of elementary processes. The first such process is stoichiometric activation, where an upstream regulator binds to, and thereby activates, a downstream target (**Figure 4.1a**)—exactly the same type of thing as happens in receptor–ligand interactions, only with intracellular components. Examples include the activation of adenylyl cyclase by a trimeric G-protein α-subunit or activation of Raf by Ras. Then there are various ways through

DOI: 10.1201/9781003124269-4

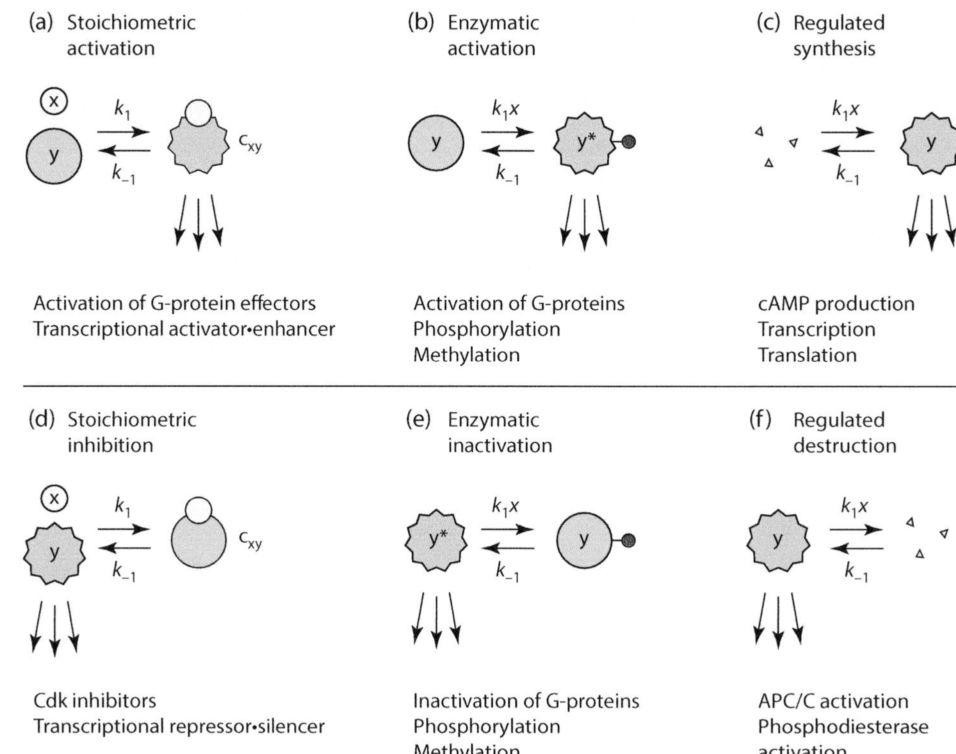

Figure 4.1 Six types of elementary processes in signal transduction.

which a protein can be activated by an upstream enzyme rather than a stoichiometric regulator. Examples include the activation of a G-protein α-subunit by the guanine nucleotide exchange activity of an activated GPCR or the activation of a MAP kinase by a MAP kinase kinase (**Figure 4.1b**). The downstream proteins are being enzymatically "marked," by GTP binding and phosphorylation, respectively, by the upstream regulatory enzyme. Finally, there is regulated production, with an upstream signaling enzyme making something rather than activating something. Examples include the regulated synthesis of cAMP, mRNAs, or proteins (**Figure 4.1c**). There are also many examples of the reverse of these schemes—stoichiometric inhibition, enzymatic inhibition, and regulated destruction (**Figure 4.1d–f**). Moreover, there are a few processes that do not fit neatly into one of these categories—we will encounter one, termed state-dependent inactivation, when we examine the phenomenon of adaptation in Chapter 13—but stoichiometric regulation, enzymatic regulation, and regulated production or destruction constitute a large fraction of the basic reactions of signal transduction.

In the next three chapters we will model each of these three elementary processes and explore the models. Along the way we will encounter the now-familiar exponential approach to equilibrium or steady state, as well as hyperbolic or Michaelian steady-state responses, but we will discover a number of other types of response as well, including ultrasensitive responses that arise from mechanisms other than cooperativity. We begin with what is probably the simplest extension of the receptor–ligand interactions we examined in Chapters 2 and 3: stoichiometric regulation, but this time occurring between intracellular components instead of through the binding of an extracellular ligand to a receptor.

STOICHIOMETRIC REGULATION INSIDE THE CELL

4.1 IN THE HIGH-AFFINITY LIMIT, DOES A HYPERBOLIC RESPONSE MAKE INTUITIVE SENSE?

For the interaction of a monomeric regulator with a monomeric target within the cell, one might think that we can just use the formulas derived in Chapter 2 for ligand–receptor interaction and change what we mean by x and y—say, make x be the Ras protein and y be its effector Raf-1. However, there is a catch, and it can be illustrated by thinking about ligand–receptor interactions in the high-affinity limit.

Suppose that we have a ligand x that regulates a receptor y and suppose they have a very high affinity for each other. Assume that a cell possesses, say, 1,000 molecules of y. What would happen as you incrementally added x to the system?

If the affinity really is very high, then the first molecule of x will yield one molecule of the ligand–receptor complex c_{xy}; 10 molecules of x will yield 10 c_{xy} complexes; 100 molecules of x will yield 100 c_{xy} complexes; and so on, until all 1,000 y molecules have been used up. Thus the expected input/output relationship will be a straight line, with slope 1, that abruptly plateaus when the amount of added x equals the total amount of y. Simple enough.

But that is not what Eq. 2.8 indicated and it is not what the plots in **Figure 2.4** showed. No matter how high the affinity, those response curves were hyperbolas that approached their maxima gradually, not straight lines that slammed into a response ceiling. What accounts for this discrepancy?

The answer is that to derive Eq. 2.8, we assumed that the ligand was in huge excess over the receptor, which means that even when the receptor was fully occupied, the depletion of x was insignificant and $x_{tot} \approx x$. But this assumption is not true in our thought experiment. In fact, until the total concentration of x exceeds y_{tot}, all of the free x is depleted by receptor binding and so $x = 0$, which is very different from $x = x_{tot}$.

This is more than just a hypothetical concern. Intracellular regulators are often similar to or even lower in abundance than the targets that they stoichiometrically regulate. So we need to derive an expression that acknowledges that both x_{tot} and y_{tot} can be depleted by complex formation.

4.2 THE EQUILIBRIUM RESPONSE CHANGES FROM HYPERBOLIC TO LINEAR WHEN DEPLETION OF THE UPSTREAM REGULATOR IS NOT NEGLIGIBLE

We start again with the rate equation for the formation of a complex between an upstream regulator x and its downstream target y—the same rate equation we used for ligand–receptor interaction (Eq. 2.1):

$$\frac{dc_{xy}}{dt} = k_1 x \cdot y - k_{-1} c_{xy}.$$

Next we use two conservation relationships to eliminate x and y, yielding:

$$\frac{dc_{xy}}{dt} = k_1\left(x_{tot} - c_{xy}\right)\cdot\left(y_{tot} - c_{xy}\right) - k_{-1}c_{xy}. \tag{4.1}$$

Expanding the right-hand side gives us:

$$\frac{dc_{xy}}{dt} = k_1 c_{xy}^2 - \left(k_1\left(x_{tot} + y_{tot}\right) + k_{-1}\right)c_{xy} + k_1 x_{tot} y_{tot}. \tag{4.2}$$

At equilibrium, the derivative must equal zero. The result is a quadratic equation in c_{xy}:

$$0 = k_1 c_{xy}^2 - \left(k_1\left(x_{tot} + y_{tot}\right) + k_{-1}\right)c_{xy} + k_1 x_{tot} y_{tot}. \tag{4.3}$$

This equation can be simplified a bit by dividing each term by k_1 and making use of the fact that $K_{eq} = k_{-1}/k_1$:

$$0 = (c_{xy})^2 - ((x_{tot} + y_{tot}) + K_{eq})c_{xy} + x_{tot} y_{to} \tag{4.4}$$

This emphasizes that it is the ratio of the rate constants—the equilibrium constant—rather than the rate constants' individual values that determines the equilibrium concentration of the c_{xy} complex. Eq. 4.4 can be solved with the quadratic formula, which yields two solutions:

$$c_{xy} = \frac{1}{2}\left(K_{eq} + x_{tot} + y_{tot} - \sqrt{\left(K_{eq} + x_{tot} + y_{tot}\right)^2 - 4x_{tot}y_{tot}}\right) \tag{4.5}$$

$$c_{xy} = \frac{1}{2}\left(K_{eq} + x_{tot} + y_{tot} + \sqrt{\left(K_{eq} + x_{tot} + y_{tot}\right)^2 - 4x_{tot}y_{tot}}\right). \tag{4.6}$$

The physically relevant solution is Eq. 4.5, which you can see by plotting both equations for some choice of K_{eq} and y_{tot} and asking whether the values of c_{xy} are physically possible or not—e.g., is c_{xy} positive as it must be and is it less than x_{tot} and y_{tot}?

If we take c_{xy} as the output and x_{tot} as the input, there are two adjustable parameters, K_{eq} and y_{tot}; we will assume that $y_{tot} = 1$ and look at a range of values for K_{eq}. As shown in **Figure 4.2a**, when $K_{eq} = 10$, the response given by Eq. 4.5 (solid red curve) is very similar to a Michaelian response (dashed black curve). This makes sense; when K_{eq} is large, it takes a large concentration (~10 units) of x_{tot} to half-maximally saturate the 1 unit of y_{tot}, and so the concentration of c_{xy} is a reasonably small fraction of x_{tot}, which means that $x_{tot} \approx x$ However, when K_{eq} is decreased to 1 and then 0.1, the approximation breaks down, and the discrepancy between the exact response (solid curves) and the Michaelian response (dashed curves) becomes greater and greater. As the assumed affinity increases further (K_{eq} decreases further), the response curve appears to approach a straight with slope = 1 that does not begin to bend over until the response is nearly maximal (**Figure 4.2b**). This is what we thought should be the behavior of the system in the high-affinity limit, and we have now analytically shown that this is in fact the case.

In the Michaelian (low affinity) limit, the stoichiometric regulation system does what a gentle audio compressor does, squeezing a range of inputs into a smaller range of outputs throughout the response range.

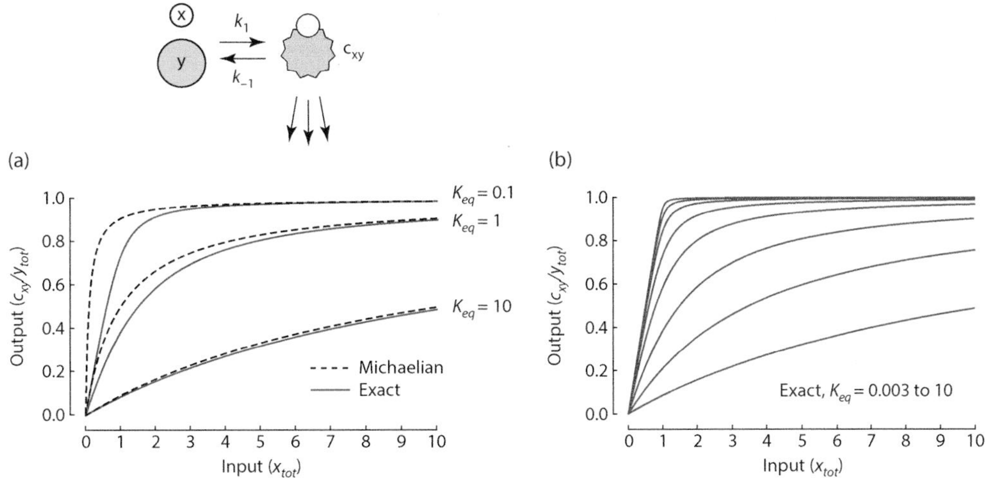

Figure 4.2 Deviation from the Michaelian response in the high-affinity limit. (a) For various values of the equilibrium constant, the exact equilibrium binding responses are shown as solid curves and the corresponding Michaelian responses are shown as dashed curves. The assumed value of y_{tot} is 1. (b) In the high-affinity limit, the response approaches a straight line of slope = 1. Responses are shown for K_{eq} = 0.003, 0.01, 0.03, 0.1, 0.3, 1, 3, and 10, and y_{tot} = 1 (left to right).

However, in the high-affinity limit, the system acts like a hard limiter, yielding a linear response up until the maximal possible response, which is set by the number of target molecules y_{tot}, is achieved.

4.3 THE DYNAMICAL RESPONSE IS SIMILAR EVEN WHEN THE DEPLETION OF THE UPSTREAM REGULATOR IS NOT NEGLIGIBLE

So the depletion of a ligand x by high-affinity binding to a target y can make the equilibrium response of the system be quite different from a Michaelian response (**Figure 4.2**). What about the dynamics of the system? Can we analytically solve the relevant rate equation (Eq. 4.2)? If so, does the time course look much different from exponential approach to equilibrium?

It turns out that it is still possible to analytically solve the rate equation, at least with the help of the differential equation solver in *Mathematica*, but the resulting analytical expression is pages long and not worth displaying here. Even if it were not possible to solve the ODE analytically, one can solve it numerically. Either way, we can plot the results and compare them to the exponential approach to equilibrium. **Figure 4.3** shows the time course for a case where the equilibrium constant K_{eq} is 10× lower than the concentrations of y_{tot} (K_{eq} = 0.1 and y_{tot} 1), with the input (x_{tot}) being stepped up from 0 to 1 and then stepped back by washing away the free x and letting c_{xy} and x_{tot} decay down to zero. With these parameters, the depletion of the added ligand x by complex formation is substantial—approximately 73% of the x_{tot} is complexed at equilibrium. The resulting time course (**Figure 4.3**, red curve) starts out like the exponential approach curve (blue curve), which makes sense since initially the depletion of x will be minimal. The system then gradually approaches an equilibrium that is somewhat lower than it would have been without the depletion. The shape of the red curve is not exactly the same as that of an exponential curve, but it is not that different either (**Figure 4.3**). When the ligand is washed away, the system returns back to c_{xy} = 0, as expected, and once again the system takes longer to respond to the step down than it did to the step up (**Figure 4.3**).

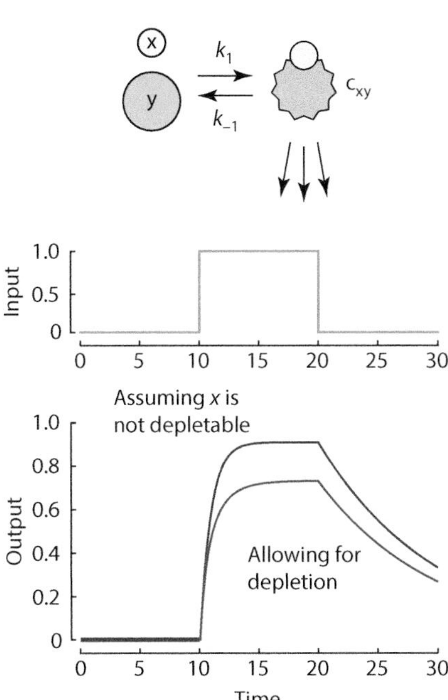

Figure 4.3 The approach to equilibrium is qualitatively similar even when the concentration of xy (c_{xy}) is comparable to the concentration of x. Here we have assumed that k_1 = 1, y_{tot} = 1, and k_{-1} = 0.1. The step up in input is an increase in x_{tot} (for the red curve) or x (for the blue curve). The step back down is a decrease in x for both.

4.4 LIGAND DEPLETION PLUS NEGATIVE COOPERATIVITY CAN PRODUCE A THRESHOLD

In Chapter 3, when we discussed the various models for cooperative binding and activation—Hill, MWC, and KNF—we also tacitly assumed that the ligand was present in great excess over the receptor. As we mentioned, this is certainly true for most experiments where dishes of cells or recombinant receptors are incubated with large volumes of buffer containing the ligand. But just as this assumption is generally not true for intracellular signaling, it is probably not true for EGFR—the receptor we used as a prototype for a multimeric receptor—either. The ligands for EGFR—epidermal growth factor (EGF), TGFα, and five other less abundant EGFR ligands (**Figure 1.3**)—are typically provided locally, released through the proteolysis of precursor proteins from the surface of one cell and then spread by diffusion or flow to allow interaction with receptors on nearby cells in, say, an epithelial sheet. So for the ligand concentration to greatly exceed [EGFR], one would expect to find at least some cell type where the ligands outnumber the receptors. So far, no such cells have been found. The same is true in the whole organism. If you tally the total number of EGFR and EGFR ligand molecules in the human, mouse, or fruit fly, there is more receptor than ligand. This means that we should consider how ligand depletion might affect the responses of cooperative receptors, as we just did for a monomeric-ligand monomeric-receptor system.

Ligand depletion would be most significant if the binding affinity were extremely high, so let us carry out a thought experiment for what we would expect in this high-affinity limit. If we start with no ligand and dial up its concentration, in the high-affinity limit we will end up with more and more ligand–receptor complexes until the concentration of the ligand exceeds that of the receptor. This is shown in both schemes in **Figure 4.4a,b**. Next, let us suppose that the receptor is activated via a KNF mechanism, and there is AND gate logic—that is, two ligand molecules must bind for the receptor to become active. If there is strong positive cooperativity in the binding, the first and second ligands will bind the same receptor dimer. Every two ligand molecules will result in the activation of one receptor (**Figure 4.4a**), yielding a linear input–output relationship (**Figure 4.4c**, blue) rather than a

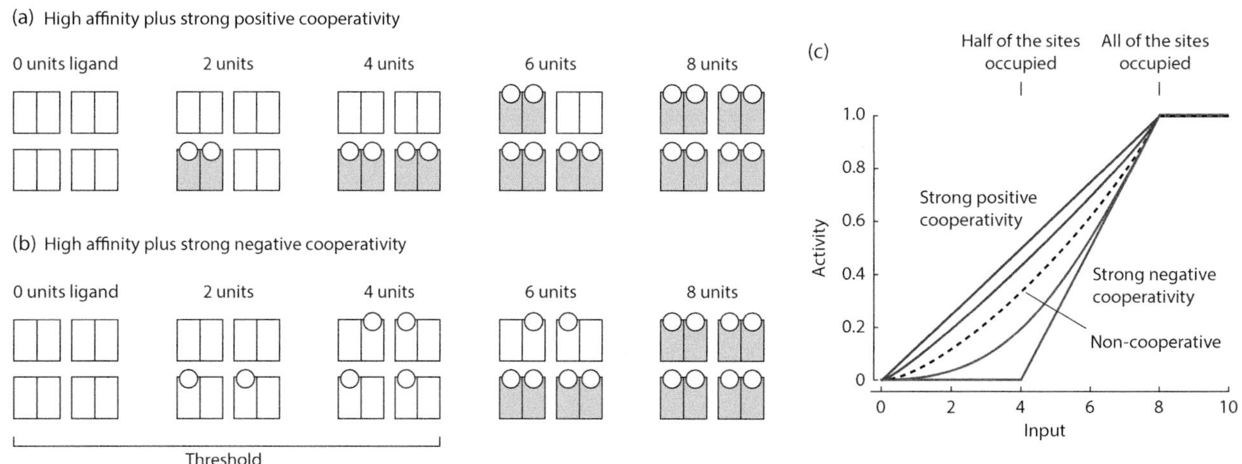

Figure 4.4 (a) Ligand depletion plus positive cooperativity can yield a linear response and (b) ligand depletion plus negative cooperativity can generate a threshold. For panel C, the activity curves assume very high-affinity binding plus cooperativity factors of ∞, 100 (blue curves), 1 (dashed black curve), 0.01, or 0 (red curves). Note that the ligand is completely depleted by receptor binding until it is present in excess of the receptor.

sigmoidal one like we saw in Chapter 3. On the other hand, if there is strong negative cooperativity in the binding, all of the receptor dimers will bind one ligand molecule before any of the receptors bind two (**Figure 4.4b**). The result is a sharp concentration threshold (**Figure 4.4c**, red), with the receptors then snapping from off to on over a tighter range of input concentrations than they do in the positive cooperativity case. Between these two extremes of cooperativity, the result is intermediate. For example, if the binding is noncooperative, the result response curve is concave-up until the maximum response is reached (**Figure 4.4c**, dashed black curve).

That is the intuitive picture for this limiting case of very high affinity. Can we derive a formula that captures this behavior and also shows the behavior when the binding affinity is not so extremely strong? Conceptually this does not seem so hard. We start again with by setting the rate equations for the net production of the empty receptor (y_0), the singly ligated receptor (y_1), and the doubly ligated receptor (y_2) equal to zero:

$$0 = -2k_1 x \cdot y_0 + k_{-1} y_1 \tag{4.7}$$

$$0 = 2k_1 x \cdot y_0 - k_{-1} y_1 - k_2 x \cdot y_1 + 2k_{-2} y_2 \tag{4.8}$$

$$0 = k_2 x . y_1 - 2k_{-2} y_2 . \tag{4.9}$$

We then add two conservation equations:

$$y_{tot} = y_0 + y_1 + y_2 \tag{4.10}$$

$$x_{tot} = x + y_1 + 2y_2 . \tag{4.11}$$

Next we would like to derive an equation for the doubly bound receptor, y_2, as a function of x_{tot}, y_{tot}, and the rate constants. We begin by eliminating the variables y_0, y_1, and x. This yields the following expression:

$$\begin{aligned} &k_2 k_{-1}^2 k_{-2} y_2 + 4k_1^2 k_{-2} y_2 \left(k_2 \left(x_{tot} - y_2 - 1\right)\left(y_2 - 1\right) + 2k_{-2} y_2\right) + \\ &k_1 k_2 k_{-1} \left(k_2 \left(x_{tot} - 2y_2\right)^2 \left(y_2 - 1\right) + 2k_{-2} y_2 \left(2 + x_{tot} - 4y_2\right)\right) = 0. \end{aligned} \tag{4.12}$$

We can simplify this a bit by using the definitions $K_1 = k_{-1}/k_1$ and $K_2 = k_{-2}/k_2$ and replacing the rate constants with equilibrium constants:

$$\begin{aligned} &K_2 y_2 \left(K_1^2 + 8K_2 y_2 + 4\left(x_{tot} - y_2 - y_{tot}\right)\left(y_2 - y_{tot}\right) + 2K_1 \left(x_{tot} - 4y_2 + 2y_{tot}\right)\right) + \\ &K_1 \left(x_{tot} - 2y_2\right)^2 \left(y_2 - y_{tot}\right) = 0. \end{aligned} \tag{4.13}$$

Equation 4.13 implicitly defines the relationship between the independent variable, x_{tot}, and the dependent variable, y_2. It is possible to solve this cubic equation to get an expression for y_2 as a function of x_{tot}, but it is much easier, and the resulting expression is much more compact, if one solves for x_{tot} as a function of y_2. Either using the quadratic formula or, more simply, using the equation solver in *Mathematica*, the result is:

$$x_{tot} = \frac{1}{2K_1\{y_2 - y_{tot}\}} \left(\begin{aligned} &-2K_1K_2y_2 + 4K_1y_2^2 - 4K_1y_2y_{tot} + 4K_2y_2y_{tot} - \\ &\sqrt{\left(\begin{aligned} &\left(2K_1K_2y_2 - 4K_1y_2^2 + 4K_2y_2^2 + 4K_1y_2y_{tot} - 4K_2y_2y_{tot}\right)^2 - \\ &4\left(K_1y_2 - K_1y_{tot}\right)\left(\begin{aligned} &K_1^2K_2y_2 - 8K_1K_2y_2^2 + 8K_2^2y_2^2 + 4K_1y_2^3 - 4K_2y_2^3 + \\ &4K_1K_2y_2y_{tot} - 4K_1y_2^2y_{tot} + 4K_2y_2y_{tot}^2 \end{aligned} \right) \end{aligned} \right)} \end{aligned} \right). \tag{4.14}$$

This looks very complicated, but still one can plot x_{tot} as a function of y_2 and then flip the axes so that the input (x_{tot}) will be on the horizontal axis and the output (y_2) on the vertical axis, and see what the input/output curves look like. As expected, the curves look like standard KNF curves when the affinities are low compared to the abundances (not shown) and like the curves we imagined in our thought experiment when the affinities are high (**Figure 4.4c**).

So is this how multimeric signaling proteins actually respond to upstream regulators? In vitro experiments using DNA oligos to stand in for the ligand and its multimeric target provide proof-of-principle for this idea (**Figure 4.5a**). The advantage here is that complementary DNA oligos of reasonable length bind each other with the requisite high affinity. Moreover, either positive or negative cooperativity can be engineered into the binding. Positive cooperativity can be produced by simply having the ligand oligos abut each other; the energetically favorable base stacking interaction between the ligand oligos makes it so that the binding of the first (whichever it is) promotes the binding of the second (**Figure 4.5a**). Noncooperativity can be achieved by putting a gap between the bind sites for the two ligand oligos, and negative cooperativity can be produced by having the ligand oligos overlap, so that each interferes with the binding of the other (**Figure 4.5a**). The equilibrium binding can be assessed by nondenaturing gel electrophoresis.

The results agree well with the theory derived above. In the case of high affinity, positively cooperative binding, the amount of doubly bound receptor increases linearly with the concentration of the two ligands (**Figure 4.5b**, blue). In the case of noncooperative binding, the binding curve is concave-up (**Figure 4.5b**, green), and in the case of negative cooperativity, there is a sharp threshold followed by a linear response (**Figure 4.5b**, yellow). Thresholds like this can allow a signaling system to filter out the first increments of stimulus and then respond decisively to suprathreshold signals, and they can be useful for generating more complex system-level behaviors such as bistability and oscillations.

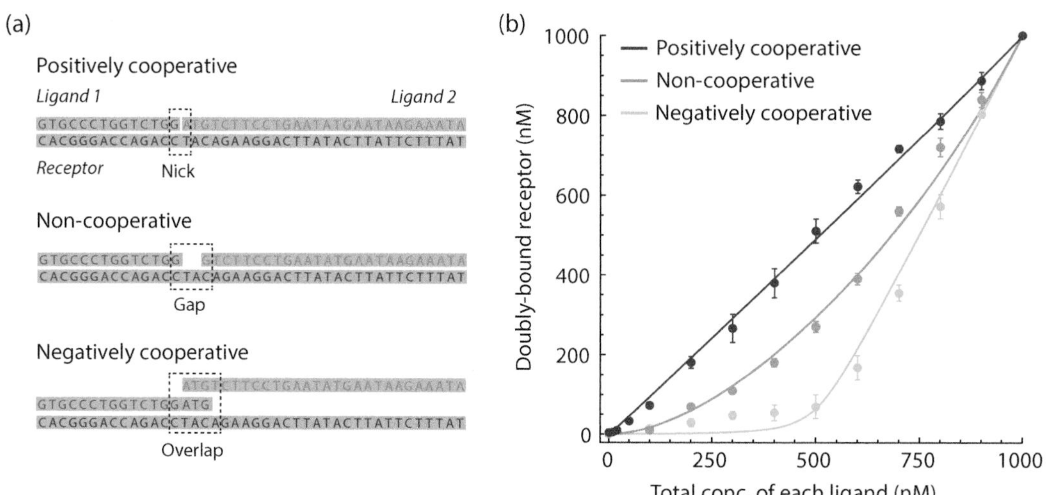

Figure 4.5 In vitro, high-affinity binding of two DNA ligand oligos to a complementary receptor strand. (a) Structures of the receptor with pairs of ligands expected to exhibit positively cooperative, noncooperative, or negatively cooperative binding. Although different in length, each ligand oligo has a similar affinity for the receptor. (b) Experimental binding data. The curves are fits to Eq. 4.14, with the cooperativity (K_1/K_2) as the only fitted parameter. (Adapted from Ha and Ferrell, *Science*. 2016.)

4.5 STOICHIOMETRIC REGULATORS MUST SOMETIMES COMPETE WITH STOICHIOMETRIC INHIBITORS

Let us take a look again at this idea that high-affinity binding with negative cooperativity can yield a switch-like, ultrasensitive response with a discrete threshold. Basically the first binding site in each receptor is acting like a buffer, soaking up the first increments of ligand, and the negative cooperativity and high-affinity binding ensure that most of these buffer sites will be filled before the ligand can bind to the receptors' second sites, which are the sites that yield receptor activation (**Figure 4.6a**). Thus, the unoccupied receptor is competing with the singly occupied receptor for access to ligand.

A similar sort of competition can happen whenever a high-affinity stoichiometric inhibitor is present in a pathway with a limited supply of upstream activator molecules, and stoichiometric inhibitors are quite common in signaling both at the level of ligand–receptor interaction and in downstream signaling. For example, the high-affinity μ-opioid receptor antagonist naloxone can compete with opioid agonists such as heroin, and so a sufficient dose of naloxone can block or reverse heroin intoxication (**Figure 4.6b**). In *Drosophila* EGF receptor signaling, the Argos protein binds to and sequesters the EGF-like protein Spitz; thus the receptor and Argos are competing for binding to Spitz/EGF (**Figure 4.6c**). Likewise, in cell cycle regulation, the cyclin-dependent kinases are regulated by a variety of high-affinity stoichiometric inhibitors, including the tumor suppressors $p21^{Cip1}$, $p27^{Kip1}$, $p16^{INK4a}$, and $p19^{INK4a}$. These can be viewed as competing with substrates for access to the active cyclin–Cdk complex (**Figure 4.6d**).

Here we will model the competition between a stoichiometric inhibitor and some other pathway component, be it a ligand or a downstream target, and see under what circumstances it can produce a threshold and ultrasensitivity in the equilibrium response. The simplest of the cases shown in **Figure 4.6** is the competition between an opioid agonist (which we will call x) and an antagonist (I) for the binding to an opioid receptor (y). We can write down the rate equations for the net production of c_{xy}—the complex of x with y—and of c_{Iy}, the complex of the inhibitor I with y:

$$\frac{dc_{xy}}{dt} = k_1 x \cdot y - k_{-1} c_{xy} \tag{4.15}$$

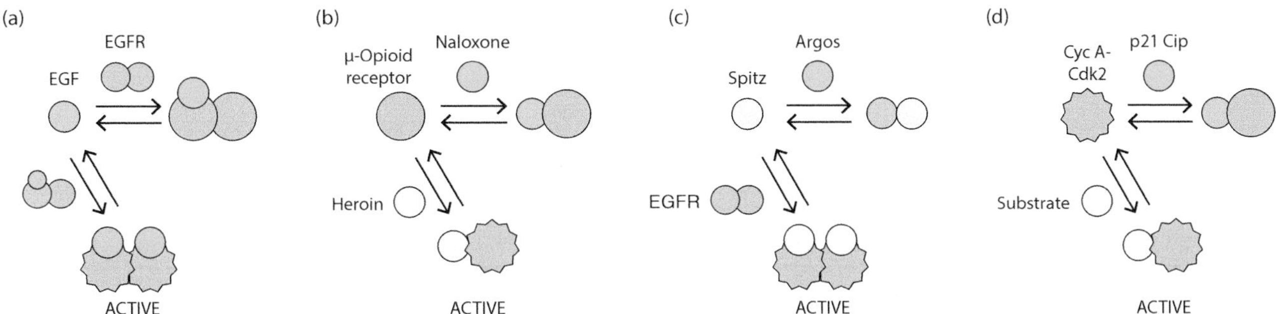

Figure 4.6 Four examples of stoichiometric competition in cell signaling. (a) Competition between unbound EGFRs and EGFRs with one bound ligand. (b) Competition between an opioid agonist (heroin) and an antagonist (naloxone) for binding to the μ-opioid receptor. (c) Competition between Argos and the *Drosophila* EGF receptor for binding to the EGF-like protein Spitz. (d) Competition between the Cdk inhibitor p21 Cip and a substrate for binding to a cyclin A–Cdk1 complex.

$$\frac{dc_{Iy}}{dt} = k_2 y \cdot I - k_{-2} c_{Iy}. \tag{4.16}$$

At equilibrium the derivatives must equal zero, and so:

$$0 = k_1 x \cdot y - k_{-1} c_{xy} \tag{4.17}$$

$$0 = k_2 y \cdot I - k_{-2} c_{Iy}. \tag{4.18}$$

In addition we have a conservation equation:

$$y_{tot} = y + c_{xy} + c_{Iy}. \tag{4.19}$$

We can assume that the ligands (x and I) are present in great excess over the receptor (y_{tot}), and so the binding of either ligand to y has a negligible effect on the total concentrations of either x or I.

We can then go ahead and solve Eqs. 4.17–4.19 to derive an expression for the output of the system, c_{xy}, as a function of the rate constants, y_{tot}, and the concentrations of x and I:

$$\left(\frac{c_{xy}}{y_{tot}}\right)_{eq} = \frac{x}{\frac{k_{-1}}{k_1}\left(1 + \frac{I}{\frac{k_{-2}}{k_2}}\right) + x} \tag{4.20}$$

$$\left(\frac{c_{xy}}{y_{tot}}\right)_{eq} = \frac{x}{K_D\left(1 + \frac{I}{K_I}\right) + x}, \tag{4.21}$$

where $K_D = k_{-1}/k_1$, the equilibrium constant for the binding of x to y, and $K_I = k_{-2}/k_2$, the equilibrium constant for the binding of I to y.

If the inhibitor concentration is zero, Eq. 4.21 reduces to the standard Langmuir equation for the binding of a non-depletable ligand x to a receptor y. Moreover, even if I is not zero, the response is still hyperbolic or Michaelian, but it has a larger $EC50$ than it would have had in the absence of inhibitor. The relationship between the equilibrium constant K_D and the observed $EC50$ is given by:

$$EC50 = K_D\left(1 + \frac{I}{K_I}\right). \tag{4.22}$$

Equation 4.22 is known by pharmacologists as the Cheng–Prusoff equation, and it can be used to extract the thermodynamic constants, K_D and K_I, from binding experiments in which the concentration of the inhibitor is varied.

The effect of a fixed concentration of a stoichiometric inhibitor on the binding of ligand x to receptor y is shown in **Figure 4.7**. Binding is still a hyperbolic function of x, just as it was in the absence of inhibitor. There is no threshold and no ultrasensitivity, despite the fact that intuition (see above) says there should be.

The reason for this is that we have not yet accounted for the possibility that either the inhibitor I or the regulator x, or both, might be in less-than-infinite supply. So now let us have x represent EGFR, y the

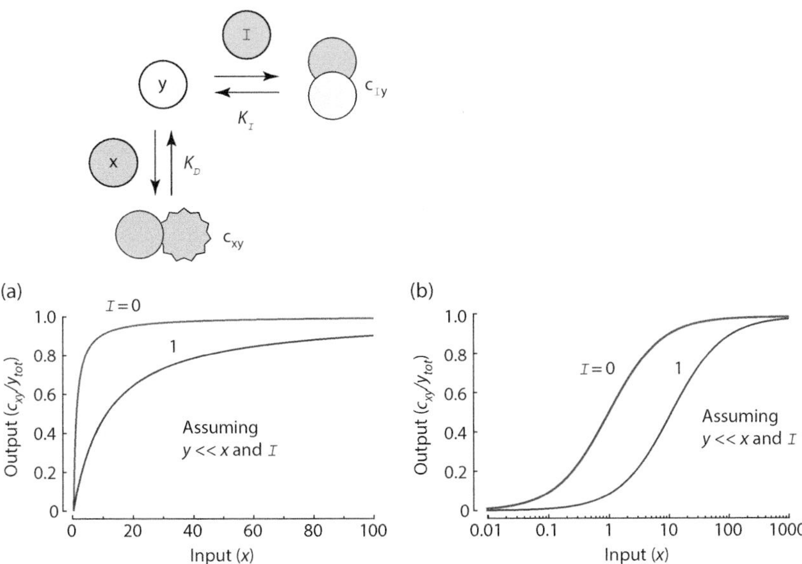

Figure 4.7 Modeling competition between a stoichiometric activator (*x*) of *y* and a stoichiometric inhibitor (*I*) of *y*, under the assumption that binding has a negligible effect on the concentrations of *x* and *I*.

EGF protein Spitz, and *I* Argos (**Figure 4.6b**), where the basic idea is that as the concentration of Spitz/EGF rises, Argos binds Spitz/EGF until all the Argos has been used up. This means that for none of the three proteins—*x*, *y*, or *I*—can we assume that the free and total protein concentrations are going to be approximately equal. Thus we write down two more conservation equations:

$$I_{tot} = I + c_{Iy} \tag{4.23}$$

$$x_{tot} = x + c_{xy}, \tag{4.24}$$

and then derive an expression for c_{xy} in terms of x_{tot}, y_{tot}, I_{tot} and the equilibrium constants. As was the case above for the treatment of negative cooperative with a depletable ligand (Eqs. 4.13 and 4.14), this requires solution of a cubic equation, which is cumbersome, but again we can make things simpler by flipping the derivation around—deriving an equation for x_{tot} as a function of c_{xy} rather than for c_{xy} as a function of x_{tot}. The result is:

$$x_{tot} = c_{xy}\left(1+K_D\right)\left(\frac{1}{y_{tot}-c_{xy}} + \frac{I_{tot}}{K_D c_{xy} + K_I\left(y_{tot}-c_{xy}\right)}\right). \tag{4.25}$$

To get some insight into what this equation means, we plot Eq. 4.25 (**Figure 4.8**). Now the binding curves agree with our intuition—there is a threshold and an ultrasensitive response. The higher the affinity of the inhibitor for the ligand *x*, the sharper the threshold is (**Figure 4.8a**). The higher the affinity of the target for the ligand, the more switch-like the approach to a maximal response is (**Figure 4.8b**). The higher the total concentration of *I* (I_{tot}), the bigger the threshold (**Figure 4.8c**). If the affinities and the concentration of I_{tot} are sufficiently high, the response approaches a step function (**Figure 4.8**). In principle, any degree of ultrasensitivity can be obtained.

Note that stoichiometric inhibition can work just as well at building a threshold into the response of an enzymatic regulator, like a protein kinase—the cyclin-dependent kinases and their many stoichiometric inhibitors are a good example (**Figure 4.8**)—or of regulated synthesis or destruction, which will be examined in Chapter 6.

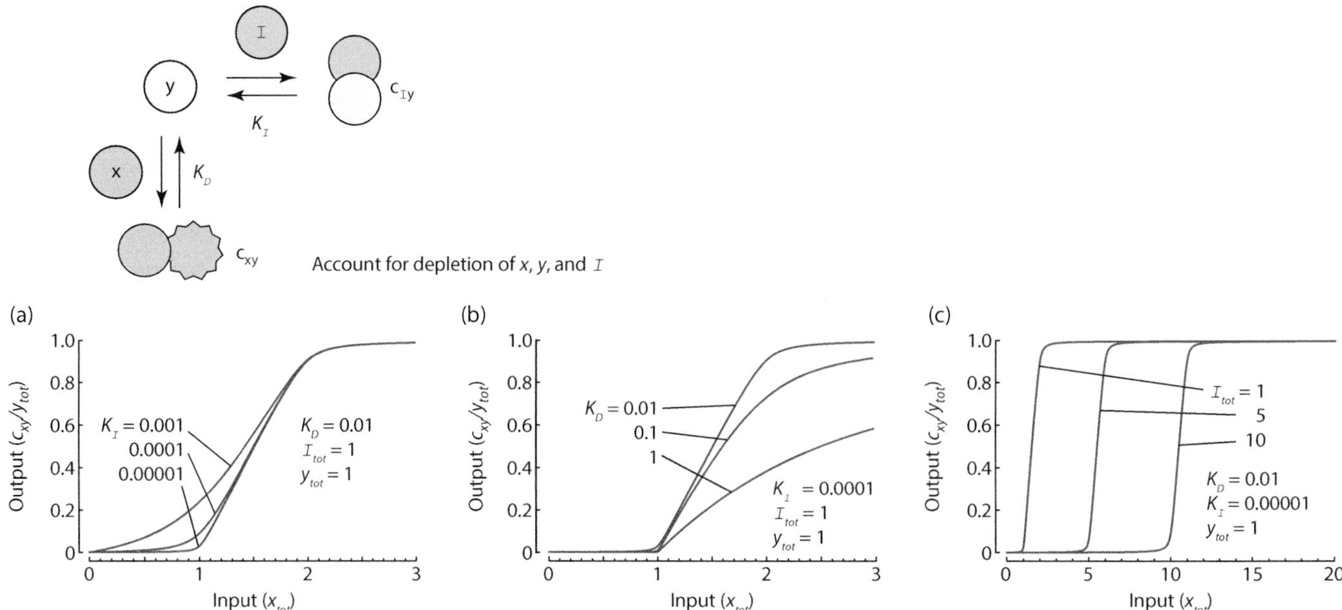

Figure 4.8 Modeling competition between a stoichiometric activator (*x*) of *y* and a stoichiometric inhibitor (*I*) of *y*, assuming that binding has a substantial effect on the concentrations of *x* and *I*.

SUMMARY

Stoichiometric regulation is commonplace in intracellular signaling, but because the upstream regulator is often substantially depleted by binding to its effector, the equations derived in Chapters 2 and 3 for the stoichiometric regulation of a receptor by a vast excess of ligand no longer apply. For the activation of a monomeric effector by a monomeric ligand, the response is more linear than that predicted by the Langmuir equation, and the dynamics also differ a bit from the exponential approach to steady state. Moreover, strangely enough, if a high-affinity upstream protein stoichiometrically regulates a multimeric effector, then negative cooperativity in the binding can result in a sharp threshold in the activation response. The same is true for a high-affinity stoichiometric inhibitor—it can build a threshold into a response, so that low concentrations of an upstream activator produce little downstream signaling, and then once the threshold is exceeded, the output of the system rises from low to high in a decisive, switch-like fashion.

This can be viewed as one simple way that intracellular regulators can generate some ultrasensitivity. We will explore several others in the next two chapters, beginning with a look at protein regulation through covalent modification as exemplified by phosphorylation and dephosphorylation.

FURTHER READING

STOICHIOMETRIC REGULATION INSIDE THE CELL

Rajalingam K, Schreck R, Rapp UR, Albert S. Ras oncogenes and their downstream targets. *Biochim Biophys Acta.* 2007 Aug;1773(8):1177–95.

Sprang SR, Chen Z, Du X. Structural basis of effector regulation and signal termination in heterotrimeric Galpha proteins. *Adv Protein Chem.* 2007;74:1–65.

COOPERATIVE BINDING WITH LIGAND DEPLETION

Ha SH, Ferrell JE Jr. Thresholds and ultrasensitivity from negative cooperativity. *Science.* 2016 May 20;352(6288):990–3.

CLASSICAL PHARMACOLOGICAL TREATMENTS OF COMPETITION

Cheng Y, Prusoff WH. Relationship between the inhibition constant (K1) and the concentration of inhibitor which causes 50 per cent inhibition (I50) of an enzymatic reaction. *Biochem Pharmacol*. 1973 Dec 1;22(23):3099–108.

Lazareno S, Birdsall NJ. Estimation of antagonist Kb from inhibition curves in functional experiments: alternatives to the Cheng-Prusoff equation. *Trends Pharmacol Sci*. 1993 Jun;14(6):237–9.

ULTRASENSITIVITY FROM COMPETITORS AND STOICHIOMETRIC INHIBITORS

Ferrell JE Jr, Ha SH. Ultrasensitivity part II: multisite phosphorylation, stoichiometric inhibitors, and positive feedback. *Trends Biochem Sci*. 2014 Nov;39(11):556–69.

DOWNSTREAM SIGNALING 2

Covalent Modification

5

IN THIS CHAPTER . . .

5.1 A MASS ACTION PHOSPHORYLATION–DEPHOSPHORYLATION CYCLE YIELDS A MICHAELIAN STEADY-STATE RESPONSE WITH EXPONENTIAL APPROACH TO THE STEADY STATE

5.2 THE STEADY-STATE RESPONSE OF A PHOSPHORYLATION–DEPHOSPHORYLATION REACTION WITH MICHAELIS–MENTEN KINETICS CAN BE ULTRASENSITIVE

5.3 RATE–BALANCE PLOTS ARE MUCH LIKE THE ECONOMIST'S SUPPLY-AND-DEMAND PLOTS

5.4 RATE-BALANCE ANALYSIS EXPLAINS THE MICHAELIAN STEADY-STATE RESPONSE

5.5 THE DYNAMICS OF THE SYSTEM CAN ALSO BE UNDERSTOOD FROM THE RATE–BALANCE PLOT

5.6 RATE-BALANCE ANALYSIS HELPS EXPLAIN ZERO-ORDER ULTRASENSITIVITY

5.7 DOES ZERO-ORDER ULTRASENSITIVITY OCCUR IN VIVO?

5.8 THE TEMPORAL DYNAMICS OF A MULTISTEP ACTIVATION PROCESS TELLS YOU THE NUMBER OF PARTIALLY RATE-DETERMINING STEPS

5.9 ASSUMING MASS ACTION KINETICS, STEADY-STATE MULTISITE PHOSPHORYLATION IS DESCRIBED BY A KNF-TYPE EQUATION

5.10 PRIMING CAN IMPART POSITIVE COOPERATIVITY ON MULTISITE PHOSPHORYLATION

5.11 DISTRIBUTIVE MULTISITE PHOSPHORYLATION IMPROVES SIGNALING SPECIFICITY

DOI: 10.1201/9781003124269-5

The second basic type of downstream signaling is the regulation of a target by a signaling enzyme rather than by a stoichiometric regulator, where the upstream regulator enzymatically tags or modifies the downstream target. This modification can and often does result in a change in the target's activity, if the target is an enzyme, but it can also regulate the protein's intracellular localization or its stoichiometric interactions with other proteins. The reversible covalent modification of a protein by phosphorylation is the most common tagging mechanism in cell signaling, so that is what we will focus on here, but the same approaches could be applied to methylation, acetylation, or ubiquitylation. Proteins can also be regulated by enzymes non-covalently, the best example being the reversible binding of a G-protein to GTP to turn it on or GDP to turn it off, with a guanine nucleotide exchange factor and a GTPase-activating protein playing similar roles in this enzymatic cycle to those played by protein kinases and phosphoprotein phosphatases in protein phosphorylation.

Protein kinases transfer the γ-phosphate of ATP to amino acid side chains in substrate proteins. In eukaryotes, most phosphorylation is carried out by the so-called classical protein kinases, which phosphorylate serine, threonine, and/or tyrosine residues, and is reversed by phosphoprotein phosphatases, with phosphorylations typically going on and coming off on a time scale of seconds to minutes. Phosphorylation can bring about a change in the overall conformation of the substrate protein, subtly or not-so-subtly affecting the folding of a protein domain. The ERK2 MAP kinase, which is a substrate of the MEK1 and MEK2 MAP kinase kinases, is a well-studied example of this: phosphorylation of a threonine residue and a tyrosine residue in the ERK2 activation loop results in a conformation change that activates the kinase and promotes its dimerization. Phosphorylation can also regulate a protein's function without inducing a conformation change. For example, the phosphorylation of cyclin B-Cdk1 at two adjacent residues in Cdk1's ATP-binding cleft, threonine 14 and tyrosine 15, interferes with the positioning of the γ-phosphate in ATP and thus decreases the activity of Cdk1, without grossly affecting the kinase's conformation. Alternatively, if the phosphorylation occurs in a poorly structured region of the substrate protein, the result may not be a conformation change but rather the production of a short phosphoepitope that can interact with "reader" domains in other signaling proteins. Given that perhaps 90% of protein phosphorylations are thought to occur in intrinsically disordered regions, this is likely to be the most common way for phosphorylation to affect protein function.

Here we will model a phosphorylation–dephosphorylation cycle, with phosphorylation activating the protein and dephosphorylation inactivating it. We will start by examining the steady-state and dynamical behaviors of a simple model of the process with mass action kinetics.

5.1 A MASS ACTION PHOSPHORYLATION–DEPHOSPHORYLATION CYCLE YIELDS A MICHAELIAN STEADY-STATE RESPONSE WITH EXPONENTIAL APPROACH TO THE STEADY STATE

A phosphorylation reaction involves the binding of two different substrates, ATP and the protein being phosphorylated, and so the production of a phosphorylated substrate is a multistep process. Likewise, dephosphorylation begins with the formation of a phosphatase–phosphosubstrate complex before catalysis occurs. But if the intermediates are low in concentration, we can reduce the system to a simple kinetic scheme where the substrate protein is phosphorylated in a one-step, mass action process and dephosphorylated in an opposing one-step, mass action process, as shown in **Figure 5.1a.** The net rate is then given by:

$$\frac{dy^*}{dt} = k_1 kinase \cdot y - k_{-1} phosphatase \cdot y^* . \tag{5.1}$$

The phosphorylated and active form of y is designated y^*; *kinase* and *phosphatase* denote the concentrations of the active enzymes; and k_1 and k_{-1} are the second-order rate constants for the forward and back reactions. We have implicitly included the concentration of ATP in the k_1 rate constant. Note that for most protein kinases, the affinity for ATP is high (~10 μM) compared to the cellular ATP concentration (~1 mM), so even if the cellular concentration of ATP varies a bit it is unlikely to affect the phosphorylation rate. We consider *kinase* to be the input to the system, and call it x. Furthermore, we assume the concentration of active phosphatase to be unchanging and lump it into the rate constant, defining a new k_{-1} to be the old k_{-1} times *phosphatase*. Equation 5.1 then becomes:

$$\frac{dy^*}{dt} = k_1 x \cdot y - k_{-1} y^* . \tag{5.2}$$

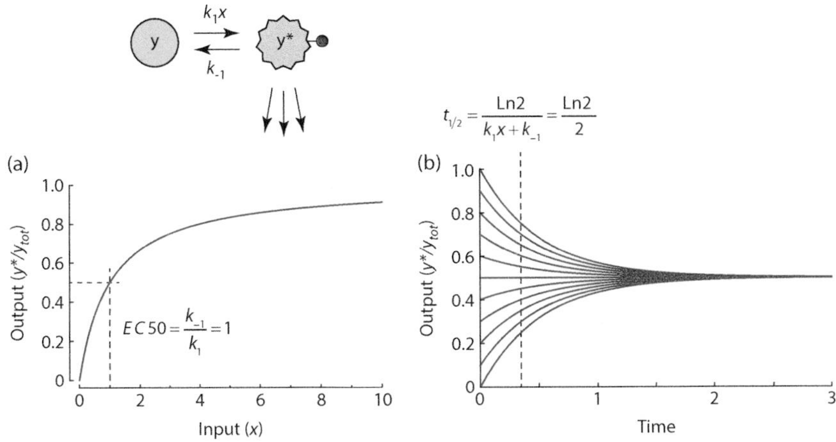

Figure 5.1 The steady-state response (a) and time course (b) for a phosphorylation–dephosphorylation reaction. In each panel we have assumed that $k_1 = k_{-1} = 1$ and in panel b we assumed $x = 1$.

In addition there is a conservation relationship:

$$y_{tot} = y + y^*. \tag{5.3}$$

Note that Eq. 5.2 is identical in form to Eq. 2.1 and the conservation relationship is identical in form to Eq. 2.2. Thus, all of the equations we derived in Sections 2.1–2.6 hold here; this model of phosphorylation is mathematically identical to our first model of ligand–receptor interaction. There are two conceptual differences worth pointing out, though. First, the system is not in equilibrium when the forward and back rates are balanced. The phosphorylation reaction consumes ATP, and the dephosphorylation reaction does not resynthesize it. The balance point is therefore termed a steady state rather than an equilibrium. Second, the ratio of the rate constants k_{-1} and k_1 is no longer an equilibrium constant. The ratio still defines the *EC*50 for the reaction, however, and it is still usually denoted by an upper case *K* just as an equilibrium constant is.

As shown in **Figure 5.1a**, the relationship between the input and the steady-state output is Michaelian:

$$\left(\frac{y^*}{y_{tot}}\right)_{ss} = \frac{x}{K_1 + x}. \tag{5.4}$$

The system responds to a step-change in input by exponentially approaching its new steady state with a halftime of $\frac{\text{Ln}2}{k_1 x + k_{-1}}$:

$$y^*[t] = y^*_{ss} + \left(y^*[0] - y^*_{ss}\right)e^{-(k_1 x + k_{-1})t} \tag{5.5}$$

Note that just as our ligand–receptor model showed a quicker response to a step up in ligand concentration than to a step back down, this phosphorylation–dephosphorylation system shows a quicker response to a step up in kinase activity than to a step back down.

If we were to take the phosphatase rather than the kinase as the input to the system—that is, we assume that the concentration of active kinase is constant and the concentration of active phosphatase varies—the steady-state response would be identical to Michaelian inhibition (Eq. 2.12), and the dynamics again would be an exponential approach to the steady state.

5.2 THE STEADY-STATE RESPONSE OF A PHOSPHORYLATION–DEPHOSPHORYLATION REACTION WITH MICHAELIS–MENTEN KINETICS CAN BE ULTRASENSITIVE

For simplicity we initially assumed that our phosphorylation and dephosphorylation reactions could be described by mass action kinetics, with the rate of each process rising linearly with the amount of substrate to be phosphorylated or dephosphorylated. But we know that that is not how enzymes work; they are saturable, and the standard way of describing that saturability is the Michaelis–Menten equation, which we first encountered in Chapter 2:

$$\frac{dP}{dt} = k_{cat}E\frac{S}{K_M + S}. \tag{5.6}$$

P is the product of the reaction, *E* is the enzyme concentration, *S* is the substrate concentration, K_M is the Michaelis constant, and the proportionality constant k_{cat} represents the maximal rate of one molecule of enzyme. What happens to the steady-state response of a phosphorylation/dephosphorylation reaction if we use Michaelis–Menten kinetics rather than mass action kinetics?

The rate of phosphorylation will be:

$$\text{Phosphorylation rate} = k_1 kin \frac{y}{K_{M1} + y} \tag{5.7}$$

where kin is the concentration of kinase, K_{M1} is the K_M value for the kinase, and k_1 is the k_{cat} value for the kinase. Note that the substrate for the kinase is the dephosphorylated form of the substrate; thus y appears where S normally would in this Michaelis–Menten expression.

Likewise, the rate of dephosphorylation will be:

$$\text{Dephosphorylation rate} = k_{-1} pase \frac{y^*}{K_{M2} + y^*} \tag{5.8}$$

where $pase$ is the concentration of phosphatase, K_{M2} is the K_M value for the phosphatase, and k_{-1} is the k_{cat} for the phosphatase. Note that phosphorylated y is the substrate of the phosphatase, and so here y^* takes the place of S. The net rate of production of y^* will therefore be:

$$\frac{dy^*}{dt} = k_1 kin \frac{y}{K_{M1} + y} - k_{-1} pase \frac{y^*}{K_{M2} + y^*}. \tag{5.9}$$

If we assume the concentrations of the kinase y and phosphatase y^* complexes are negligible, we can invoke the conservation relationship $y_{tot} = y + y^*$ and eliminate y from Eq. 5.9:

$$\frac{dy^*}{dt} = k_1 kin \frac{y_{tot} - y^*}{K_{M1} + y_{tot} - y^*} - k_{-1} pase \frac{y^*}{K_{M2} + y^*}. \tag{5.10}$$

At steady state, the derivative must be equal to zero. The resulting algebraic equation can be solved for y^*. The result is formidable:

$$y^*_{ss} = \frac{\begin{array}{c} -K_{M2}kin - K_{M1}K_1 pase + kin \cdot y_{tot} - K_1 pase \cdot y_{tot} \\ \pm\sqrt{4K_{M2}kin(kin - K_1 pase)y_{tot} + (K_{M2}kin + K_{M1}K_1 pase - kin \cdot y_{tot} + K_1 pase \cdot y_{tot})^2} \end{array}}{2(kin - K_1 pase)}, \tag{5.11}$$

where $K_1 = k_{-1}/k_1$. Only the solution with the +sign is biologically relevant, and it is called the **Goldbeter–Koshland equation**. The Goldbeter–Koshland equation describes the steady-state output of a phosphorylation–dephosphorylation system, or any covalent modification cycle, where the enzyme-mediated reaction rates are described by the Michaelis–Menten equation.

The next question is how to understand such a behemoth of an equation. A plausible first step is to plot it. There are four parameters in the equation (y_{tot}, K_{M1}, K_{M2}, and K_1) and two "knobs" to turn—the kinase (kin) and phosphatase ($pase$) concentrations. For a start, let us take all of the K values equal to 1, fix the phosphatase concentration at 1, and look at the output as a function of the kinase concentration for different values of y_{tot} and hence different degrees of saturation of the two opposing enzymes. As shown in **Figure 5.2b**, when y_{tot}=1 the response resembles that of a Michaelian system, though a bit

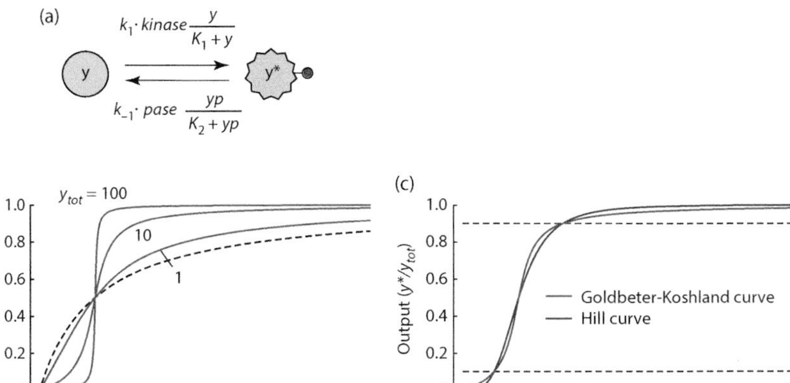

Figure 5.2 Zero-order ultrasensitivity. (a) Schematic view of a phosphorylation–dephosphorylation cycle with saturable enzymes. (b) Steady-state phosphorylation for different *kinase* concentrations and different assumed total concentrations of the substrate y (1, 10, or 100, as indicated). In each case it is assumed that $K_{M1}=K_{M2}=K_1=1$. The black dashed curve is a Michaelian response, for comparison. (c) Similarity between a Goldbeter–Koshland curve, assuming $K_{M1}=K_{M2}=K_1=1$ and $y_{tot}=10$, and a Hill curve with the same effective Hill exponent ($n=3.7$). The $EC50$ and Hill exponent for the Hill curve were chosen so that the Hill and Goldbeter–Koshland curves would have the same $EC10$ and $EC90$ values. The 10% and 90% responses are indicated by the horizontal dashed lines.

less graded. When $y_{tot}=10$, so that the enzymes' substrates (y and y^* respectively) will be present at up to 10× the K_M values, the result is a markedly switch-like, sigmoidal response curve (**Figure 5.2b**). The effective Hill exponent for the response, calculated with Eq. 3.53, is about 3.7, and the overall response resembles that of a Hill function, though it is a bit steeper in the middle of the response range and less steep at the top and bottom of the range (**Figure 5.2c**). When $y_{tot}=100$, the response is nearly a step function (**Figure 5.2b**), with an effective Hill exponent of about 26. Koshland and Goldbeter termed this phenomenon **zero-order ultrasensitivity**, because it is an ultrasensitive response that occurs when the opposing enzymes have a nearly flat, zero-order dependence on substrate concentration.

In fact it was the discovery of this phenomenon that led them to coin the term ultrasensitivity. With the KNF and MWC models one only gets a sigmoidal response if one assumes there is cooperativity in the binding. The same is almost true for multisite phosphorylation, which we will examine in Sections 5.8–5.11, although if the phosphorylation is ordered, the response is slightly sigmoidal even when the phosphorylation and dephosphorylation reactions are noncooperative. So prior to the discovery of zero-order ultrasensitivity, cooperativity was almost synonymous with sigmoidal response. But in the case of zero-order ultrasensitivity, nothing even remotely resembling cooperativity is involved, and yet a sigmoidal response is still obtained. The term ultrasensitivity implicitly acknowledges that cooperativity is not the only type of mechanism that can yield a sigmoidal response.

5.3 RATE–BALANCE PLOTS ARE MUCH LIKE THE ECONOMIST'S SUPPLY-AND-DEMAND PLOTS

At this point we have seen that enzyme saturation can produce an ultrasensitive response—even an extremely ultrasensitive, highly switch-like response—but it is not clear why. One good approach to this question is **rate-balance analysis**, a way of depicting rates and steady states graphically. Rate–balance plots can deepen one's understanding of dynamical systems in cases where the algebra is simple, like the simple Michaelian phosphorylation response we got when we assumed mass action kinetics (Eq. 5.4), and in cases where the algebra is not so simple, like zero-order ultrasensitivity (Eq. 5.11), the rate–balance plots are often still easy to analyze and understand, and thus are especially helpful.

A rate-balance plot is similar to the supply-and-demand curves seen in introductory economics. Economists plot the supply of some commodity—let us say widgets—as a function of price. Typically the supply goes up monotonically with price (**Figure 5.3**, blue curve). Then they plot the demand for the commodity as a function of price—typically demand goes down as price goes up (**Figure 5.3**, green curve). Where the two curves cross, supply equals demand, and the price at which this intersection occurs is the steady-state market price for the commodity.[1] For the curves shown in **Figure 5.3**, the steady-state price for a widget is \$2.40, and the number of widgets produced and sold per year is 3 million.

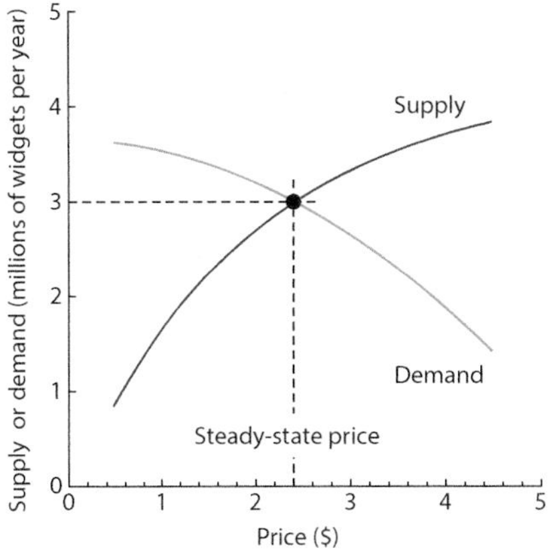

Figure 5.3 Supply-and-demand curves. Supply (blue curve) typically goes up with price and demand (green curve) typically goes down. The steady-state market price, and the quantity of widgets produced and bought at steady state, is determined by where the curves cross (filled circle).

5.4 RATE-BALANCE ANALYSIS EXPLAINS THE MICHAELIAN STEADY-STATE RESPONSE

We can do the same thing for any signal transduction reaction that involves just one time-dependent variable. Here we will start with the mass action model of a phosphorylation–dephosphorylation cycle, and then work our way up to the Goldbeter–Koshland model with its Michaelis–Menten rate expressions.

We begin with the rate equation (Eq. 5.2) for the phosphorylation of y by kinase x and plug in the conservation relationship (Eq. 5.3) so that we have an equation with a single time-dependent variable:

$$\frac{dy^*}{dt} = k_1 x\left(y_{tot} - y^*\right) - k_{-1} y^*. \tag{5.12}$$

We break down the right-hand side of this expression into one term for the rate of the forward reaction and one term for the back reaction. Both of these rates can be thought of as functions of y^*. We plot the forward and back reaction rates vs. y^* on one set of axes, and the point where the two rate curves cross is the steady state for the system. For Eq. 5.12, the forward reaction rate, which is the rate of phosphorylation, is:

$$\text{Phosphorylation rate} = k_1 x\left(y_{tot} - y^*\right), \tag{5.13}$$

and the rate of the back reaction, which is the dephosphorylation reaction, is:

$$\text{Dephosphorylation rate} = k_{-1} y^* \tag{5.14}$$

We can explain these two rate equations in intuitive terms. The dephosphorylation reaction rate will be maximal if all of the substrate y is phosphorylated and therefore available to be dephosphorylated ($y^* = y_{tot}$); the rate will be zero if none of the substrate is phosphorylated; and if the concentration of y^* is somewhere between these two extremes, the rate will be intermediate. Because we have assumed mass action kinetics, the result is a straight line with a positive slope of k_{-1} (**Figure 5.4**, blue line). For simplicity we have arbitrarily taken $k_{-1} = 1$.

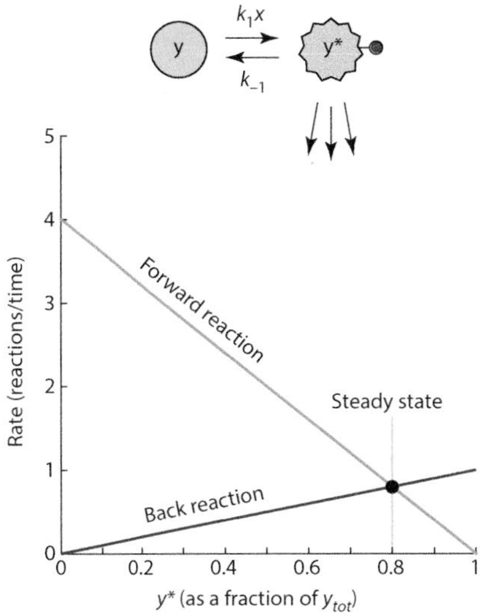

Figure 5.4 Rate-balance analysis for a phosphorylation–dephosphorylation cycle with mass action kinetics. The back reaction rate (blue line) goes up as the concentration of phosphorylated y^* goes up, and the forward reaction rate (green line) goes down. The system's steady state is determined by where the green and blue lines cross; at this concentration of y^*, the forward and back reactions are balanced. For the particular choice of rate constants and x concentration assumed here ($k_1 = k_{-1} = 1$; $x = 4$), the steady state occurs when 80% of the substrate y is phosphorylated.

We can explain the forward reaction rate curve similarly. The phosphorylation rate will be maximal when the concentration of unphosphorylated y is the largest, which occurs when $y^* = 0$.

1 Economists plot price on the y-axis and supply or demand on the x-axis. We have put the independent variable (price, since we are considering how supply and demand vary as the price is changed) on the x-axis, as is more common in science and mathematics.

Conversely, the forward rate will be zero when $y^*=y_{tot}$ so that there are no substrate molecules available for phosphorylation. In between there is again a straight-line relationship because we have assumed mass action kinetics, this time with a negative slope of $-k_1x$ (Figure 5.4, green line). We have arbitrarily assumed that $k_1=1$ and the kinase $x=4$. The blue and green lines cross at a single point, at which $y^*=0.8$. This means that the equilibrium for the system occurs when 80% of the substrate molecules are phosphorylated. This makes sense; the forward reaction is "stronger" than the reverse reaction since its proportionality constant is bigger (4 vs. 1), so the steady state favors the forward reaction.

No matter what slopes we choose for the green and blue curves, there will always be a single intersection point—a single steady state for the system. Moreover, the steady state will always be stable. By stable, we mean that if we were to displace the system from its steady state, the association and dissociation reaction rates would change in such a way as to make the system return toward the steady state. For example, for the system shown in **Figure 5.4**, suppose you were to increase y^* from 0.8 to 0.9 without changing the rate curves. At this value of y^*, the blue curve is higher than the green curve—the dephosphorylation rate is larger than the phosphorylation rate—and there will be net dephosphorylation, returning the system toward its stable steady state. Conversely, suppose you were to decrease y^* to 0.7. Now the green curve is higher than the blue curve, so the phosphorylation reaction predominates, returning the system toward its stable steady state.

It might at first seem obvious that a steady state should be stable, but actually it is not always going to be the case. The lowest point in a valley is a steady state, and it is stable, but the peak of a mountain is a steady state too, and it is unstable. If you are perfectly balanced you can rest on the mountaintop indefinitely, but any tiny perturbation will send you plummeting down one side or another. Unstable steady states are involved in some of the most interesting behaviors in signal transduction, including bistability, excitability, and oscillations, and we will encounter them in Chapters 8, 9, 13, and on.

So far we have chosen one particular value for the kinase x in our rate-balance analysis. What happens if we vary x? Varying x has no effect on the back reaction curve (the blue lines in Figures 5.4 and 5.5a) since the rate of y^* dephosphorylation does not depend on x. However, it does affect the forward reaction curve; the rate of phosphorylation is directly proportional to x. Increasing x increases the slope of the green line, and decreasing it decreases the slope (Figure 5.5a). As the slope of the line increases, the point at which it intersects the blue line—the steady-state concentration of y^*—shifts to the right, but it does so with a law-of-diminishing-returns quality; the geometry of the situation is such that a change from 0 to 1 units of x shifts the equilibrium point further than a change from 1 to 2, which shifts it further than a change from 2 to 3, and so on.

One can extract and reconstruct the Michaelian steady state response curve from the rate–balance plot. On the x-axis we plot the assumed value of x; on the y-axis we plot the corresponding steady state value of y^*, which we read from the rate–balance plot (**Figure 5.5**). By repeating this for a few additional assumed values of x, we can sketch out the whole response curve (**Figure 5.5b**). Not surprisingly,

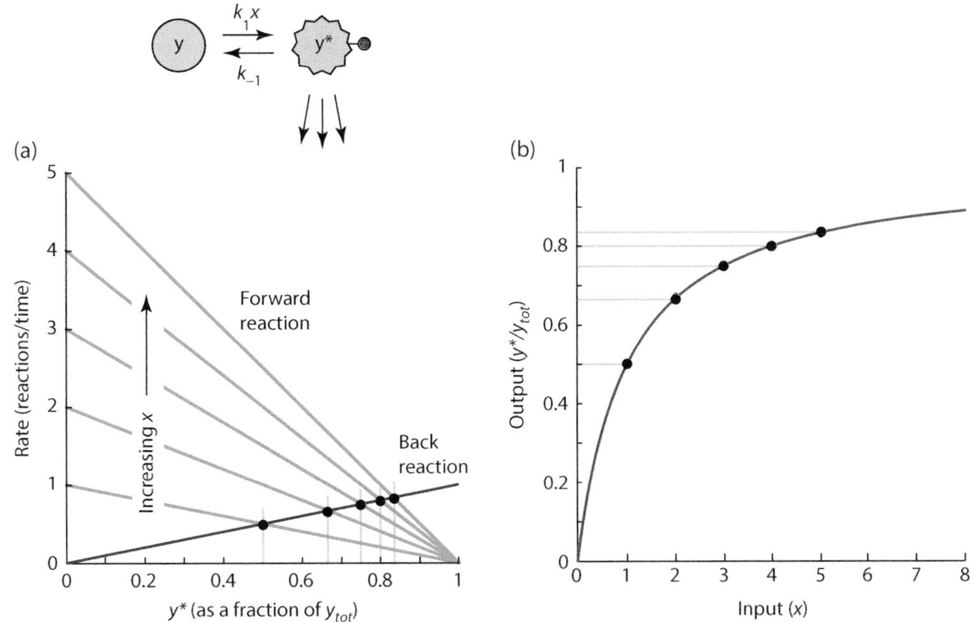

Figure 5.5 Extracting a Michaelian steady-state response from the rate–balance plot. (a) The rate–balance plot for a mass action phosphorylation–dephosphorylation reaction. The parameter values are $k_1 = k_{-1} = 1$ and $x = 1, 2, 3, 4$, or 5. The steady-state value of y^* for each assumed value of x can be read off the plot from the intersections between the green lines and the blue line. (b) The steady-state response curve can be constructed by plotting the values of x on the x-axis and the graphically determined steady-state values of y^* on the y-axis.

the points fall exactly on the curve defined by a Michaeilian response equation (Eq. 5.4).

5.5 THE DYNAMICS OF THE SYSTEM CAN ALSO BE UNDERSTOOD FROM THE RATE–BALANCE PLOT

So far we have focused on what the rate–balance plot can tell us about the steady state of the system. But actually the plot contains more information than that. For any value of the concentration of y^*, the plot shows us whether there will be net phosphorylation or dephosphorylation, and it shows us how fast or slow the phosphorylation or dephosphorylation will be. This is emphasized in **Figure 5.6a**. To the left of the steady state, the forward rate (green line) is larger than the back rate (blue line), and so there is net association; to the right of the steady state the back rate is larger than the forward rate, and so there is net dissociation. The net rate of phosphorylation or dephosphorylation is simply the vertical distance between the rate curves.

Note that for each value of y^*, the net rate can be viewed as a vector whose direction is either to the right (positive) or the left (negative) and whose magnitude is the absolute value of the rate. The collection of all such vectors is the one-dimensional vector field of the system's reaction rates. We can display a sampling of the vector field to show us in which direction and at what rate the system will move as a function of y^* (**Figure 5.6b**). We have depicted the steady state—that is, the point where the magnitude of the vector field is zero—as a filled circle, which is customary for stable steady states. The steady state is stable because all of the vectors point toward it.

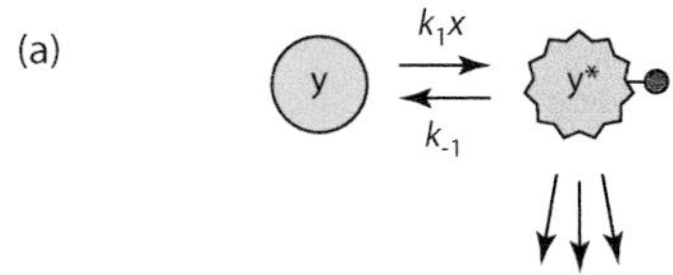

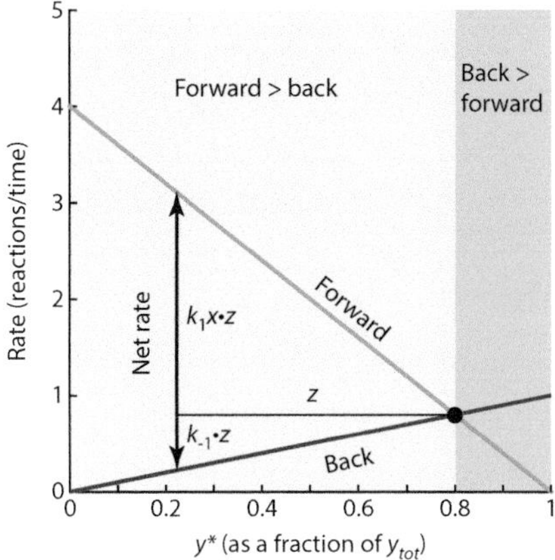

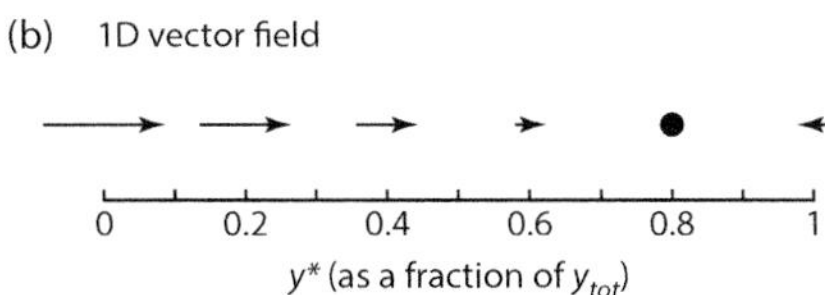

Figure 5.6 Extracting dynamics from the rate–balance plot. (a) The magnitude of the net rate can be calculated geometrically; it is simply the negative slope of the forward reaction line (k_1x) times the distance z to the steady state, plus the slope of the back reaction line (k_{-1}) times z. (b) A one-dimensional vector field, where the vector length is proportional to the net reaction rate. The stable steady state is denoted by the solid circle.

From either the rate–balance plot (**Figure 5.6a**) or the vector field plot (**Figure 5.6b**) we can see that the rate at which the system moves toward the equilibrium is linearly proportional to how far the system is from equilibrium. This is exactly what it takes for a system to exponentially approach its equilibrium. If we define the variable $z[t]$ to represent the distance from the present value of $y*$ to the equilibrium and if we know that the velocity of the system, $\frac{dz[t]}{dt}$, is linearly proportional to $z[t]$ and opposite in sign (the velocity opposes the displacement from steady-state), this means that:

$$\frac{dz[t]}{dt} = -k_{apparent}z^{[t]} \tag{5.15}$$

and it immediately follows that:

$$z[t] = z[0]e^{-k_{apparent}t}. \tag{5.16}$$

Two things contribute to $k_{apparent}$: the (positive) slope of the back reaction line and the (negative) slope of the forward reaction line. By simple geometry (**Figure 5.6**), it follows that at any point in time, the net rate $\frac{dz[t]}{dt} = -(k_1x + k_1)z[t]$, and so $k_{apparent} = k_1x + k_{-1}$. We have re-derived Eq. 5.5 and in a way that is arguably simpler and more physically motivated than the approach taken in Section 5.1.

Anytime a one-dimensional system's vector field has its magnitude increase linearly with the distance from equilibrium or steady state, the system will approach the steady state exponentially. Likewise, if the system approaches its steady state exponentially, the magnitude of the vector field must increase linearly with the distance from the steady state. Even if we cannot solve the ODE for some complicated kinetic system exactly, we may be able to tell by examining the vector field whether it will approach its steady state with a time course that falls off more or less abruptly than an exponential approach does.

The rate–balance plot shows us geometrically why the halftime for the response to a step up in the input ligand will always be smaller (faster) than the halftime for the response to a step back down. Suppose that we step the quantity x up from 0 to 4, so that the concentration of $y*$ will rise from 0 to 0.8 (**Figure 5.7a**). The net rate at which $y*$ approaches its steady state are relatively large quantities, depicted by the vertical distances between the green and the blue rate curves (**Figure 5.7a**). On the way back down, the net rate is the distance between the blue curve and the x-axis, which will be smaller than the on-the-way-up rate was (**Figure 5.7b**). Thus the response to a step up is faster than the response to the step back down.

5.6 RATE-BALANCE ANALYSIS HELPS EXPLAIN ZERO-ORDER ULTRASENSITIVITY

With rate-balance analysis now in our armamentarium, we can take on the challenge of trying to understand why enzyme saturation gives rise to ultrasensitivity. Let us saturate one enzyme at a time: first the phosphatase (**Figure 5.8b**), then the kinase (**Figure 5.8c**), and finally both (**Figure 5.8d**), and see what happens. For comparison, **Figure 5.8a** shows the now-familiar rate–balance plot for a phosphorylation–dephosphorylation reaction with no saturation.

If the phosphatase is saturable, we know that the blue dephosphorylation curve will be a hyperbola rather than a straight line. Arbitrarily

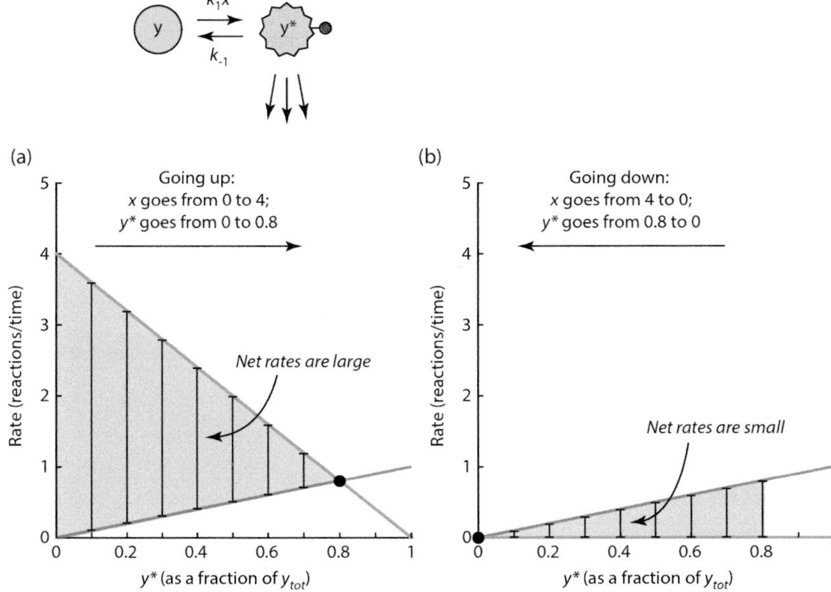

Figure 5.7 The rate–balance plot rationalizes why the response to a step up in kinase activity (a) is always faster than the response to a step back down (b).

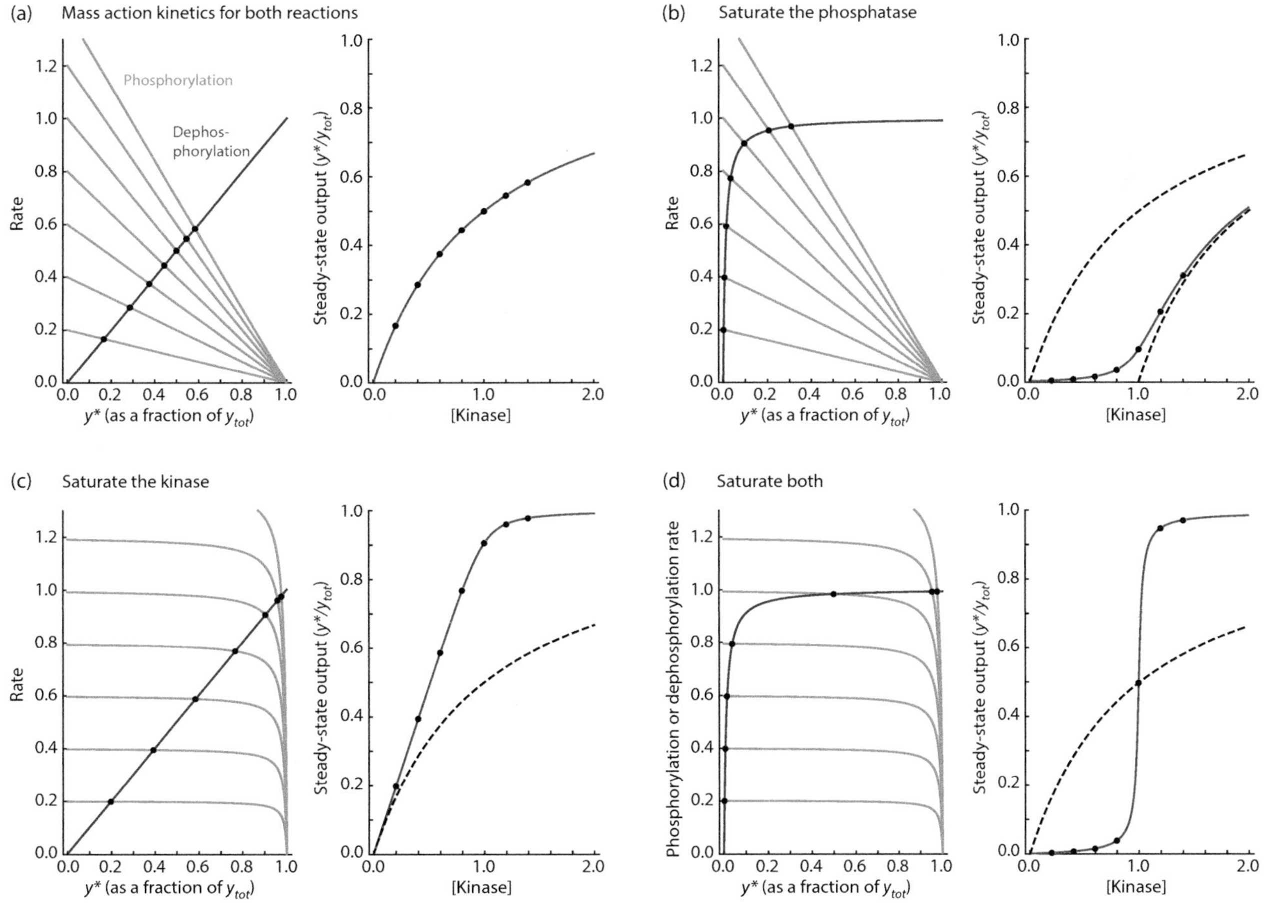

Figure 5.8 Rate-balance analysis for a phosphorylation–dephosphorylation system with saturable kinase and/ or phosphatase. Each panel shows a rate–balance plot with a range of kinase concentrations (left) and a steady-state response curve (right) for a phosphorylation/dephosphorylation system. The black dashed curves in panels b–d are Michaelian response curves (or in one instance in panel b, a Michaelian response curve shifted to the right by one unit), shown for comparison. (Adapted from Ferrell and Ha, *Trends Biochem Sci.* 2014.)

we will assume the K_M value for dephosphorylation (K_{M2}) to be 100-fold lower than the total concentration of the substrate y; i.e., the phosphatase is running close to saturation for most of the range of $y*$ concentrations. The result is shown in **Figure 5.8b** (blue curve); the dephosphorylation rate rises steeply initially and then levels off once $y*$ reaches about 0.1 or so. As the kinase activity increases in 0.2 unit increments (**Figure 5.8b**, green lines), there is very little production of $y*$ until the intersection points get to where the blue curve starts to level off, which happens at a kinase concentration of about 1. As a result, the steady-state response curve (**Figure 5.8b**, right, red curve) acquires a threshold. At kinase concentrations beyond the threshold, the system responds in a graded way to additional increments of kinase. We are halfway toward generating a full sigmoidal response curve.

Next, let us assume that the phosphatase is not saturable, but the kinase is, and again let us take its K_M value (K_{M1}) to be 100-fold lower than the total concentration of substrate y. Now the dephosphorylation rate curve is a straight line (**Figure 5.8c**, blue line) and the phosphorylation rate curves are a family of hyperbolas with different maximum values (**Figure 5.8c**, green curves). The result is that as the kinase activity increases, the nearly horizontal green curves march their way up the blue line (**Figure 5.8c**, left), yielding a linear response until the response is very close to maximal (**Figure 5.8c**, right). The steady-state response curve does not have a threshold but does have an abrupt approach to maximum.

Finally, if both the kinase and phosphatase are assumed to be close to saturation, the response shows both a threshold and an abrupt leveling off at maximal response (**Figure 5.8d**). There is virtually no response until a kinase concentration of about 1 is obtained, and beyond that the response is nearly maximal. The result is a steeply sigmoidal input–output relationship (**Figure 5.8d**, right). Thus, saturation of the kinase and phosphatase synergize to produce a highly switch-like response, and the reason is because both rate curves are nearly flat over most of the range of $y*$ concentrations. Note that the flatness of the rate curves not only produces a switch-like steady-state response but also makes it so that the steady states are approached relatively sluggishly. The vertical distance between the forward and back rates when the system is away from steady state—which is the rate at which the system heads back toward the steady state—is substantially smaller than it would be for a mass action system with the same steady-state and the same steady-state phosphorylation turnover rate.

5.7 DOES ZERO-ORDER ULTRASENSITIVITY OCCUR IN VIVO?

Does this simple, powerful mechanism for generating a switch-like response actually occur in vivo? Ultrasensitivity does seem to be fairly common in cellular regulation (**TABLE 5.1**), and there are a couple of classic examples of zero-order ultrasensitivity from experiments done in the 1980s. For example, reconstitution studies showed that the steady-state phosphorylation of muscle glycogen phosphorylase by phosphorylase kinase in the presence of phosphorylase phosphatase in vitro is sigmoidal when the substrate is present at a concentration of 70 μM, which is a few-fold higher than the observed K_M values for the two enzymes, but not at 20 μM. This demonstrates that zero-order ultrasensitivity can be made to happen in vitro. Moreover, the substrate protein, glycogen phosphorylase, is highly abundant, perhaps as high as 100 μM in skeletal muscle, so the concentrations required to produce zero-order ultrasensitive in vitro are probably relevant to the situation in vivo.

Sadly there are few other examples where we know with reasonable certainty that a protein kinase substrate is present in vivo at a concentration several-fold above its K_M values for phosphorylation and dephosphorylation. Thus, zero-order ultrasensitivity remains a triumph of theory that seems likely, but not certain, to be of broad significance in cell signaling systems.

MULTISITE PHOSPHORYLATION

5.8 THE TEMPORAL DYNAMICS OF A MULTISTEP ACTIVATION PROCESS TELLS YOU THE NUMBER OF PARTIALLY RATE-DETERMINING STEPS

Most proteins that are phosphorylated at all are phosphorylated at multiple sites. Sometimes several kinases and/or phosphatases are involved, and sometimes a single kinase will phosphorylate a target protein at several or even dozens of sites. Under the right circumstances, a multisite phosphorylation process can be mathematically equivalent to a KNF mechanism and so can yield ultrasensitive steady-state responses. Here we will work through such a case.

To make things as simple as possible, let us assume that our target protein y is phosphorylated by a kinase x at two specific sites, yielding a singly phosphorylated species y_1 and a doubly phosphorylated species y_2. Next let us make a series of simple but plausible mechanistic assumptions:

- Both the phosphorylation and dephosphorylation reactions can be described as mass action processes.
- Only the doubly phosphorylated species is active. This means that protein y acts like an AND gate; phosphorylation of site 1 AND phosphorylation of site 2 yields activation.
- The phosphorylation and the dephosphorylation reactions are distributive rather than processive; that is, both the kinase and phosphatase release their protein substrate after each phosphorylation or dephosphorylation reaction, so that the dual phosphorylation of y requires two productive collisions between kinase and substrate, and the complete dephosphorylation of y_2 requires two productive collisions between phosphatase and substrate.
- The phosphorylation and dephosphorylation reactions are strictly ordered, so that only one mono-phosphorylated form is ever produced.
- The kinase x can be considered the input to the system, and the phosphatase(s) can be lumped into the two dephosphorylation rate constants.

These assumptions yield the mechanism shown schematically in **Figure 5.9a**.

We can write down three rate equations for the system:

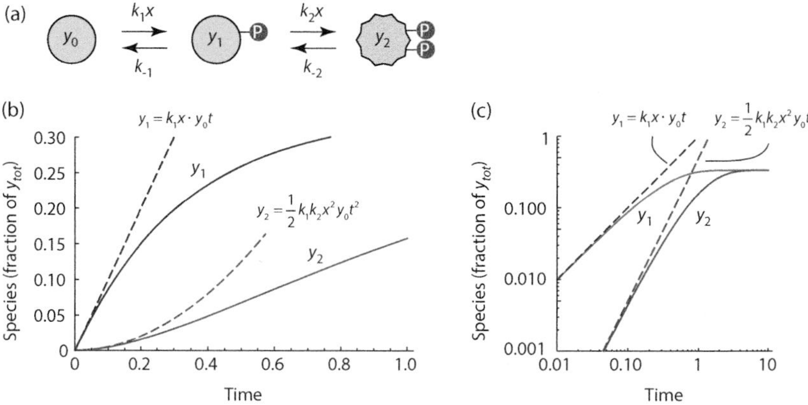

Figure 5.9 The initial time course of distributive dual phosphorylation. (a) Schematic of the process. (b) The time course for the formation of singly phosphorylated y_1 (blue) and doubly phosphorylated y_2 (red). For comparison, the small t limits are shown as dashed curves. (c) The same data as in B but plotted on a log–log plot. The slope of the curve is its polynomial order. For both B and C, we have assumed that $k_1 = k_2 = k_{-1} = k_{-2} = y_0 = x = 1$.

$$\frac{dy_0}{dt} = -k_1x \cdot y_0 + k_{-1}y_1 \tag{5.17}$$

$$\frac{dy_1}{dt} = k_1x \cdot y_0 - k_{-1}y_1 - k_2x \cdot y_1 + k_{-2}y_2 \tag{5.18}$$

$$\frac{dy_2}{dt} = k_2x \cdot y_1 - k_{-2}y_2. \tag{5.19}$$

These three rate equations are almost identical to those for the binding of two ligands to a dimeric receptor through a KNF mechanism (**Eqs. 3.41–3.43**); the only difference is a factor of two that arose from the assumption that a ligand could bind to either of two equivalent sites in the KNF case, which is lost here because we are considering the two phosphorylation sites to be nonequivalent and the two phosphorylations to be strictly ordered.

Let us begin with a look at the initial dynamics of this system. Suppose we carried out an experiment where we added a fixed amount of x to some totally unphosphorylated y, and then monitored the time course of appearance of singly phosphorylated y_1 and doubly phosphorylated y_2. Furthermore, let us assume that the two phosphorylation reactions are similar in speed. For the initial increments of time, y_1 and y_2 are negligible, which means that **Eq. 5.18** reduces to:

$$\frac{dy_1}{dt} = k_1x \cdot y_0. \tag{5.20}$$

We can solve this by integrating both sides:

$$y_1 = k_1x \cdot y_0 t + C. \tag{5.21}$$

As we have assumed that $y_1[0] = 0$, the constant of integration C must equal 0. Thus, initially y_1 increases linearly with time:

$$y_1 \propto t. \tag{5.22}$$

This is shown in **Figure 5.9**.

For the doubly phosphorylated y_2 species, at very early times Eq 5.19 reduces to:

$$\frac{dy_2}{dt} = k_2 x \cdot y_1 = k_1 k_2 x^2 y_0 t, \quad (5.23)$$

which means that y_2 is initially proportional to t^2:

$$y_2 = \frac{1}{2} k_1 k_2 x^2 y_0 t^2. \quad (5.24)$$

If there is AND gate logic, and y_2 is the only active form of y, then there will be a time lag in the appearance of active y_2—the time course curve will be concave-up and parabolic in shape (**Figure 5.9b**). Likewise, in a triple-phosphorylation process, where all steps are partially rate determining (i.e. none one is hugely faster than the others), the concentration of the triply phosphorylated species will increase like t^3, and for four, t^4, and so on.

In general, if you have an input connected to an output through some incompletely understood mechanism, you should be able to estimate the number of partially rate-determining steps between the input and the output by plotting the data on a log–log plot and extracting the initial slope, which is the polynomial order of the relationship (Figure 5.9c). It requires having a sensitive, accurate way of obtaining time course data, but in principle it can be done.

Note that if we were to remove x after y had been maximally phosphorylated, the logic would be reversed: y_1 would increase proportionally with t, and y_0 would go up like t^2. Thus, if y_1 is an inactive species (as we assumed), then there will be no time lag in the inactivation response; only in the activation response.

5.9 ASSUMING MASS ACTION KINETICS, STEADY-STATE MULTISITE PHOSPHORYLATION IS DESCRIBED BY A KNF-TYPE EQUATION

To analyze the steady-state response of our multisite phosphorylation model, we set the three derivatives equal to zero:

$$0 = -k_1 x.y_0 + k_{-1} y_1 \quad (5.25)$$

$$0 = -k_1 x.y_0 - k_{-1} y_1 - k_2 x.y_1 + k_{-2} y_2 \quad (5.26)$$

$$0 = -k_2 x.y_1 - k_{-2} y_2. \quad (5.27)$$

Note that we can use Eq. 5.25 to eliminate the first two terms from the right-hand side of Eq. 5.26; just as we found with equilibrium binding to multisubunit receptors, this system is in steady state only when all of the individual phosphorylation–dephosphorylation reactions are in steady state.

There is also a conservation equation:

$$y_{tot} = y_0 + y_1 + y_2. \quad (5.28)$$

From these four algebraic equations (5.25–5.28) we can derive expressions for the fraction of the total y in each of its three phosphorylation states, y_0, y_1, and y_2, as a function of the kinase concentration x and the rate constants. In particular, the fraction in the active, doubly phosphorylated y_2 form is:

$$\left(\frac{y_2}{y_{tot}}\right)_{ss} = \frac{k_1 k_2 x^2}{k_{-1}k_{-2} + k_1 k_{-2} x + k_1 k_2 x^2}. \tag{5.29}$$

If we define $K_1 = k_{-1}/k_1$ and $K_2 = k_{-2}/k_2$, we can write:

$$\left(\frac{y_2}{y_{tot}}\right)_{ss} = \frac{x^2}{K_1 K_2 + K_2 x + x^2}. \tag{5.30}$$

Keep in mind that these K's are not equilibrium constants but instead are ratios of rate constants for two opposing reactions. From past experience (**Eqs. 3.48–3.50**) and the form of the denominator in **Eq. 5.30**, you can probably correctly guess what the corresponding expressions for y_0 and y_1 are—they have K_1K_2 (for y_0) or K_2x (for y_1) in the numerator in the place of x^2.

Let us examine the response function defined by **Eq. 5.30** in more detail. If we assume that the two K values are equal, meaning that no matter what the individual values are for the phosphorylation and dephosphorylation of the two sites, the balance between the forward and back reactions is no more favorable for the first phosphorylation than for the second, the result is a slightly sigmoidal response curve (**Figure 5.10b,c**, dashed black curve). We can extract the $EC10$ and $EC90$ values from this curve, and then use **Eq. 3.54** to calculate an effective Hill coefficient for the response of 1.36. If we assume that K_2 is smaller than K_1, either because the phosphorylation of y_1 is more favorable than the phosphorylation of y_0, or the dephosphorylation of y_1 is more favorable than the dephosphorylation of y_2, or both—i.e., there is positive cooperativity in the phosphorylation and/or dephosphorylation reactions—the resulting curves are more switch-like. In the limit where $K_2 / K_1 \to 0$, the response approaches a Hill curve with $n=2$ (**Figure 5.10b,c**, blue curves). And if K_2 is larger than K_1, meaning that there is negative cooperativity in the phosphorylation and/or dephosphorylation, then the curves will be intermediate between a Michaelian curve and the $n=1.36$ curve (**Figure 5.10b,c**, red curves).

These ultrasensitivities are fairly modest compared to the largest Hill exponents in **TABLE 5.1**. Keep in mind, though, that some substrates are phosphorylated at many sites—sometimes dozens—which could yield higher degrees of ultrasensitivity. In general, if a protein's activity depends on the ordered, mass action, non-processive phosphorylation of n sites, the resulting steady-state activity is:

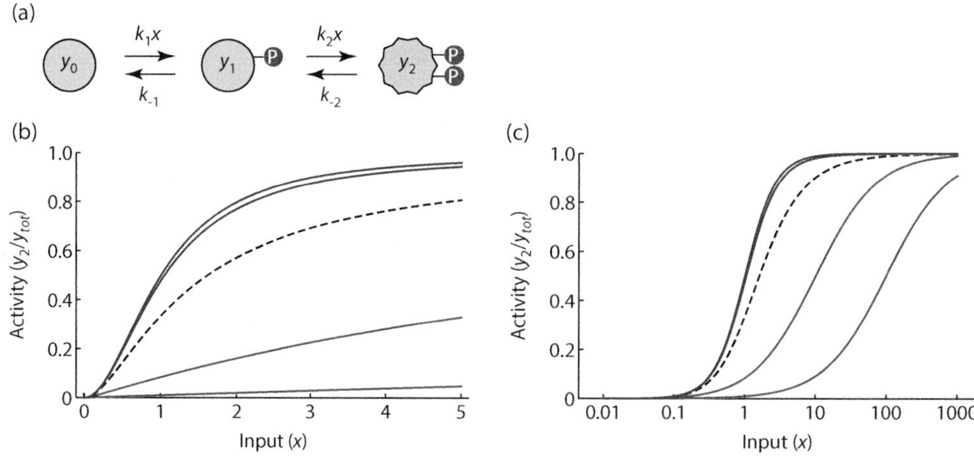

Figure 5.10 Multisite phosphorylation. (a) Schematic view. Here we have assumed that substrate y can be phosphorylated at two sites by kinase x; that the phosphorylation and dephosphorylation reactions are distributive and ordered; and that only the doubly phosphorylated form of y (y_2) is active. (b, c) Steady-state activities are plotted for 5 choices of K_1 and K_2: 100 and 0.01 (red); 10 and 0.1 (red); 1 and 1 (dashed black); 0.1 and 10 (blue); and 0.01 and 100 (blue). The activity curves are shown on linear plots (b) and on semilog plots (c).

TABLE 5.1 **Examples of Subsensitive and Ultrasensitive Responses**

Stimulus	Response	Effective Hill Exponent	Experimental System	Reference
EGF (Spitz)	EGFR binding	0.24	Purified recombinant *Drosophila* proteins	Alvarado et al. (2010)
Insulin	Insulin receptor binding	0.46	Purified recombinant human proteins	Whittaker et al. (2008)
Aspartate	Aspartate receptor (Tar) binding	0.6	Purified *Salmonella typhimurium* Tar protein	Biemann and Koshland (1994)
Acetylcholine	Nicotinic cholinergic receptor conductance	1.3	Chicken neuronal homomeric σ7 receptors	Galzi et al. (1996)
Delta (in trans only)	Notch production	1.7	CHO cells	Sprinzak et al. (2010)
Mos	MEK1 activity	1.7	*Xenopus laevis* oocyte extracts	Huang and Ferrell (1996)
Phosphorylase kinase/ phosphatase	Glycogen phosphorylase activity	2	Reconstituted mammalian muscle enzymes	Meinke et al. (1986)
RsbQP	σ^B activity	2.1	*Bacillus subtilis*	Locke et al. (2011)
AICAR	AMPK activity	2.5	Rat INS-1 cells	Hardie et al. (1999)
Ca^{2+}	Calmodulin-dependent cAMP phosphodiesterase activity	2.7	Purified beef heart proteins	Teo and Wang (1973)
IP_3	Calcium release	3	Permeablized rat basophilic leukemia cells	Meyer et al. (1988)
Cdk1	Wee1A hyperphosphorylation	3.5	*Xenopus laevis* egg extracts	Kim and Ferrell (2007)
Anisomycin or sorbitol	JNK activity	3–10	HeLa, HEK293, and Jurkat cells	Bagowski et al. (2003)
Mos	Erk2 activity	5	*Xenopus laevis* oocytes	Huang and Ferrell (1996)
Cln2	Cln2 synthesis	5	*Saccharomyces cerevisiae*	Charvin et al. (2010)
KinA	σ^E and σ^F activities	10	*Bacillus subtilis*	Narula et al. (2012)
CheY-P	Flagellar motor output	~10–20	*Escherichia coli*	Cluzel et al. (2000); Yuan and Berg (2013); Yuan et al. (2012)
Cdk1	Cdc25C hyperphosphorylation	11	*Xenopus laevis* egg extracts	Trunnell et al. (2011)
Delta (cis and trans)	Notch production	12	CHO cells	Sprinzak et al. (2010)
Cdk1	APC/C^{Cdc20} activity	≥17	*Xenopus laevis* egg extracts and embryos	Tsai et al. (2014); Yang and Ferrell (2013)
HGF	HGF-inducible mRNAs	>1,000 mRNAs showed ultrasensitive responses with Hill exponents ranging from just above 1 to 76	MDCK cells	Senthivel et al. (2016)

$$\left(\frac{y_n}{y_{tot}}\right)_{ss} = \frac{x^n}{(K_1K_2K_3\cdots K_n)+(K_2K_3\cdots K_n)x+(K_3\cdots K_n)x^2+\cdots+K_nx^{n-1}+x^n}. \tag{5.31}$$

With this model the effective Hill exponent will always be less than or equal to the number of phosphorylation sites, with the greatest ultrasensitivity achieved when there is substantial cooperativity in the phosphorylation and/or dephosphorylation.

So far we have assumed AND gate logic for the activation of the substrate by phosphorylation—so for a substrate with two phosphorylation sites, only the doubly phosphorylated species are active, and the result is an ultrasensitive activity curve. By analogy with the KNF model, we should be able to achieve a subsensitive response as well.

For this to occur, we would need (1) negative cooperativity, with the first phosphorylation making the second more difficult, and, importantly (2) an activation mechanism where the singly phosphorylated form is half as active as the doubly phosphorylated form. I do not know of an example where this latter assumption is true, but it is probably worth bearing in mind that an ultra-graded, subsensitive response is a theoretical possibility. It is also possible that the singly and doubly phosphorylated forms are comparably active; in other words, there is OR gate logic in the activation. This is in fact true for the activation of the MAP kinase MEK1 by Raf or Mos. This means that MEK should respond promptly when Raf is turned on but only after a time lag when it is turned back off—the opposite of what we found for dual phosphorylation when both sites must be phosphorylated for the substrate to change in activity.

5.10 PRIMING CAN IMPART POSITIVE COOPERATIVITY ON MULTISITE PHOSPHORYLATION

Let us return now to a multisite phosphorylation system with AND gate logic. As seen in **Figure 5.10**, the most switch-like responses are obtained when K_2 is small compared to K_1, which could arise because of positive cooperativity in the phosphorylation reaction, the dephosphorylation reaction, or both. Is positive cooperativity ever actually found in such reactions, and, if so, how does it arise?

It is not hard to imagine how the positive cooperativity arises a multisubunit protein like hemoglobin. The hemoglobin tetramer consists of globular subunits with their amino acids packed in an orderly fashion (**Figure 3.1**), so the binding of oxygen to one heme group could cause a local conformational change that would then be transmitted through a network of energetically coupled residues to the other heme groups. Likewise with the negative cooperativity of the epidermal growth factor receptor (EGFR), the binding of the first epidermal growth factor (EGF) molecule appears to cause a gross rearrangement of one receptor subunit, which then pushes on the other receptor subunit, putting into a conformation that makes the binding of the second EGF molecule less favorable.

This kind of allosteric regulation might well contribute cooperativity to some instances of multisite protein phosphorylation, with, say, a first phosphorylation inducing a conformation change that makes a second phosphorylation site more accessible to the kinase. But, as mentioned earlier, most phosphorylations occur at residues that reside in intrinsically disordered regions of proteins. It is not so easy to see how conformation changes and a traditional allosteric mechanism could link the phosphorylation of two residues in intrinsically disordered regions.

However, a simpler type of coupling can still take place. There are a number of examples now where the phosphorylation of one residue, even in an intrinsically disordered region, produces a phosphoepitope that can serve as a docking site for a protein kinase—sometimes, but not always, the same kinase that carried out the first priming phosphorylation. In this way, a priming phosphorylation can promote subsequent phosphorylations through enforced proximity, a simple but powerful alternative to allostery. The cyclin–Cdk–Cks complex is one nice example of a kinase that makes use of priming phosphorylations, as shown schematically in **Figure 5.11**. In this case, the Cks subunit (called Suc1 in *S. pombe*, Cks1 in *S. cerevisiae*, and Cks1 or 2

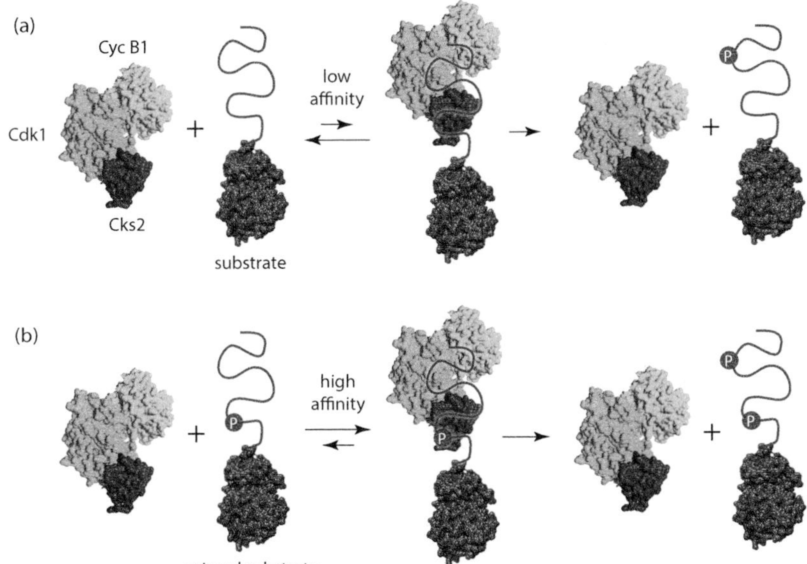

Figure 5.11 Positive cooperativity in multisite phosphorylation through the interaction of a phosphoepitope-binding domain with an unstructured priming site. (a) The unprimed substrate interacts relatively weakly with the kinase. (b) The primed substrate interacts more strongly through interaction with the blue Cks2 subunit. (Based on work from McGrath et al., *Nat Struct Mol Biol*. 2013.)

in vertebrates) serves as the phosphoepitope-binding subunit, binding phosphothreonine residues in a particular primary sequence context and facilitating the phosphorylation of other nearby serine and threonine residues. This priming-and-multisite-phosphorylation theme is probably relevant to the regulation of scores of substrate proteins by the cell cycle Cdk complexes. Glycogen synthase kinase 3 (GSK3) and casein kinase 1 (CK1) are two other kinases that depend on a priming phosphorylation to carry out multisite phosphorylation and regulate their substrates. In these cases, the priming phosphorylation interacts with a region of the kinase's catalytic domain rather than a separate subunit, but otherwise the basic phenomenon is similar.

5.11 DISTRIBUTIVE MULTISITE PHOSPHORYLATION IMPROVES SIGNALING SPECIFICITY

In any signaling system, specificity is an important issue, and in the case of protein phosphorylation, it is a particularly challenging one. There are probably approximately 10,000 different proteins in a typical human cell, with an average length of ~400 amino acids. This means that there are about 4,000,000 different amino acid residues. About 17% of these are Ser (8.5%), Thr (5.7%), or Tyr (3.0%) residues and hence are potential phosphorylation sites, which means that there are ~700,000 possible phosphorylation sites for a protein kinase to choose among. Yet despite the fact that all of the 500+ classical protein kinases are evolutionarily related and have similar structures, each one is somehow able to focus in on a few to a few hundred of the ~700,000 possible target sites. Maximizing on-target phosphorylation and minimizing off-target phosphorylation is a formidable challenge.

To achieve this specificity, protein kinases often make use of multiple binding interactions to recognize their on-target substrates. Amino acids adjacent to the phosphorylation site interact with residues near the kinase's active site, and this interaction contributes to specificity, but kinases also often make use of a separate docking site that binds to residues far from the phosphorylation site to increase the binding energy. Only substrates that have the right primary sequence motif around the phosphorylation site and the right docking motif will be phosphorylated efficiently.

In addition, multisite phosphorylation can amplify whatever specificity is inherent in the kinase–substrate interaction. We can illustrate this with a specific example: consider the MAP kinase (MAPK) cascades that crop up over and over again in eukaryotic signaling and which were introduced in Section 1.5. Most eukaryotic cells possess several classes of MAPK cascades that operate in parallel, with little cross talk. In human cells, one cascade starts with four MAPKKKs (MAP kinase kinase kinases; the Raf proteins and Mos) followed by two MAPKKs (MAP kinase kinases; MEK1/2) and two MAPKs (ERK1 and 2); another starts with one of at least 13 JNKKKs followed by two JNKKs (MKK4 and 7) and three JNKs (JNK1, 2, and 3). The last two levels of these cascades are shown schematically in **Figure 5.12a**. How do the MEKs manage to activate ERKs, and not the structurally related JNKs, and how do the JNKKs manage to activate JNKs without activating ERKs?

Let us first consider the issue of specificity in a hypothetical system where only one phosphorylation was required for activation of JNK and ERK. For simplicity let us also assume that mass action kinetics describes the phosphorylation and dephosphorylation reactions. Let us suppose that MEK phosphorylates its on-target substrate ERK with a rate constant of k_1 and the off-target substrate JNK with a rate constant of ϕk_1, where the factor ϕ is less than 1. The steady-state level of

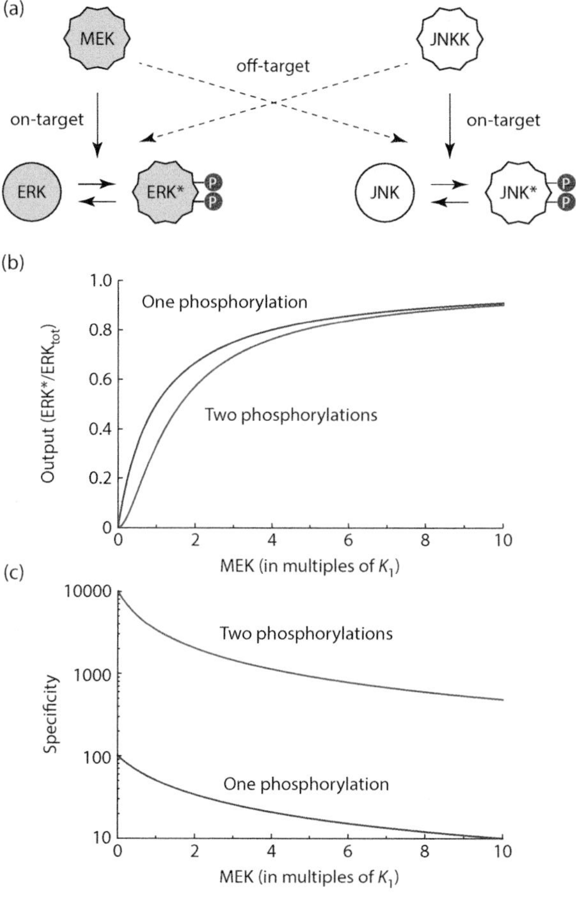

Figure 5.12 Distributive multisite phosphorylation can enhance signaling specificity. (a) Two parallel MAP kinase cascades that operate independently. (b, c) On-target steady-state response (ERK*) (b) and signaling specificity (taken to be the ratio of ERK*/JNK* at steady state) (c), assuming one (blue) or two (red) mass action phosphorylation reactions are required for ERK or JNK activation, with $\phi = 0.01$.

active ERK (ERK*) in the presence of some concentration x of the right MEK will be Michaelian:

$$\left(\frac{ERK^*}{ERK_{tot}}\right)_{ss} = \frac{x}{K_1 + x}. \tag{5.32}$$

Next consider the off-target phosphorylation of JNK by MEK. If we assume for simplicity that the rate constants for JNK and ERK dephosphorylation (the k_{-1} values) are equal, then since $K_1 = \frac{k_{-1}}{\phi k_1}$, the steady-state level of off-target activation of JNK* will be:

$$\left(\frac{JNK^*}{JNK_{tot}}\right)_{ss} = \frac{x}{\frac{K_1}{\phi} + x} \tag{5.33}$$

We can define the specificity of the process to be the ratio of the resulting steady-state activities of ERK and JNK:

$$Specificity = \frac{\frac{K_1}{\phi} + x}{K_1 + x}. \tag{5.34}$$

Equation 5.34 is plotted as the blue curve in **Figure 5.12c** for one value of ϕ, ϕ=0.01. We used a semilog plot, because it facilitates the comparison of this specificity curve to the one we will obtain for the dual phosphorylation case, and we show the steady-state response in **Figure 5.12** as well (on a regular linear plot). Specificity is maximal, approaching $1/\phi = 100$, when the stimulus x is infinitesimal, and it falls quickly as x is increased. When $x = K_1$ and the ERK activation is half-maximal, the specificity falls to 50.5, and when $x = 9K_1$ so that the ERK activation is 90% maximal, the specificity falls to 10.9. As the input approaches infinity and the output approaches 100%, the specificity approaches 1; in other words, the MEK is producing as much off-target JNK activation as on-target ERK activation. Thus, the system is very good at distinguishing on-target from off-target MAPKs when the output of the system is very low but not so good at high output levels.

Now consider the case where we have two distributive phosphorylations, each described by mass action kinetics. Furthermore, assume that each of the two phosphorylations is a factor of ϕ slower for JNK than for ERK and again assume the dephosphorylation rate constants for the two substrates are equal. From Eq. 5.30, the steady-state activation of ERK by MEK (x) is given by:

$$\left(\frac{ERK^*}{ERK_{tot}}\right)_{ss} = \frac{x^2}{K_1K_2 + K_2x + x^2}, \tag{5.35}$$

and the steady-state activation of JNK is:

$$\left(\frac{JNK^*}{JNK_{tot}}\right)_{ss} = \frac{x^2}{\frac{K_1K_2}{\phi^2} + \frac{K_2}{\phi}x + x^2}. \tag{5.36}$$

The specificity of ERK activation is the ratio of these two expressions:

$$Specificity = \frac{\frac{K_1K_2}{\phi^2} + \frac{K_2}{\phi} + x}{K_1K_2 + K_2x + x}. \tag{5.37}$$

This specificity function is plotted as the red curve in **Figure 5.12c**, taking $K_1 = K_2 = 1$ and ϕ=0.01. Now the specificity approaches $1/\phi^2$ = 10,000 for small levels of activation. The specificity still falls off as the input (x) and the output (ERK*) increase (and, in fact, in relative terms it falls off more abruptly), but because the specificity starts at such a high value, even when the output is 90% maximal (which occurs when $x=9.9K_1$) the specificity is still a whopping 102, higher than the specificity for activation by a single phosphorylation ever was. Thus, distributive multisite phosphorylation can greatly enhance the specificity of a signal transduction process, and in the limit of low inputs and outputs, the specificity of an n-site activation process is equal to $1/\phi^n$.

So far we have assumed that the phosphorylation and dephosphorylation reactions are distributive. This is true for the ERK2 MAPK in vitro, but there is evidence that ERK2 phosphorylation may be processive in vivo, possibly because of scaffold proteins or macromolecular crowding. If phosphorylation and dephosphorylation are strictly processive, would the dual phosphorylation still enhance specificity? The simple answer is no. However, if the process is only semi-processive—for example, if some fraction of the time the kinase (MEK1/2) dissociates from the mono-phosphorylated form of pMAPK substrate, which is then rapidly dephosphorylated, and the off-target MEK-pJNK complexes are more likely to dissociate than the on-target MEK–pERK complexes—then some specificity enhancement will still occur.

This sort of mechanism has been called kinetic proofreading, and it was proposed by John Hopfield to explain the high fidelity of amino acid incorporation during mRNA translation and nucleotide incorporation during DNA synthesis. Kinetic proofreading is now regarded as a critically important contributor to specificity in a wide variety of biological contexts.

5.12 INESSENTIAL PHOSPHORYLATION SITES CAN CONTRIBUTE TO ULTRASENSITIVITY

So far we have assumed that our doubly phosphorylated protein substrate is active, and the non-phosphorylated and singly phosphorylated species are inactive. To a first approximation, this is true for the ERK2 MAP kinase—the singly phosphorylated form does possess a measurable activity, but it is 10- to 100-fold lower than the doubly phosphorylated form. But this is not always the case for proteins regulated by multisite phosphorylation. For example, as mentioned above, MEK1, the immediate upstream activator of ERK2, undergoes the ordered, distributive phosphorylation of two sites during its activation, but it appears that the protein is maximally activated by the first phosphorylation. This means that ERK2 is relatively hard to turn on but easy to turn off, whereas MEK1 is easy to turn on but hard to turn off. It also means there will be a time lag in ERK2 activation and a time lag in MEK1 inactivation, which could act as a temporal filter. And, finally, it means that the shapes of the steady-state response curves will be a bit different from each other. This raises the question of which requirement produces the more switch-like, ultrasensitive response. Or, more generally, if a protein is phosphorylated at n sites, how many phosphorylations should it take to activate the protein to produce the maximal degree of ultrasensitivity?

Let us begin with the dual phosphorylation case. If both phosphorylation sites must be phosphorylated to achieve activation (AND gate logic), the fraction of the protein activated is, as discussed above:

$$Activity = \frac{x^2}{K_1K_2 + K_2x + x^2}. \tag{5.38}$$

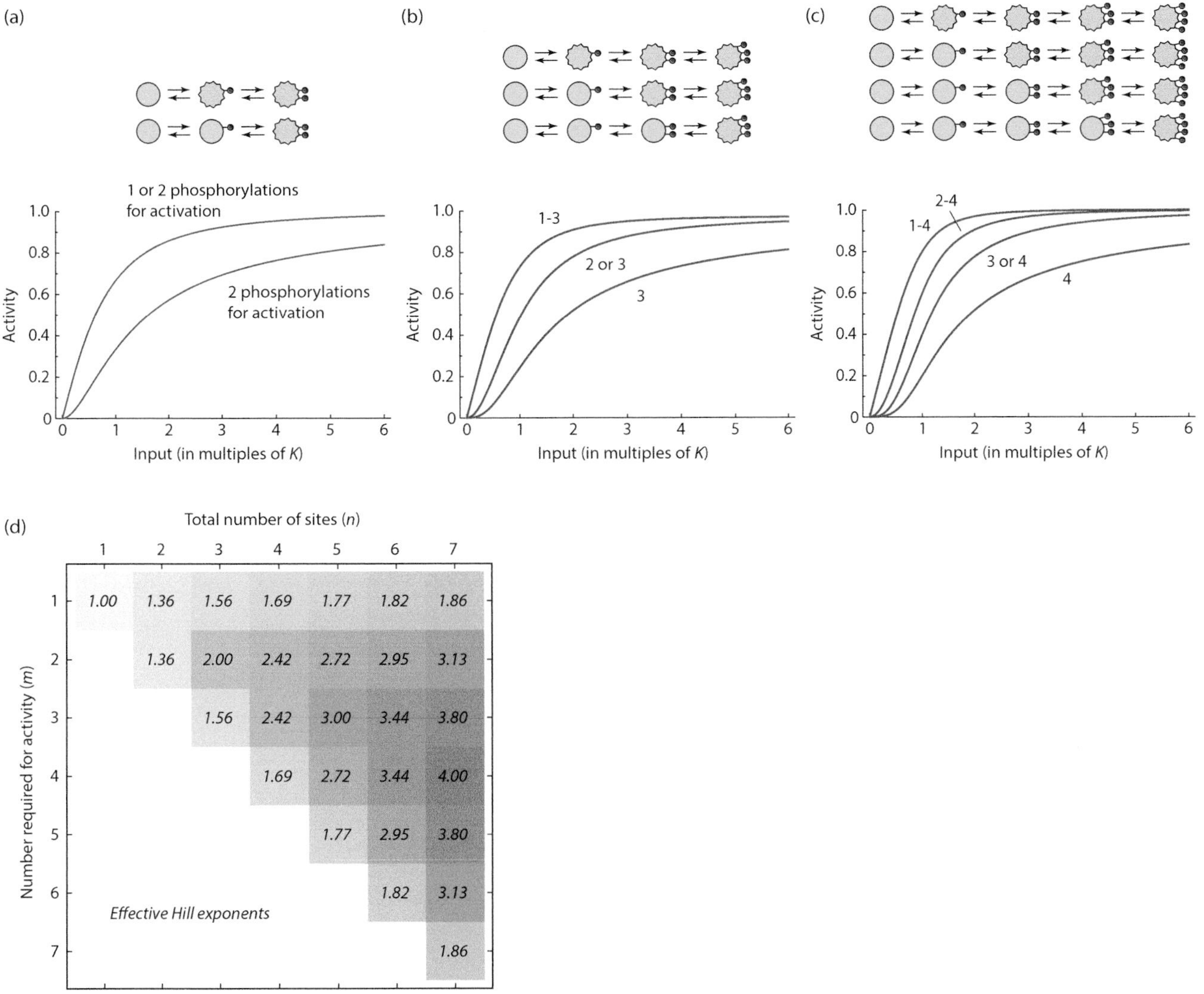

Figure 5.13 Steady-state activity curves for a phosphorylation–dephosphorylation system with different numbers of phosphorylations required for activation. Examples shown are for a protein with one (a), two (b), or three (c) ordered phosphorylations. The calculated effective Hill exponents for systems with *n* ordered phosphorylations where *m* phosphorylations are required for activity are shown in panel d. All of the parameters (the total concentration of the substrate y_{tot} and the *K* values) have been taken to be equal to 1. Note that if it is assumed that all sites must be phosphorylated for the substrate to become activated, the result is a prominent threshold, but the steady-state response approaches maximum very gradually. Conversely, if a single phosphorylation is assumed to suffice for activation, the steady-state response approaches maximum abruptly, but there is no threshold. The maximum ultrasensitivity is achieved when the number of sites required (*m*) is about halfway between these two extremes ($m = \frac{n+1}{2}$ if *n* is odd, and $m = \frac{n}{2}$ or $\frac{n}{2}+1$ if *n* is even).

On the other hand, if both the singly and doubly phosphorylated forms are active, the activity is:

$$Activity = \frac{K_2 x + x^2}{K_1 K_2 + K_2 x + x^2}. \tag{5.39}$$

We can gauge the overall ultrasensitivity of these two activation responses by calculating the *EC*10 and *EC*90 values and plugging them into Eq. 3.53. To start with, let us assume that the two *K*'s are equal; i.e., that there is no cooperativity, positive or negative, in the phosphorylation. Equation 5.33 yields a sigmoidal curve with an effective Hill exponent of 1.36. Equation 5.34 yields a response that is shaped differently—it starts out linear and then approaches a maximum relatively abruptly—but, strangely enough, its effective Hill exponent is exactly the same, 1.36. The requirement for two phosphorylations makes the activity curve more switch-like at low levels of input and output; the requirement for one phosphorylation makes the activity curve less graded at high levels of input and output; and, based on effective Hill exponents, overall the two curves are equally ultrasensitive (**Figure 5.13a,d**).

So then what happens if there are three phosphorylation sites? We can imagine three activation models, one where one phosphorylation suffices for activation and two are "extras;" one where two phosphorylations are required and one is an extra; and one where all three phosphorylations are required (**Figure 5.13b**). The first model yields the most linear response at the low end of the response range and the most abrupt response at the high end (**Figure 5.13b**, blue curve); the third model yields the most non-linear response at the low end and the most gradual response at the high end (**Figure 5.13b**), and the second model is intermediate at both ends. Models 1 and 3 yield identical effective Hill exponents, 1.56 (**Figure 5.13d**), and model 2 yields a higher n value (exactly 2).

And four phosphorylation sites? The same pattern is seen (**Figure 5.13c**). In general, for a system with n phosphorylation sites, if n is an odd number, requiring $m = \frac{n+1}{2}$ phosphorylations for activation yields the highest effective Hill exponent, and if n is even, then $m = \frac{n}{2}$ and $\frac{n}{2}+1$ yield identical, maximal Hill exponents (**Figure 5.13d**). Somehow having extra inessential phosphorylation sites contributes to the switch-like character of the response.

So far we have assumed that all of the K values are equal. What if, say, the final K were 100-times smaller than the others, so that there is substantial positive cooperativity to the phosphorylation–dephosphorylation reactions. What would happen to the ultrasensitivities? From **Figure 5.13d**, we know that for the equal K value case, the effective Hill exponents for a 5-site phosphorylation reaction are 1.77, 2.72, 3, 2.72, and 1.77, and the maximum value is obtained when 3 out of 5 sites must be phosphorylated for activation. If we now assume that K_5=0.01 and the other K's are equal to one, the resulting effective Hill exponents are 2.44, 4.11, 4.35, 4.29, and 4.18. Overall the Hill exponents are higher, but the middle value is still the highest. At the opposite extreme, if we assume that the first K is 100 times smaller than the others so that there is substantial negative cooperativity, the resulting effective Hill exponents are 1.02, 1.74, 2.43, 2.01, and 1.69. Again, the middle value is the highest. It is possible to select K values so that some number of sites other than 3 gives the highest effective Hill exponent, but for a wide range of K values this general up-then-down trend holds.

5.13 INESSENTIAL BINDING SITES CAN CONTRIBUTE TO ULTRASENSITIVE RECEPTOR ACTIVATION

Note that these results apply equally well to the regulation of a multimeric receptor by ligands. Suppose we have a KNF model of a pentameric receptor, like the nicotinic cholinergic receptor, and for simplicity, let us suppose that all of the binding constants are equal. From the approach laid out in Chapter 3.6 and Eq. 3.61, the fraction of the receptor (y_{tot}) in the 6 different ligand-bound states (0 through 5 molecules bound) as a function of the free ligand concentration x is:

$$\frac{y_0}{y_{tot}} = \frac{K^5}{K^5 + 5K^4x + 10K^3x^2 + 10K^2x^3 + 5Kx^4 + x^5} \quad (5.40)$$

$$\frac{y_0}{y_{tot}} = \frac{5K^4x}{K^5 + 5K^4x + 10K^3x^2 + 10K^2x^3 + 5Kx^4 + x^5} \quad (5.41)$$

$$\frac{y_0}{y_{tot}} = \frac{10K^3x^2}{K^5 + 5K^4x + 10K^3x^2 + 10K^2x^3 + 5Kx^4 + x^5} \quad (5.42)$$

$$\frac{y_0}{y_{tot}} = \frac{10K^2x^3}{K^5 + 5K^4x + 10K^3x^2 + 10K^2x^3 + 5Kx^4 + x^5} \quad (5.43)$$

$$\frac{y_0}{y_{tot}} = \frac{5Kx^4}{K^5 + 5K^4x + 10K^3x^2 + 10K^2x^3 + 5Kx^4 + x^5} \quad (5.44)$$

$$\frac{y_0}{y_{tot}} = \frac{1}{K^5 + 5K^4x + 10K^3x^2 + 10K^2x^3 + 5Kx^4 + x^5}. \quad (5.45)$$

If we now consider 5 activation schemes, where 1 bound ligand is sufficient for activation, 2 ligands are required, and so on, and calculate the effective Hill exponents, we get values of 1.33, 1.83, 1.97, 1.82, and 1.33. These are all a bit smaller than they were for ordered phosphorylation (1.77, 2.72, 3, 2.72, and 1.77), but still the effective Hill exponents go up and then down, and the most ultrasensitive activation is obtained when the binding of ligand molecules to 3 out of the 5 receptor subunits is required for activation.

5.14 VARIATION: COHERENT FEED-FORWARD REGULATION

The regulation of an oligomeric receptor by multiple ligand molecules and the regulation of a protein via multisite phosphorylation are examples of processes where an input feeds into the production of an output more than once. Both of these processes can be viewed as variations on what is termed coherent feed-forward regulation, a signaling motif mentioned back in Chapter 1. There are other more explicit examples of coherent feed-forward regulation in cell signaling as well, and, like oligomeric receptors and multiply phosphorylated proteins, they can yield ultrasensitive responses. Feed-forward regulation is commonplace in *Escherichia coli* transcriptional networks. One well-studied example is the arabinose-utilization system (**Figure 5.14a**). The upstream transcription factor CRP (for cAMP receptor protein) can, in the presence of its activating ligand cAMP, stimulate the transcription of various downstream genes, including those of the araBAD and araFGH operons.

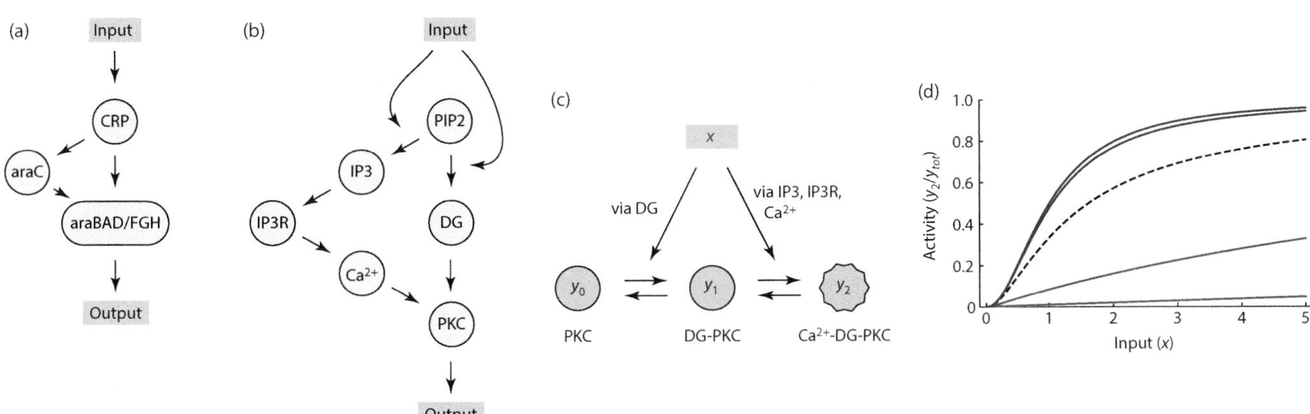

Figure 5.14 Coherent feed-forward regulation. (a) A transcriptional feed-forward system from *Escherichia coli*. (b) A post-translation feed-forward system in the activation of PKC by the PIP2-derived second messengers Ca^{2+} and diacylglycerol (DG). (c) Schematic view of a simple model of the activation of PKC by the sequential binding of DG and Ca^{2+}. (d) Steady-state response curves. Activities are plotted for 5 choices of K_1 and K_2: 100 and 0.01 (red); 10 and 0.1 (red); 1 and 1 (dashed black); 0.1 and 10 (blue); and 0.01 and 100 (blue).

CRP also induces the transcription of araC, and the araC protein, when bound to its activating ligand arabinose, stimulates the transcription of the araBAD and araFGH operons. Therefore when arabinose is present, the system functions as a coherent feed-forward system, with CRP feeding into to the transcription of araBAD/FGH via two routes.

Another famous example of coherent feed-forward regulation, this time from eukaryotic signaling, is the activation of classical protein kinase C (**Figure 5.14b**). The activation of a phospholipase C (PLC) brings about the conversion of the inner leaflet phospholipid PIP_2 into diacylglycerol (DG) and IP_3. DG acts as a stoichiometric regulator of PKC, helping to recruit it to the plasma membrane. IP_3 acts as a regulator of Ca^{2+} release from the endoplasmic reticulum, and cytosolic Ca^{2+} in turn acts as a stoichiometric activator of DG-bound PKC.

Here we will construct a simple model for coherent feed-forward regulation based on the PKC system. As shown in **Figure 5.14c**, we assume that the output protein (PKC) can exist in three states: the apo-PKC (y_0), PKC bound to DG (DG-PKC, y_1), and PKC bound to both DG and Ca^{2+} (Ca^{2+}-DG-PKC, y_2). For simplicity we assume the binding is ordered and that the Ca^{2+}-PKC complexes are negligible. We can write down three rate equations:

$$\frac{dy_0}{dt} = -k_1 DG \cdot y_0 + k_{-1} y_1 \tag{5.46}$$

$$\frac{dy_1}{dt} = k_1 DG \cdot y_0 - k_{-1} y_1 - k_2 Ca^{2+} y_1 + k_{-2} y_2 \tag{5.47}$$

$$\frac{dy_2}{dt} = k_2 Ca^{2+} y_1 - k_{-2} y_2. \tag{5.48}$$

We can simplify this further by assuming that the concentrations of both DG and Ca^{2+} are directly proportional to the phospholipase C activity, which we denote as x. It follows that:

$$\frac{dy_0}{dt} = -k_1 x \cdot y_0 + k_{-1} y_1 \tag{5.49}$$

$$\frac{dy_1}{dt} = k_1 x \cdot y_0 - k_{-1} y_1 - k_2 x \cdot y_1 + k_{-2} y_2 \tag{5.50}$$

$$\frac{dy_2}{dt} = k_2 x \cdot y_1 - k_{-2} y_2. \tag{5.51}$$

Note that we have redefined the rate constants to include the proportionality factors that relate x to DG and Ca^{2+}. At steady state all of these derivatives must equal zero:

$$0 = -k_1 x \cdot y_0 + k_{-1} y_1 \tag{5.52}$$

$$0 = k_1 x \cdot y_0 - k_{-1} y_1 - k_2 x \cdot y_1 + k_{-} \tag{5.53}$$

$$0 = k_2 x \cdot y_1 - k_{-2} y_2. \tag{5.54}$$

We also have a conservation relationship:

$$y_{tot} = y_0 + y_1 + y_2. \tag{5.55}$$

These four equations, which define the steady-state response of this stoichiometric feed-forward system, are identical in form to those for

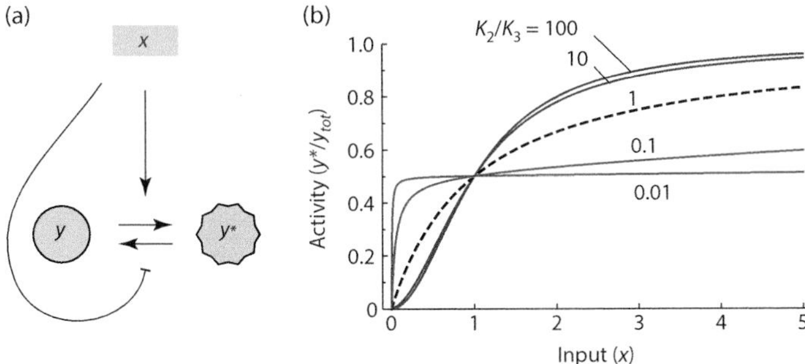

Figure 5.15 Reciprocal regulation. (a) Schematic view of a reciprocal regulation where a stimulus *x* both activates a kinase and inhibits a phosphatase. (b) Steady-state activity curves. The response can be ultrasensitive (blue curves), Michaelian (dashed black curve), or subsensitive (red curve) depending on which enzyme, the kinase or phosphatase, is more responsive to *x*.

the enzymatic activation of a protein through two-site phosphorylation (Eqs. 5.25–5.28). It follows then that the output of the system is given by:

$$\left(\frac{y_2}{y_{tot}}\right)_{ss} = \frac{x^2}{K_1K_2 + K_2x + x^2}. \tag{5.56}$$

As usual, $K_1 = k_{-1} / k_1$ and $K_2 = k_{-2} / k_2$. Thus, the result is a response approaching an n=2 Hill function when the cooperativity in the two activation steps is very high (i.e. $K_1 >> K_2$) and approaching an n=1 Michaelian curve when there is strong negative cooperativity ($K_1 << K_2$) (**Figure 5.14d**).

5.15 VARIATION: RECIPROCAL REGULATION

One final variation on feed-forward regulation is probably worth examining: reciprocal regulation. Sometimes an input stimulus will not only turn on a process but also turn off the corresponding reverse reaction (**Figure 5.15a**). For example, stresses like DNA damage result in the phosphorylation and activation of Wee1, a protein kinase that blocks mitotic entry by phosphorylating the master mitotic regulator cyclin B-Cdk1. These same stresses result in the phosphorylation and inhibition of Cdc25, the protein phosphatase that undoes the phosphorylations done by Wee1. Thus DNA damage feeds into Cdk1 regulation in two ways, by reciprocally regulating the opposing enzymes that determine the phosphorylation state of Cdk1. This is not exactly the same as a traditional feed-forward system, where an input affects two different proteins, nor is it exactly the same as multisite phosphorylation or multisubunit receptors, where an input affects two different protein states. But, like these processes, reciprocal regulation can generate some ultrasensitivity.

To show that this is the case, we begin by writing down the rate equation for the phosphorylation and dephosphorylation of Cdk1 by Wee1 and Cdc25:

$$\frac{dyp}{dt} = k_1 Wee1(1 - yp) - k_{-1} Cdc25 \cdot yp, \tag{5.57}$$

where *yp* represents the fraction of the Cdk1 molecules that are phosphorylated. At steady state, the derivative is zero:

$$0 = k_1 Wee1_{ss}(1 - yp) - k_{-1} Cdc25_{ss} \cdot yp. \tag{5.58}$$

The activities of both Wee1 and Cdc25 are functions of the upstream stress stimulus *x*. We could make the simple assumption that the steady-state responses are Michaelian activation and inactivation, respectively:

$$Wee1_{ss} = \frac{x}{K_2 + x} \tag{5.59}$$

$$Cdc25_{ss} \frac{K_3}{K_3 + x}. \tag{5.60}$$

If we plug Eqs. 5.59 and 5.60 into Eq. 5.58 and solve for yp, we get:

$$yp_{ss} = \frac{K_3 x + x^2}{K_1 K_2 K_3 + (K_1 K_3 + K_3)x + x^2}. \tag{5.61}$$

We can simplify this a little by choosing to measure the concentration of x in multiples of K_1, so that $K_1 = 1$. This yields:

$$yp_{ss} = \frac{K_3 x + x^2}{K_2 K_3 + 2K_3 x + x^2}. \tag{5.62}$$

This equation is identical in form to the KNF saturation equation (Eq. 3.51). If K_2 equals K_3, it reduces to a Michaelian response (**Figure 5.15b**, dashed black curve). If K_2 is smaller than K_3—i.e., it is easier for the stimulus to activate the forward reaction than it is to inactivate the reverse reaction—the result is a subsensitive response (**Figure 5.15b**, red curves), like what one would get from negatively cooperative binding in the KNF case. If K_2 is larger than K_3, the result is an ultrasensitive response (**Figure 5.15b**, blue curves), like what one would get from positive cooperativity.

SUMMARY

For an enzymatic regulation system, such as a phosphorylation–dephosphorylation reaction, the system approaches a steady state rather than an equilibrium. In the simplest models of this type of process—with mass action kinetics for both the phosphorylation and dephosphorylation reactions—the steady-state response is Michaelian and the system responds to a step-change in input by exponentially approaching its new equilibrium with a halftime of $\frac{\text{Ln}2}{k_1 x + k_{-1}}$. If, instead, the enzymes are operating close to saturation, so that Michaelis–Menten kinetics rather than mass action kinetics applies, the system can give rise to zero-order ultrasensitivity. This is a sigmoidal, switch-like steady-state response that arises from enzyme saturation rather than cooperativity. The conceptual basis of zero-order ultrasensitivity—and particularly why saturating the phosphatase and saturating the kinase have different effects on the steady-state response curve—can be understood through rate-balance analysis, a powerful graphical method for analyzing the steady states and the dynamics of systems with one time-dependent variable.

Even if the kinase and phosphatase are not operating close to saturation, ultrasensitivity can arise if a substrate's regulation depends on multisite phosphorylation, especially if the first phosphorylations promote the subsequent phosphorylations through a mechanism such as priming. Multisite phosphorylation is very common in cell signaling, and there are at least a few examples of priming as well. Multisite phosphorylation has the potential for improving the specificity of regulation; multistep systems can accomplish kinetic proofreading. Finally, we examined two related types of signaling circuits—coherent feed-forward regulation and reciprocal regulation—that can, under the right circumstances, generate ultrasensitive responses.

FURTHER READING

PROTEIN PHOSPHORYLATION

Canagarajah BJ, Khokhlatchev A, Cobb MH, Goldsmith EJ. Activation mechanism of the MAP kinase ERK2 by dual phosphorylation. *Cell*. 1997 Sep 5;90(5):859–69.

Holt LJ, Tuch BB, Villén J, Johnson AD, Gygi SP, Morgan DO. Global analysis of Cdk1 substrate phosphorylation sites provides insights into evolution. *Science*. 2009 Sep 25;325(5948):1682–6.

Hunter T. Why nature chose phosphate to modify proteins. *Philos Trans R Soc Lond B Biol Sci*. 2012 Sep 19;367(1602):2513–6.

Iakoucheva LM, Radivojac P, Brown CJ, O'Connor TR, Sikes JG, Obradovic Z, Dunker AK. The importance of intrinsic disorder for protein phosphorylation. *Nucleic Acids Res*. 2004 Feb 11;32(3):1037–49.

Johnson LN. The regulation of protein phosphorylation. *Biochem Soc Trans*. 2009 Aug;37(Pt 4):627–41.

Khokhlatchev AV, Canagarajah B, Wilsbacher J, Robinson M, Atkinson M, Goldsmith E, Cobb MH. Phosphorylation of the MAP kinase ERK2 promotes its homodimerization and nuclear translocation. *Cell*. 1998 May 15;93(4):605–15.

Mann M, Jensen ON. Proteomic analysis of post-translational modifications. *Nat Biotechnol*. 2003 Mar;21(3):255–61.

McGrath DA, Balog ER, Kõivomägi M, Lucena R, Mai MV, Hirschi A, Kellogg DR, Loog M, Rubin SM. Cks confers specificity to phosphorylation-dependent CDK signaling pathways. *Nat Struct Mol Biol*. 2013 Dec;20(12):1407–14.

Pawson T, Nash P. Assembly of cell regulatory systems through protein interaction domains. *Science*. 2003;300(5618):445–52.

Ubersax JA, Ferrell JE Jr. Mechanisms of specificity in protein phosphorylation. *Nat Rev Mol Cell Biol*. 2007 Jul;8(7):530–41. Review. Erratum in: *Nat Rev Mol Cell Biol*. 2007 Aug;8(8): 665.

Welburn JP, Tucker JA, Johnson T, Lindert L, Morgan M, Willis A, Noble ME, Endicott JA. How tyrosine 15 phosphorylation inhibits the activity of cyclin-dependent kinase 2-cyclin A. *J Biol Chem*. 2007 Feb 2;282(5):3173–81.

ULTRASENSITIVITY

Ferrell JE Jr. Tripping the switch fantastic: how a protein kinase cascade can convert graded inputs into switch-like outputs. *Trends Biochem Sci*. 1996 Dec;21(12):460–6.

Ferrell JE Jr, Ha SH. Ultrasensitivity part I: Michaelian responses and zero-order ultrasensitivity. *Trends Biochem Sci*. 2014 Oct;39(10):496–503.

Ferrell JE Jr, Ha SH. Ultrasensitivity part II: multisite phosphorylation, stoichiometric inhibitors, and positive feedback. *Trends Biochem Sci*. 2014 Nov;39(11):556–69.

Ferrell JE Jr, Ha SH. Ultrasensitivity part III: cascades, bistable switches, and oscillators. *Trends Biochem Sci*. 2014 Dec;39(12):612–8.

Goldbeter A, Koshland DE Jr. An amplified sensitivity arising from covalent modification in biological systems. *Proc Natl Acad Sci U S A*. 1981 Nov;78(11):6840–4.

Gomez-Uribe C, Verghese GC, Mirny LA. Operating regimes of signaling cycles: statics, dynamics, and noise filtering. *PLoS Comput Biol*. 2007 Dec;3(12):e246.

LaPorte DC, Koshland DE Jr. Phosphorylation of isocitrate dehydrogenase as a demonstration of enhanced sensitivity in covalent regulation. *Nature*. 1983 Sep 22–28;305(5932):286–90.

Wang L, Nie Q, Enciso G. Nonessential sites improve phosphorylation switch. *Biophys J*. 2010;99(6):L41–L43.

EXAMPLES OF ULTRASENSITIVITY

Alvarado D, Klein DE, Lemmon MA. Structural basis for negative cooperativity in growth factor binding to an EGF receptor. *Cell*. 2010 Aug 20;142(4):568–79.

Bagowski CP, Besser J, Frey CR, Ferrell JE Jr. The JNK cascade as a biochemical switch in mammalian cells: ultrasensitive and all-or-none responses. *Curr Biol*. 2003 Feb 18;13(4):315–20.

Biemann HP, Koshland DE Jr. Aspartate receptors of Escherichia coli and Salmonella typhimurium bind ligand with negative and half-of-the-sites cooperativity. *Biochemistry*. 1994 Jan 25;33(3):629–34.

Charvin G, Oikonomou C, Siggia ED, Cross FR. Origin of irreversibility of cell cycle start in budding yeast. *PLoS Biol*. 2010 Jan 19;8(1):e1000284.

Cluzel P, Surette M, Leibler S. An ultrasensitive bacterial motor revealed by monitoring signaling proteins in single cells. *Science*. 2000 Mar 3;287(5458):1652–5.

Galzi JL, Bertrand S, Corringer PJ, Changeux JP, Bertrand D. Identification of calcium binding sites that regulate potentiation of a neuronal nicotinic acetylcholine receptor. *EMBO J*. 1996 Nov 1;15(21):5824–32.

Hardie DG, Salt IP, Hawley SA, Davies SP. AMP-activated protein kinase: an ultrasensitive system for monitoring cellular energy charge. *Biochem J*. 1999 Mar 15;338 (Pt 3)(Pt 3):717–22.

Huang CY, Ferrell JE Jr. Ultrasensitivity in the mitogen-activated protein kinase cascade. *Proc Natl Acad Sci U S A*. 1996 Sep 17;93(19):10078–83.

Kim SY, Ferrell JE Jr. Substrate competition as a source of ultrasensitivity in the inactivation of Wee1. *Cell*. 2007 Mar 23;128(6):1133–45.

Locke JC, Young JW, Fontes M, Hernández Jiménez MJ, Elowitz MB. Stochastic pulse regulation in bacterial stress response. *Science*. 2011 Oct 21;334(6054):366–9.

Meinke MH, Bishop JS, Edstrom RD. Zero-order ultrasensitivity in the regulation of glycogen phosphorylase. *Proc Natl Acad Sci U S A*. 1986 May;83(9):2865–8.

Meyer T, Holowka D, Stryer L. Highly cooperative opening of calcium channels by inositol 1,4,5-trisphosphate. *Science*. 1988 Apr 29;240(4852):653–6.

Narula J, Devi SN, Fujita M, Igoshin OA. Ultrasensitivity of the Bacillus subtilis sporulation decision. *Proc Natl Acad Sci U S A*. 2012 Dec 11;109(50):E3513–22.

Senthivel VR, Sturrock M, Piedrafita G, Isalan M. Identifying ultrasensitive HGF dose-response functions in a 3D mammalian system for synthetic morphogenesis. *Sci Rep*. 2016 Dec 16;6:39178.

Sprinzak D, Lakhanpal A, Lebon L, Santat LA, Fontes ME, Anderson GA, Garcia-Ojalvo J, Elowitz MB. Cis-interactions between Notch and Delta generate mutually exclusive signalling states. *Nature*. 2010 May 6;465(7294):86–90.

Teo TS, Wang JH. Mechanism of activation of a cyclic adenosine 3':5'-monophosphate phosphodiesterase from bovine heart by calcium ions. Identification of the protein activator as a Ca2+ binding protein. *J Biol Chem*. 1973 Sep 10;248(17):5950–5.

Trunnell NB, Poon AC, Kim SY, Ferrell JE Jr. Ultrasensitivity in the Regulation of Cdc25C by Cdk1. *Mol Cell*. 2011 Feb 4;41(3):263–74.

Tsai TY, Theriot JA, Ferrell JE Jr. Changes in oscillatory dynamics in the cell cycle of early Xenopus laevis embryos. *PLoS Biol*. 2014 Feb 11;12(2):e1001788.

Whittaker L, Hao C, Fu W, Whittaker J. High-affinity insulin binding: insulin interacts with two receptor ligand binding sites. *Biochemistry*. 2008 Dec 2;47(48):12900–9.

Yang Q, Ferrell JE Jr. The Cdk1-APC/C cell cycle oscillator circuit functions as a time-delayed, ultrasensitive switch. *Nat Cell Biol*. 2013 May;15(5):519–25.

Yuan J, Berg HC. Characterization of the adaptation module of the signaling network in bacterial chemotaxis by measurement of step responses. *Biophys J*. 2012 Sep 19;103(6):L31–3.

Yuan J, Berg HC. Ultrasensitivity of an adaptive bacterial motor. *J Mol Biol*. 2013 May 27;425(10):1760–4.

DOWNSTREAM SIGNALING 3

Regulated Production or Destruction

6

IN THIS CHAPTER . . .

6.1 STIMULATED PRODUCTION YIELDS A LINEAR STEADY-STATE RESPONSE WITH EXPONENTIAL APPROACH TO THE STEADY STATE

The third basic type of signal transduction process is the regulated production or destruction of some downstream signaling molecule. Famous examples include the second messenger cyclic AMP (cAMP), and the best-studied (but not the only) way of bringing about its production is through the stoichiometric activation of an adenylyl cyclase protein, which can be considered the input stimulus, by a trimeric G-protein's GTP-bound $G\alpha_S$ subunit. The result is that some of the cell's

DOI: 10.1201/9781003124269-6

stockpile of ATP is converted into cAMP, which is considered to be the output of this process. The cAMP can then stoichiometrically activate downstream targets like protein kinase A and cAMP-regulated ion channels.

This type of regulation is not confined to second messenger signaling. The regulated transcription of a gene is, in broad strokes, the same type of process. An input stimulus—in this case an activated transcription factor—brings about the production of one or more mRNAs from a stockpile of ribonucleotide bases. Likewise for translation: an input mRNA brings about the production of a protein, the output, from a stockpile of amino acids via charged tRNAs.

A stimulus can also bring about a decrease, rather than an increase, in the concentration of a downstream signaling molecule through stimulated destruction. This occurs during photoreception in the vertebrate retina. The activation of an opsin photoreceptor protein brings about the activation of the G-protein transducin, which then stoichiometrically activates a cyclic GMP (cGMP) phosphodiesterase, resulting in a decrease in the concentration of cGMP and a closure of cGMP-regulated sodium channels.

Here we will model a stimulated production system. We assume that the rate of the synthesis of y is proportional to an input stimulus x and that the rate of destruction of y is proportional to the amount of y present—mass action kinetics, the simplest case scenario. This yields the following rate equation:

$$\frac{dy}{dt} = k_1x - k_{-1}y. \tag{6.1}$$

To obtain an expression for the steady-state concentration of y in response to a constant concentration of the input x, we set the time derivative equal to zero:

$$0 = k_1x - k_{-1}y \tag{6.2}$$

$$y_{ss} = \frac{k_1}{k_{-1}}x. \tag{6.3}$$

This shows that there is a simple linear relationship between the input (x) and the steady-state output (y), as depicted in **Figure 6.1a**, rather than the Michaelian relationship seen with the simplest models of stoichiometric regulation (Chapter 2, Eq. 2.8) and with mass action enzymatic regulation (Chapter 5, Eq. 5.4).

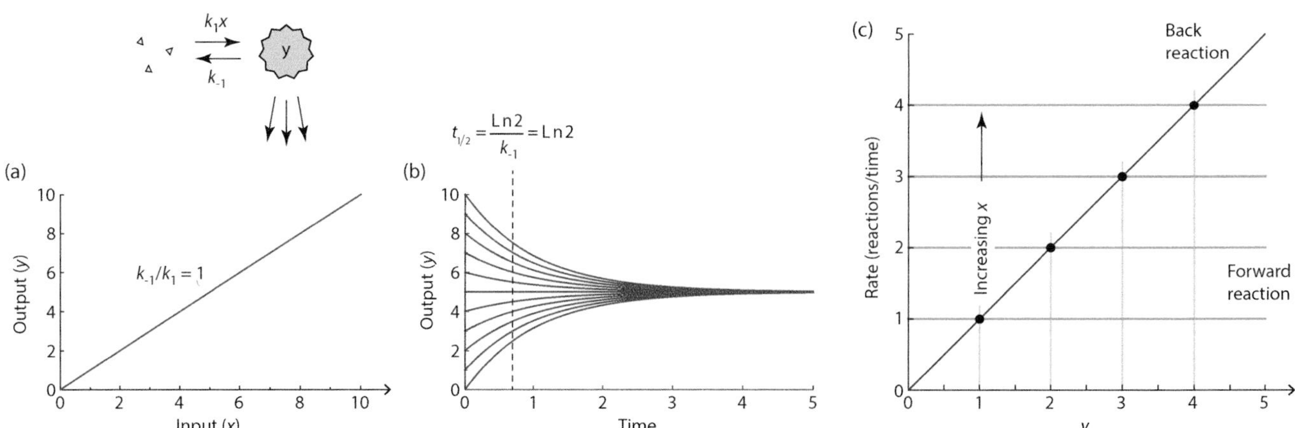

Figure 6.1 The steady-state response (a), time course (b), and rate–balance plot (c) for a stimulated production reaction. In each panel we have assumed that $k_1 = k_{-1} = 1$, and for panel c we have assumed $x = 1, 2, 3,$ or 4.

An equation for the dynamical response (**Figure 6.1b**) is suggested by the rate–balance plot (**Figure 6.1c**). We define a new variable $z[t] = y[t] - y_{SS}$, which is a measure of how far y is from its steady-state value. The rate–balance plot shows that the rate at which the system approaches the steady state, $\frac{dz[t]}{dt}$, is linearly proportional to $z[t]$ and opposite in sign (the rate opposes the displacement from steady state), which means that:

$$Dz[t] = -k_{apparent} z[t] \tag{6.4}$$

In the case of our phosphorylation–dephosphorylation rate–balance plot (**Figure 5.4**), two factors contributed to the net rate back toward the steady state: a change in the forward reaction rate and a change in the back reaction rate. However, now only the back reaction rate changes when the system is pushed out of steady state; the forward reaction rate lines are flat (**Figure 6.1c**). Therefore, only one slope figures into the value of the apparent rate constant, the slope of the back reaction line:

$$k_{apparent} = k_{-1}. \tag{6.5}$$

Therefore,

$$z[t] = z[0]e^{-k_{-1}t}. \tag{6.6}$$

And, in terms of y,

$$y[t] - y_{ss} = (y[0] - y_{ss})e^{-k_{-1}t} \tag{6.7}$$

$$y[t] = y_{ss} - y_{ss}e^{-k_{-1}t} + y[0]e^{-k_{-1}t} \tag{6.8}$$

$$y[t] = \frac{k_1 x}{k_{-1}}\left(1 - e^{-k_{-1}t}\right) + y[0]e^{-k_{-1}t}. \tag{6.9}$$

The output exponentially approaches steady state, with a halftime that depends only on the speed of the reverse reaction:

$$t_{1/2} = \frac{\text{Ln}2}{k_{-1}}. \tag{6.10}$$

This seems a little counterintuitive. Should not the system approach its steady state faster if the synthesis rate, which is determined by k_1

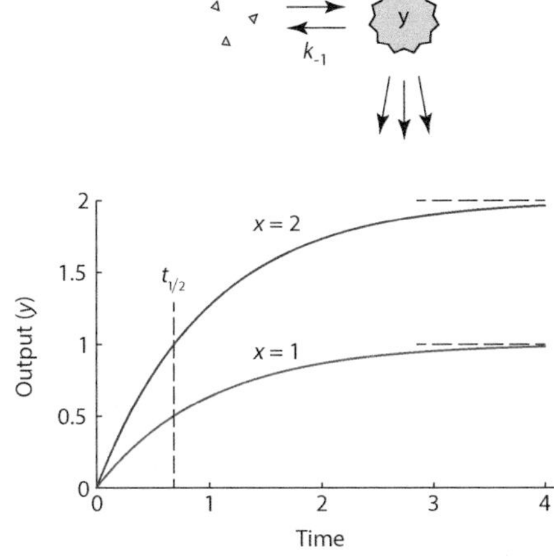

Figure 6.2 Changing the rate of production does not affect the halftime.

and x, is faster? The answer is no, and we can see why from the time courses shown in **Figure 6.2**. We have assumed that $y[0] = 0$ and have calculated the time courses for $x = 1$ and $x = 2$, with the rate constants kept constant ($k_1 = k_{-1} = 1$). Doubling x doubles the initial rate at which y approaches steady state, but, as seen in Eq. 6.3, it also doubles the final steady-state level of y. Thus, when x is bigger, y initially increases faster, but it has farther to go. The same is true if k_1 rather than x is varied. Therefore the halftime does depend only on the k_{-1} term and is independent of k_1 and x.

6.2 THE STABILITY OF THE STEADY STATE CAN BE QUANTIFIED BY THE EXPONENT IN THE EXPONENTIAL APPROACH EQUATION

Suppose we have two signaling processes operating at steady state—a phosphorylation–dephosphorylation reaction that activates 500 of a cell's 1,000 molecules of y_1 and a synthesis–destruction reaction that produces 500 molecules of active y_2. Suppose also that the two reactions have the same flux, so that at steady state 500 molecules of y_1 or y_2 turns over every second. Now suppose that some fluctuation transiently pushes both processes out of steady state, so that there are 600 active molecules of y_1 and y_2. Both systems will return toward their stable steady state, but the phosphorylation reaction will return faster than the synthesis reaction (**Figure 6.3a**). The reason for this can be seen from the two rate–balance plots. In the phosphorylation reaction, when the system is pushed away from its steady state, the back reaction rate increases and the forward reaction rate decreases, and both of these factors contribute (equally in this case) to the driving force that restores the system back toward its steady state (**Figure 6.3b**). However, in the synthesis reaction, only the back reaction changes when the system is out of steady state (**Figure 6.3c**). The driving force that restores the system back toward steady state is smaller, and so the return to steady state is slower. The steady state is still stable, but less stable, in some sense,

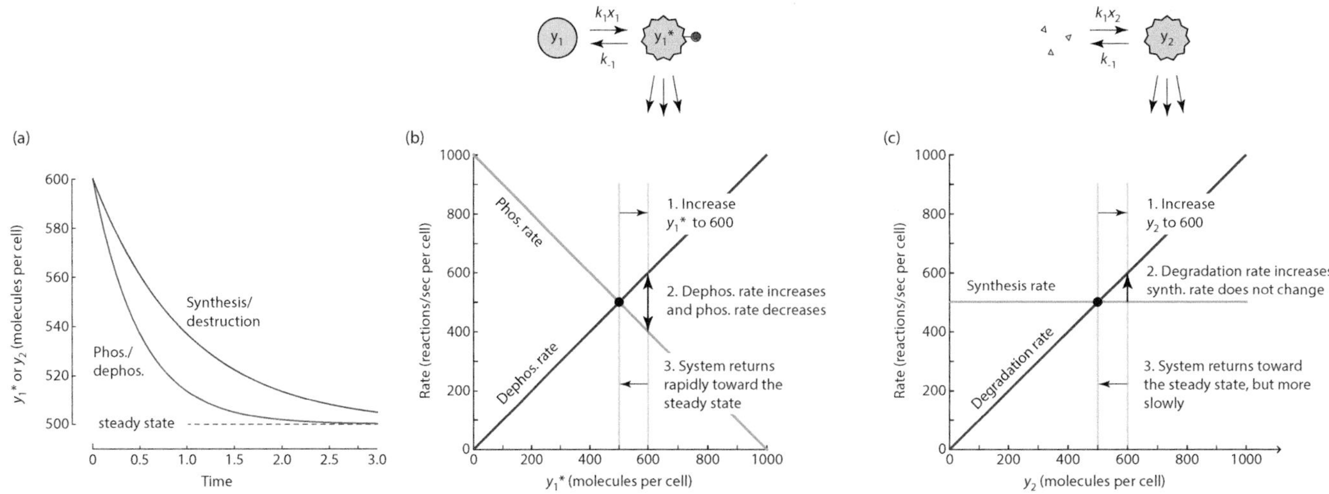

Figure 6.3 The steady state of a phosphorylation–dephosphorylation reaction is more stable than that of a synthesis–destruction reaction. (a) The time course for the return to a steady-state output of 500 molecules of y_1* or y_2 per cell after transiently increasing the output to 600 molecules per cell. (b) Rate–balance plot for the phosphorylation–dephosphorylation reaction. (c) Rate–balance plot for the synthesis–destruction reaction. In panel b we have assumed that $k_1 = k_{-1} = 1$ and $y_{tot} = 1{,}000$ molecules per cell. In panel c we have assumed that $k_1 = 500$ and $k_{-1} = 1$. In both panels the flux at steady state is 500 reactions/sec per cell.

than the steady state was in the phosphorylation–dephosphorylation system.

The crux of the matter is that the forward reaction in the phosphorylation system not only produces an active y^* molecule, it also consumes an activatable y molecule. The same is true for stoichiometric regulation; the binding event both produces an active c_{xy} complex, but also consumes an activatable free receptor (y) molecule. It is this target depletion that makes the forward reaction rate decrease with increasing y^* or c_{xy}, and it is this decrease in the forward rate that makes the system return to steady state so fast.

To make the synthesis–destruction reaction return to steady state as quickly as the phosphorylation–dephosphorylation reaction does, one would have to double the phosphorylation and dephosphorylation rate constants, and thus double the flux through the system.

A simple way to assess whether a steady state is stable, and to quantify how stable steady state is, is from the sign and the magnitude of the exponential factor in the time course equation (Eq. 6.6). If the exponential factor, usually designated λ, is negative, the steady state is stable, and the larger λ is in magnitude, the more stable the steady state is. For the phosphorylation system shown in **Figure 6.3b**, $\lambda = -2$, and for the synthesis reaction shown in **Figure 6.3c**, $\lambda = -1$. Thus, the steady state in the phosphorylation–dephosphorylation system is twice as stable, by this measure, as the steady state of a comparable synthesis–destruction system.

6.3 SATURATING THE BACK REACTION BUILDS A THRESHOLD INTO THE STEADY-STATE RESPONSE

So far we have assumed that both the forward and the back reactions of our synthesis–destruction system are described by mass action kinetics. What if the back reaction was saturable and was running close to saturation? Would we get something akin to zero-order ultrasensitivity out of the system?

In this case, Eq. 6.1 would become:

$$\frac{dy}{dt} = k_1 x - k_{-1}\frac{y}{K_M + y}, \tag{6.11}$$

and the steady-state concentration of y as a function of the input x and the parameters would be:

$$y = \frac{K_M x}{\frac{k_{-1}}{k_1} - x}. \tag{6.12}$$

Note that this equation implies that as x approaches a critical value, k_{-1}/k_1, the steady-state value of y increases without bound. This makes sense; if the synthesis rate ($k_1 x$) exceeds the maximum possible degradation rate (k_{-1}), the amount of y will blow up.

The steady-state response curves defined by Eq. 6.12 are shown in **Figure 6.4** for $k_1 = k_{-1} = 1$ and various assumed values of K_M. When x is small, the response curves are not too different from the linear responses predicted by Eq. 6.3, but as x gets close to 1, the curves turn upward, and a tiny change in x results in a large change in y.

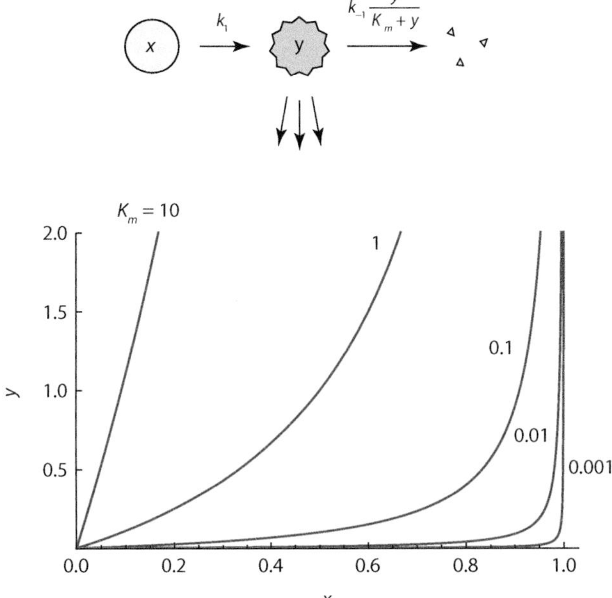

Figure 6.4 Saturating the back reaction adds a threshold to the steady-state response.

Thus, saturating the back reaction adds a threshold to the response, and the higher the saturation (the lower the K_M value), the sharper the threshold. In terms of the local definition of ultrasensitivity discussed in Section 3.7 and defined in Eq. 3.57, $S = \frac{d \ln Output}{d \ln Input}$, the response is highly ultrasensitive near the critical point.

6.4 ZERO-ORDER DEGRADATION MAKES DRUG DOSING DICEY

The synthesis–degradation models we have been examining can also be applied to models of drug levels in patients treated with drugs. If the drug is administered by constant infusion or constant-release pills, the rate accumulation of the drug in the blood will be a linear function of the drug dosage, just as the production of second messenger was assumed to be a linear function of the input stimulus. Likewise, the rate of elimination of the drug, by excretion and metabolism, corresponds to the degradation of a second messenger. If the elimination of the drug is first order in drug concentration—i.e., the kidney's excretion capacity and the liver's metabolic capacity are far from being maxed out—then Eq. 6.3 describes the blood level of the drug as a function of dose. If the dose is doubled, the steady-state blood concentration doubles too. This is the case for most drugs in clinical use.

But if the elimination of the drug is saturable and is close to saturation at therapeutic blood levels, then Eq. 6.12 applies, and this can spell trouble. A small change in the daily dose of one of these drugs can produce a big change in its blood concentration.

I used to tell the following story to our medical students to drive home the distinction between zero-order and first-order elimination. The story is not true (and I admitted as much to the students), but it could be—the basic facts are right—and it seemed to make the importance of degradation kinetics stick. Here goes:

Our medical student, who has a knack for quantitative biology, sees a newly diagnosed patient with epilepsy in clinic. The patient has been treated with carbamazepine (trade name Tegretol) for two weeks and had a seizure yesterday. His carbamazepine blood concentration is measured, and it comes back right at the bottom of the therapeutic range; it needs to be boosted by 50%. Our medical student knows that carbamazepine is eliminated with first-order kinetics and decides to increase the dosage by 50%. One week later the patient returns to clinic. His carbamazepine level is smack in the middle of the therapeutic range, and he has not had any more seizures. The medical student is a hero. Everybody loves her.

The medical student's intern, who does not have such a good grasp of quantitative biology, sees another newly diagnosed seizure patient in clinic the next day. This patient has been treated with a different antiseizure drug, phenytoin (trade name Dilantin), and, wouldn't you know it, he had a seizure the day before and his phenytoin levels came back at the bottom of the therapeutic range (10 μg/mL). Phenytoin is one of those rare drugs whose elimination is close to saturation at therapeutic levels—a typical daily dose might be 300 mg and a typical maximum elimination rate might be only 400 mg per day. Nevertheless, the intern figures a 50% increase in dose should do it, because that worked for the medical student's patient, and so the patient is sent home with a prescription for 450 mg phenytoin per day. Four days later the patient is brought to the emergency room by his partner with signs and symptoms of phenytoin intoxication—unsteady gait, weakness, and drowsiness. His serum phenytoin concentration comes back at 42 μg/mL—way too high. The intern is a goat.

The attending physician draws **Figure 6.5** to explain why a small change in phenytoin dosage can lead to big changes in the blood concentration of phenytoin, and she gently points out that since 450 mg of phenytoin is more than many patients can clear in a day, this patient would maybe eventually turn into a pillar of solid phenytoin if he had continued with taking a 450-mg dose for long enough.

The long and the short of it is, most drugs do not come close to saturating the body's elimination mechanisms, but a few do, and for those that do, small changes in dosing can result in big changes in steady-state blood concentrations. Phenytoin is one such drug. The other two are drugs most of us have some experience with: aspirin and ethanol. For all of these drugs, one needs to be particularly cautious with the dosage.

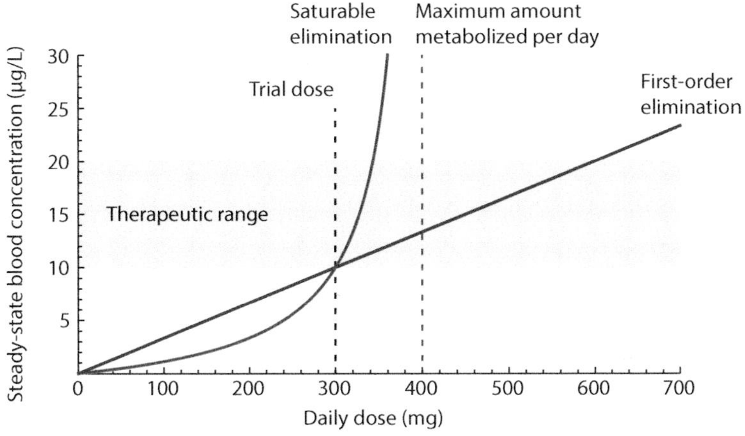

Figure 6.5 Steady-state blood concentrations of two drugs as a function of dose. For one (in blue), its elimination rate is directly proportional to its blood concentration. This is typical of most therapeutic agents. The other (in red) has saturable elimination. This is true of a few drugs, notably the anticonvulsant phenytoin, aspirin, and ethanol.

SUMMARY

For regulated production or destruction, the input–output relationship is simple; if the destruction is first order, then the steady-state concentration of the product is proportional to the input stimulus and the proportionality constant is simply $\frac{k_1}{k_{-1}}$. The system responds to a step-change in input by exponentially approaching its new steady state with a halftime of $\frac{\text{Ln}2}{k_{-1}}$; that is, the halftime is determined solely by the kinetics of the destruction reaction, which might seem counterintuitive but is true. If the destruction process is saturable rather than first order, then the steady-state response will have a threshold, and the system will blow up if the synthesis rate exceeds the maximal destruction rate.

Finally, we introduced a way for quantifying the stability of a stable steady state. For a system that exponentially approaches steady state, we simply look at the proportionality constant in the exponential term. If the system approaches (rather than being repelled by) the steady state, the constant (traditionally designated λ) will be negative, and the larger the magnitude of λ, the higher the stability of the steady state. We will return to this way of quantifying stability in Chapters 8 and 9, when we will investigate the local stability of multiple steady states in bistable systems.

FURTHER READING

THE RATE–BALANCE PLOT

Ferrell JE Jr. Tripping the switch fantastic: how a protein kinase cascade can convert graded inputs into switch-like outputs. *Trends Biochem Sci*. 1996 Dec;21(12):460–6.

LaPorte DC, Koshland DE Jr. Phosphorylation of isocitrate dehydrogenase as a demonstration of enhanced sensitivity in covalent regulation. *Nature*. 1983 Sep 22–28;305(5932): 286–90.

STOICHIOMETRIC REGULATION INSIDE THE CELL

Rajalingam K, Schreck R, Rapp UR, Albert S. Ras oncogenes and their downstream targets. *Biochim Biophys Acta*. 2007 Aug;1773(8):1177–95.

Sprang SR, Chen Z, Du X. Structural basis of effector regulation and signal termination in heterotrimeric Galpha proteins. *Adv Protein Chem*. 2007;74:1–65.

ENZYMATIC REGULATION

Hunter T. Why nature chose phosphate to modify proteins. *Philos Trans R Soc Lond B Biol Sci*. 2012 Sep 19;367(1602):2513–6.

Johnson LN. The regulation of protein phosphorylation. *Biochem Soc Trans*. 2009 Aug;37(Pt 4):627–41.

Mann M, Jensen ON. Proteomic analysis of post-translational modifications. *Nat Biotechnol*. 2003 Mar;21(3):255–61.

Ubersax JA, Ferrell JE Jr. Mechanisms of specificity in protein phosphorylation. *Nat Rev Mol Cell Biol*. 2007 Jul;8(7):530–41. Review. Erratum in: *Nat Rev Mol Cell Biol*. 2007 Aug;8(8):665.

STIMULATED PRODUCTION OR DESTRUCTION

Covert MW. *Fundamentals of Systems Biology: From Synthetic Circuits to Whole-Cell Models*. CRC Press, Taylor and Francis Group, Boca Raton, FL, 2015.

CASCADES AND AMPLIFICATION

7

IN THIS CHAPTER . . .

INTRODUCTION

A few eukaryotic signaling systems are about as simple as prokaryotic two-component systems are (**Figure 7.1a**), with a receptor directly regulating one or two transcription factors. One such system mediates cytokine signaling. Various cytokines and hormones, including growth hormone, the interferons, and tissue necrosis factor α (TNFα), function by binding to what are termed cytokine receptors and they thereby activate an associated Janus-family (Jak) tyrosine kinase. The active Jak protein then directly phosphorylates and activates STAT family transcription factors (**Figure 7.1b**). Another simple example is TGFβ signaling. TGFβ and its relatives function by binding to and activating a multimeric receptor serine/threonine kinase complex, which then phosphorylates and activates SMAD family transcription factors. Once again, the terminal effector, here the SMAD complex, is

DOI: 10.1201/9781003124269-7

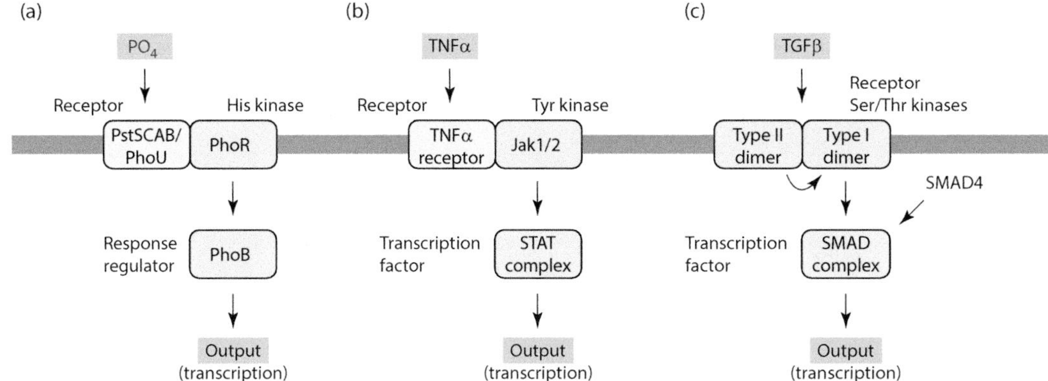

Figure 7.1 Cytokine and TGFβ signaling, like two-component signaling in bacteria, directly link a kinase to a transcription factor. (a) Phosphate detection in *E. coli*. Phosphate is sensed by the PstACAB/PhoU receptor/transport complex, which brings about activation of the associated PhoR histidine kinase, which phosphorylates and activates the PhoB transcription factor. (b) TNFβ signaling in mammalian cells. TNFβ binds to the trimeric TNFβ receptor, which activates the associated Jak family tyrosine kinases, which phosphorylate and activate STAT family transcription factors. (c) TGFβ signaling in mammalian cells. TGFβ binds to a tetrameric receptor complex composed of type I and type II dimers. The type II dimer phosphorylates and activates the type I dimer, which phosphorylates SMAD proteins. The phosphorylated SMAD proteins form an active transcription factor complex with SMAD4.

regulated by a receptor without a long chain of intermediary signaling proteins (**Figure 7.1c**).

However, many eukaryotic signaling systems interpose at least a few intermediaries between the receptor and the terminal regulator. For example, as we saw in Chapter 1, there are six proteins between the epidermal growth factor receptor and the Ets family transcription factors: the Shc and/or Grb2 adaptors, the Sos guanine nucleotide exchange factor, the Ras protein, and the three sequential protein kinases of the evolutionarily ancient MAP kinase cascade (**Figures 1.3 and 1.4**). This raises the question of what the advantages (and disadvantages) of signaling via so many intermediaries are. Here we begin by examining how a cascade like the MAP kinase cascade can function as an amplifier, turning small signals into large ones.

SIGNALING CASCADES AND TWO TYPES OF AMPLIFICATION

7.1 CASCADES CAN DELIVER SIGNALS FASTER THAN SINGLE SIGNAL TRANSDUCERS

As a thought experiment, suppose that nature dispensed with the three-kinase cascade and instead had the first kinase—e.g., Raf in the Raf/MEK/ERK cascade—directly regulate some terminal effectors. There would be a number of advantages to this stripped-down arrangement. Only one protein kinase would need to somehow be insulated from off-target upstream regulators and only one protein kinase would need to be prevented from regulating off-target downstream substrates. If every component in a cascade possesses some vulnerabilities, then the smaller the number of levels in the cascade, the fewer the vulnerabilities.

But there are potential advantages to the three-kinase system too; one such advantage is the speed with which it can regulate abundant terminal effectors. To see why this is so, suppose that in our hypothetical one-kinase cascade, some modest concentration of active

Raf—say 10 nM—must regulate substrates whose total concentration is 1,000-fold higher (10 μM). Furthermore, let us assume that the phosphatases acting on these substrates are low in activity compared to the fully active Raf protein, so that the phosphorylation reaction goes to completion as rapidly as possible. The maximum speed (k_{cat}) of a protein kinase is typically about one phosphorylation reaction per molecule per second, and if Raf was functioning at this maximum speed, it would take 1,000 s, or about 17 min, for Raf to phosphorylate all of its substrate molecules (**Figure 7.2a**).

Next, let us consider the actual system, where Raf sits at the top of a three-kinase cascade. For simplicity, assume that each kinase in the cascade is regulated by a single phosphorylation. In the first 10 s, 10 nM Raf operating at maximal velocity could activate 100 nM MEK. In the next 10 s, those activated MEK proteins could activate a total of 1 μM ERK—less than 10 s, actually, since some of the MEK molecules will be active before the first 10 s is up. The active ERK could regulate 10 μM substrate proteins in the next (less than) 10 s. Hence in less than 30 s the three-kinase cascade could accomplish what it would take 17 min for a single kinase to do (**Figure 7.2b**).

Of course you could imagine a system with 10 μM active Raf directly phosphorylating 10 μM substrate proteins. This would be a fast process, but it just pushes back the problem upstream; there is not enough Ras to stoichiometrically activate that much Raf. And even if there was, it might take a long time for the super-abundant Ras to become activated.

These arguments assume constant rates of phosphorylation to make the calculations easy. What if we try a more realistic kinetic model, using realistic abundances and K_M values? Let us assume again that the k_{cat} values for all of the kinases are $1\,s^{-1}$. Let us take the abundance of active Raf again to be 10 nM and the abundances of MEK and ERK to each be 1 μM, which is the approximate concentration of these kinases in many vertebrate cell lines and tissues. Let us assume the K_M

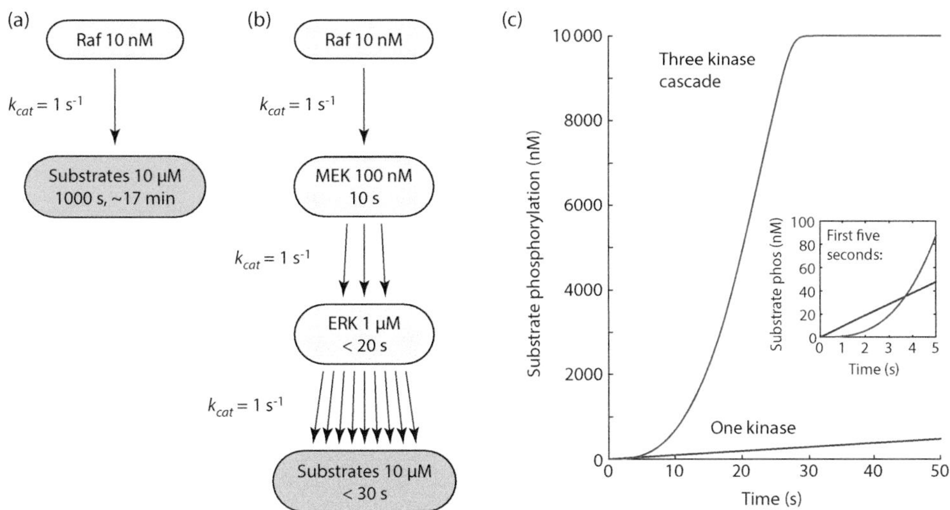

Figure 7.2 Magnitude amplification from a protein kinase cascade. (a) A single protein kinase could eventually phosphorylate even very abundant substrate proteins, but it would take some time. At a constant speed of 1 phosphorylation per kinase molecule per second, it would take ~17 min for 10 nM Raf to phosphorylate a 1000× excess of substrates. (b) A three-kinase cascade can phosphorylate abundant substrates much faster. At a speed of 1 phosphorylation per kinase molecule per second, it would take the cascade less than 30 s to phosphorylate the same number of substrate molecules. (c) Time courses for substrate phosphorylation by a single kinase (blue) or a three-kinase cascade (red). Here the assumed abundances are Raf, 10 nM; MEK, 1,000 nM; and ERK, 1,000 nM. The assumed K_M values for all three kinases are 500 nM and the assumed k_{cat} values for all three kinases are $1\,s^{-1}$. The inset highlights the first 5 s of the time course.

values are all equal to 500 nM, which is in line with the admittedly limited in vitro kinetics data available for the pathway, so that the kinases are somewhat but not completely saturated by their substrates, and let us again assume that the dephosphorylation reactions are insignificant compared to the phosphorylation reactions, so that all of the phosphorylation reactions will go to completion as quickly as possible. Under these assumptions, the single kinase system takes a little longer to phosphorylate its 10 μM substrates—about 1,200 s, or 20 min, for the phosphorylation to reach 99% of completion (**Figure 7.2c**, blue curve, compared to **Figure 7.2a**). Again, the three-kinase cascade is much quicker, now taking about 28 s to reach 99% of completion (**Figure 4.2c**, red curve). From the inset to the graph in **Figure 7.2c**, one can see that the single kinase system is actually faster than the three-kinase cascade for the first 3 s or so (**Figure 7.2c**)—interposing additional kinases adds a time lag to the process. However, very quickly the three-kinase system outpaces the one-kinase system.

Note that with a single kinase in the cascade, substrate phosphorylation initially increases linearly with respect to time (**Figure 7.2c**). If there were two comparably slow kinases, the substrate phosphorylation would increase like time squared, and for the three-kinase case shown here, substrate phosphorylation is initially proportional to t^3 (**Figure 7.2c**). Just as we found for multisite phosphorylation (Chapter 5), the initial polynomial order of a cascade-mediated signal transduction process provides an estimate of how many slow, rate-determining steps connect the input to the output.

The phenomenon demonstrated here, where a cascade of signaling molecules quickly converts a small signal at the top of the cascade into a large signal at the bottom, is termed **magnitude amplification**. Note though that it is not the case that a single kinase could not produce just as large of a signal; it is just that it would take longer to do so.

There are a number of striking examples of magnitude amplification in cell signaling—for instance, signal transduction in the vertebrate retina. A single photon activates one molecule of rhodopsin, which activates hundreds of molecules of the G-protein transducin. Each active transducin molecule can stoichiometrically activate a phosphodiesterase molecule, and the active phosphodiesterase can cleave up to 4,200 cGMP molecules per second. Finally, the cGMP molecules bind to a multimeric cation channel, whose activity is a highly ultrasensitive function of the cGMP concentration. All told, the amplification achieved is estimated to be a few hundred thousand-fold. The activation of T cells, which can be initiated by the binding of a few peptides to T-cell receptors, is another good example of magnitude amplification.

In the MAPK cascade, the current evidence is that the Raf and Mos MAPKKKs are relatively scarce—in HeLa cells, the B-Raf and C-Raf proteins together are ~50 nM and in *Xenopus* oocytes the Mos protein is ~5–10 nM—whereas the MEK and ERK proteins are relatively abundant (nearly micromolar for both). In both systems, physiological stimuli result in the activation of most of the MEK and ERK. This means that there is at least 20-fold magnitude amplification at the MAPKKK-to-MAPKK step. However, there is probably little, if any, magnitude amplification at the MAPKK-to-MAPK step. Thus overall the magnitude amplification achieved by the MAPK cascade appears to be pretty modest.

But magnitude amplification is not the only useful systems-level behavior that can be accomplished by a cascade of signaling proteins; a cascade can also generate what is termed **sensitivity amplification**, converting slightly ultrasensitive responses into highly ultrasensitive ones. However, before we discuss sensitivity amplification, we will examine how signals can, in principle, be severely degraded when they propagate down a cascade.

7.2 A CASCADE OF MICHAELIAN RESPONSES LEADS TO SIGNAL DEGRADATION

So far we have been discussing the speed of signaling down a cascade; here we will turn to the steady-state response, taking both the phosphorylation reactions and dephosphorylation reactions into account.

Let us suppose again that active Raf, at the top of the cascade, relays signals to MEK, and then ERK, and then on to some terminal effectors like transcription factors. Typically the transcription factors act on a time scale of tens of minutes, so there should be enough time for the phosphorylation–dephosphorylation reactions to approach their steady states.

MEK and ERK are both activated through the phosphorylation of two sites that reside in the kinases' activation loops. From Chapter 5, we would expect that the steady-state responses of MEK to Raf and ERK to MEK should be something between a Michaelian response and a Hill response with a Hill exponent of 2. For now, though, let us assume that they are both Michaelian responses, for simplicity (**Figure 7.3a**).

The fraction of the MEK that is activated (*MEK**) as a function of the fraction of Raf that is activated (*Raf**) would therefore be:

$$(MEK*)_{ss} = \frac{Raf*}{K_1 + Raf*}. \tag{7.1}$$

The fraction of the ERK that is activated (*ERK**) as a function of *MEK** is:

$$(ERK*)_{ss} = \frac{MEK*}{K_2 + MEK*}. \tag{7.2}$$

Likewise, if the response of an ERK substrate is Michaelian, the fraction activated at steady state will be:

$$(Substrate*)_{ss} = \frac{ERK*}{K_3 + ERK*}. \tag{7.3}$$

We can combine these three equations and eliminate two of the variables (*MEK** and *ERK**) to obtain an expression for substrate activation as a function of *Raf**:

$$(Substrate*)_{ss} = \frac{1}{1+K_3+K_2K_3} \frac{Raf*}{\frac{K_1K_2K_3}{1+K_3+K_2K_3} + Raf*}. \tag{7.4}$$

We can simplify this equation a bit by letting:

$$Substrate*_{max} = \frac{1}{1+K_3+K_2K_3} \tag{7.5}$$

and

$$EC50 = \frac{K_1K_2K_3}{1+K_3+K_2K_3}. \tag{7.6}$$

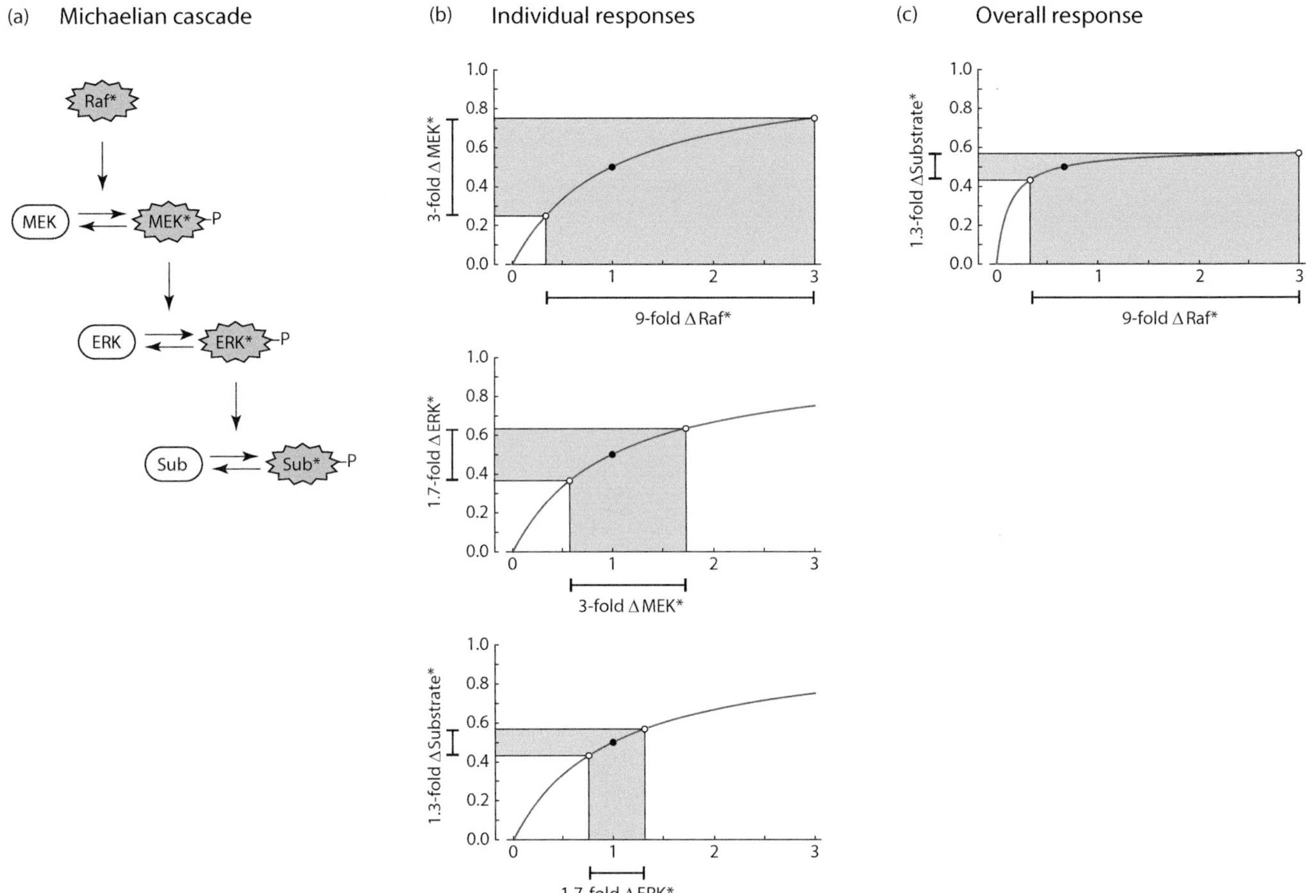

Figure 7.3 Signal degradation on a Michaelian protein kinase cascade. (a) Schematic view of the cascade. (b) Responses of each level to its immediate upstream activator. In all cases the input stimulus, on the *x*-axis, is expressed in multiples of the corresponding *K* value. The input to each level is geometrically centered on the *K* value so as to maximize the resulting change in output. For example, *Raf** goes from 3-fold below the *K* value to 3-fold above it, so that MEK activity increases from 0.25 to 0.75 (Δ=0.5). (c) Overall the cascade compresses a 9-fold change in input into a ~32% (1.3-fold) change in substrate activation.

Plugging these definitions into **Eq. 7.4** yields:

$$(Substrate^*)_{ss} = Substrate^*_{max} \frac{Raf^*}{EC50 + Raf^*} \tag{7.7}$$

Equation 7.7 shows that the steady-state response of a cascade whose individual levels exhibit Michaelian responses is Michaelian, although the maximum response is less than 100% since $\frac{1}{1+K_3+K_2K_3}<1$. The $EC50$ for this Michaelian response (**Eq. 7.6**) is no greater than the product of the three K values.

To get a better idea of what this would mean for signal propagation down the cascade, we can look at the steady-state responses at each level. First, consider the response of MEK to some change in Raf activity. We could assume that the system steps up from zero input (*Raf**=0) to some constant level of input, but instead let us assume something more realistic—that the net level of Raf activity goes from some small but non-zero level to a higher level. The initial level of Raf activity could be due either to some small amount of active Raf even when there is no upstream input or to "inactive" Raf having some small but non-zero level of activity. Let us say that a signal impinging upon the system causes the Raf activity to increase 9-fold (where this number is chosen in part because the resulting arithmetic is fairly simple). The resulting change in the steady-state activities of the downstream components—MEK, ERK, and ERK's substrates—can be regarded as the amplitude of the signal as it passes down the cascade, and this amplitude can be expressed in either absolute terms (numbers or concentrations of active molecules) or in fold-change terms (the ratio of the stimulated activity to the basal activity).

The change in the response of MEK to this 9-fold change in *Raf** will depend on how the initial and final values of *Raf** compare to K_1, the *EC50* value for the Michaelian response. If we want to maximize the change in MEK in absolute terms ($MEK^*[\infty] - MEK^*[0]$), it turns out we need to choose the initial value of *Raf** to be 3-fold below K_1 and the final value of *Raf** to be 3-fold above it. That is, we geometrically center the initial and final Raf activities about the *K* value. From Eq. 4.1, when $Raf^* = \frac{1}{3}K_1$, the steady-state value of *MEK** will be:

$$(MEK^*)_{ss} = \frac{\frac{1}{3}K_1}{K_1 + \frac{1}{3}K_1} = \frac{1}{4}. \tag{7.8}$$

Likewise, when $Raf^* = 3K_1$:

$$(MEK^*)_{ss} = \frac{3K_1}{K_1 + 3K_1} = \frac{3}{4}. \tag{7.9}$$

Thus, the 9-fold change in Raf activity has been squashed into a 3-fold change in *MEK**. This is shown in **Figure 7.3b**.

If we feed this 3-fold change in *MEK** into Eq. 7.2, we get the largest change in *ERK** if we choose K_2 to be $\sqrt{3}$-fold above the initial value of *MEK** (1/4) and $\sqrt{3}$-fold below the final value (3/4); this means $K_2 = \frac{\sqrt{3}}{4}$. This results in a basal *ERK** value of $\frac{1}{1+\sqrt{3}}$ and a final *ERK** value of $\frac{3}{3+\sqrt{3}}$. The 9-fold change in Raf activity is now down to a meager $\sqrt{3}$-fold change in *ERK** (**Figure 7.3b**).

If we feed this optimally into Eq. 4.3, the $\sqrt{3}$-fold change in *ERK** yields a $\sqrt[4]{3}$-fold change in the steady-state value of *Substrate**. All in all, our 9-fold change in Raf activity has been whittled down to about a 32% change in substrate activation (**Figure 7.3c**). The fold-change amplitude of a signal becomes severely degraded after even a few levels of propagation down a Michaelian signaling cascade. The floor of the response gets higher, and the ceiling of the response gets lower. This seems like a big problem.

7.3 FOLD-SENSITIVITY DECREASES AS A SIGNAL DESCENDS A CASCADE OF MICHAELIAN RESPONSES

One way to see why this phenomenon happens is to consider the sensitivity function for a Michaelian response. As defined in Chapter 5, the local sensitivity is given by:

$$S = \frac{dOutput}{dInput}\frac{Input}{Output}, \tag{7.10}$$

or, equivalently:

$$S = \frac{d\ln Output}{d\ln Input}, \tag{7.11}$$

which emphasizes that the sensitivity represents the slope of the input/output curve plotted on a log–log plot.

For a Michaelian response, **Eq. 7.10** becomes:

$$S = \frac{K}{K + Input}. \tag{7.12}$$

This sensitivity function S is a maximum of 1 when *Input* is zero (i.e. the response is at most a first-order function of *Input*) and approaches zero when *Input* approaches infinity. For intermediate values of *Input*, $0<S<1$.

Now, by the chain rule, the sensitivity for the whole signaling cascade is equal to the product of the sensitivities of the individual levels:

$$S = \frac{d \ln Substrate^*}{d \ln Raf^*} \tag{7.13}$$

$$S = \left(\frac{d \ln Substrate^*}{d \ln ERK^*}\right)\left(\frac{d \ln ERK^*}{d \ln MEK^*}\right)\left(\frac{d \ln MEK^*}{d \ln Raf^*}\right). \tag{7.14}$$

Each multiplicative term in **Eq. 7.14** is less than 1, no matter what the value of the input *Raf**, so each level of the cascade decreases the value of S and makes the response less switch-like. Even if the amount of signal goes up as the cascade is descended (magnitude amplification), the fold-change sensitivity of the response goes down.

This decrease can be partially mitigated by using the low end of the Michaelian response curve, where the response is closest to first order, rather than the middle of the curve. For example, if instead of geometrically centering the inputs above and below the *EC*50 values for each level (which maximizes the change in response), as we did in **Figure 7.3**, suppose we start with the basal Raf activity at 1/9 of K_1 and the final Raf activity equal to K_1 (**Figure 7.4**). Then the output (*MEK**) would rise from 1/10 to 1/2. This is a smaller absolute increase in *MEK** than we obtained in **Figure 4.3b** (0.40 vs. 0.80), but it is a larger fold-change (5-fold vs. 3-fold) (**Figure 4.4b**). If we feed this change in *MEK** into *ERK** the same way, *ERK** changes from 1/6 to 1/2 (0.33), an increase of 3-fold. If we feed *ERK** into *Substrate**, *Substrate** will change from 1/4 to 1/2 (0.25), an increase of 2-fold. So the response has still become less switch-like—a 9-fold change in *Raf** translates to a 2-fold change in *Substrate**—but not to the extent it did in **Figure 7.3**.

This improvement has come at a cost. We had to assume that nature uses only the lower halves of the cascade's Michaelian response curves, so the cell is effectively carrying around a 2-fold excess of each of the kinases. In addition, since we are not activating as many signaling molecules as we might at each step of the cascade, we are slowing down signal propagation. Nevertheless, we have substantially decreased the degradation of the signal, which is a plus.

But we can do better than this.

7.4 ULTRASENSITIVITY CAN RESTORE OR INCREASE THE DECISIVENESS OF A SIGNAL

So far we have assumed that each level of the cascade exhibits a Michaelian response. But both MEK and ERK are activated through the phosphorylation of two sites, and, as we saw in Chapter 5, that can result in an ultrasensitive steady-state response. In particular, if there is strong positive cooperativity in the phosphorylation/dephosphorylation reactions—i.e., the first phosphorylation promotes the

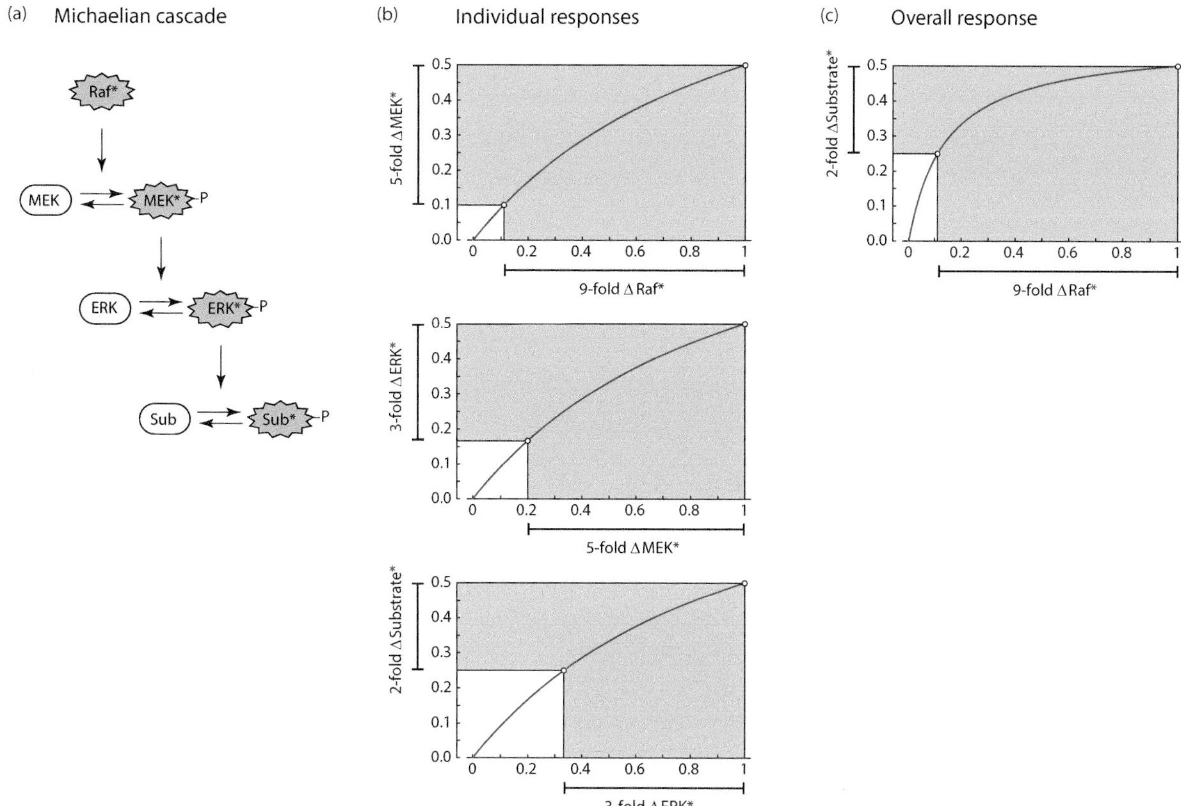

Figure 7.4 Signal degradation can be partially mitigated by using only the lower half of the Michaelian response curve. (a) Schematic view of the cascade. (b) Responses of each level to its immediate upstream activator. In all cases the input stimulus, on the *x*-axis, is expressed in multiples of the corresponding *K* value. The input is assumed to never exceed the *K* value. Thus, *Raf** goes from 9-fold below the *K* value to the *K* value. (c) Overall the cascade still compresses the response, but not as much as in Figure 7.3; a 9-fold change in input becomes a 2-fold change in substrate activation.

second or the first dephosphorylation promotes the second—then the steady-state response can approach a Hill function with a Hill exponent of 2. What happens if we stack $n=2$ responses one on top of another in a signaling cascade?

We can approach this question algebraically. Suppose we now have the following equations for the individual steady-state responses:

$$(MEK*)_{ss} = \frac{(Raf*)^2}{K_1^2 + (Raf*)^2} \tag{7.15}$$

$$(ERK*)_{ss} = \frac{(MEK*)^2}{K_2^2 + (MEK*)^2} \tag{7.16}$$

$$(Substrate*)_{ss} = \frac{(ERK*)^2}{K_3^2 + (ERK*)^2}. \tag{7.17}$$

We can combine these three equations to obtain an expression for the fraction of the substrate activated as a function of *Raf**. The result is complicated:

$$Substrate*_{ss} = \frac{(Raf*)^8}{K_1^8K_2^4K_3^2 + 4K_1^6K_2^4K_3^2(Raf*)^2 + (2K_1^4K_2^2K_3^2 + 6K_1^4K_2^4K_3^2)(Raf*)^4 + (4K_1^2K_2^2K_3^2 + 4K_1^2K_2^4K_3^2)(Raf*)^6 + (1 + K_3^2 + 2K_2^2K_3^2 + K_2^4K_3^2)(Raf*)^8} \tag{7.18}$$

However, complicated or not, one thing is clear—for small values of *Raf**, the response $(Substrate^*)_{SS}$ is no longer a linear function of *Raf**, as it was with a cascade of Michaelian signaling responses; it is now 8th-order in *Raf**, a hugely nonlinear response that would start out nearly flat and then explode upward as the stimulus increases. With this high degree of nonlinearity, perhaps we can not only maintain the fold-change difference between input and output but even increase it.

Let us examine how a 9-fold increase in active Raf would propagate down a cascade built out of $n=2$ responses. As was the case with a Michaelian cascade, the maximum amount of activation of *MEK** will be achieved if we arrange the initial and final values of *Raf** to be 3-fold below and 3-fold above K_1. From **Eq. 7.13**, the initial steady-state value of *MEK** will be:

$$(MEK^*)_{ss} = \frac{\left(\frac{1}{3}K_1\right)^2}{K_1^2 + \left(\frac{1}{3}K_1\right)^2} = \frac{1}{10}. \tag{7.16}$$

Likewise, the final steady-state value of *MEK** will be:

$$(MEK^*)_{ss} = \frac{(3K_1)^2}{K_1^2 + (3K_1)^2} = \frac{9}{10}. \tag{7.17}$$

Thus, the 9-fold change in *Raf** has yielded a 9-fold change in *MEK**. The fold-change in the signal has neither decreased nor increased; it has been transmitted with perfect fidelity. The same would be true for any fold-change in *Raf**, provided it is geometrically centered on K_1, and the same is true for the subsequent levels of the cascade, as shown in **Figure 7.5**. Thus, our cascade with $n=2$ responses at each level has preserved the fold-change amplitude of the input signal.

In addition, the overall response curve has become more switch-like (**Figure 7.5c**). Maximal substrate activation in the presence of infinite Raf* turns out to be ~90.3%; to get from 10% of this maximum output (0.10×0.903) to 90% of the maximum output (0.90×0.903), *Raf** must go from ~0.33 units to ~0.80 units. From **Eq. 3.53**, this corresponds to an effective Hill exponent of 5.0, a pretty big number as Hill exponents go.

What would happen if we were to use only the first half of the response range, starting with an input *Raf** that increases from $\frac{1}{9}K_1$ to K_1? In this case, *MEK** would change from a basal level of $\frac{1}{82}$ to $\frac{1}{2}$, which is a 41-fold change (**Figure 7.6b**). If this 41-fold change in y is fed into the next level of the cascade the same way, *ERK** turns out to change by 841-fold (**Figure 7.6b**). And if this change is fed into the next level, the fold-change in substrate activity is a whopping 353,641-fold (**Figure 7.6b**). The basal input to the cascade is almost completely suppressed, and the overall response is highly switch-like, with an effective Hill exponent of 3.7 (**Figure 7.6c**).

One way to understand why the fold-change of the signal increases is to examine the sensitivity functions. When we assumed the cascade was composed of Michaelian responses (above), the sensitivity function for each level were always less than 1, irrespective of the input stimulus, and so they combined multiplicatively to be an even smaller number. But here we have assumed the individual steady-state responses are Hill functions with $n=2$. From **Eq. 4.10**, the local sensitivity is now:

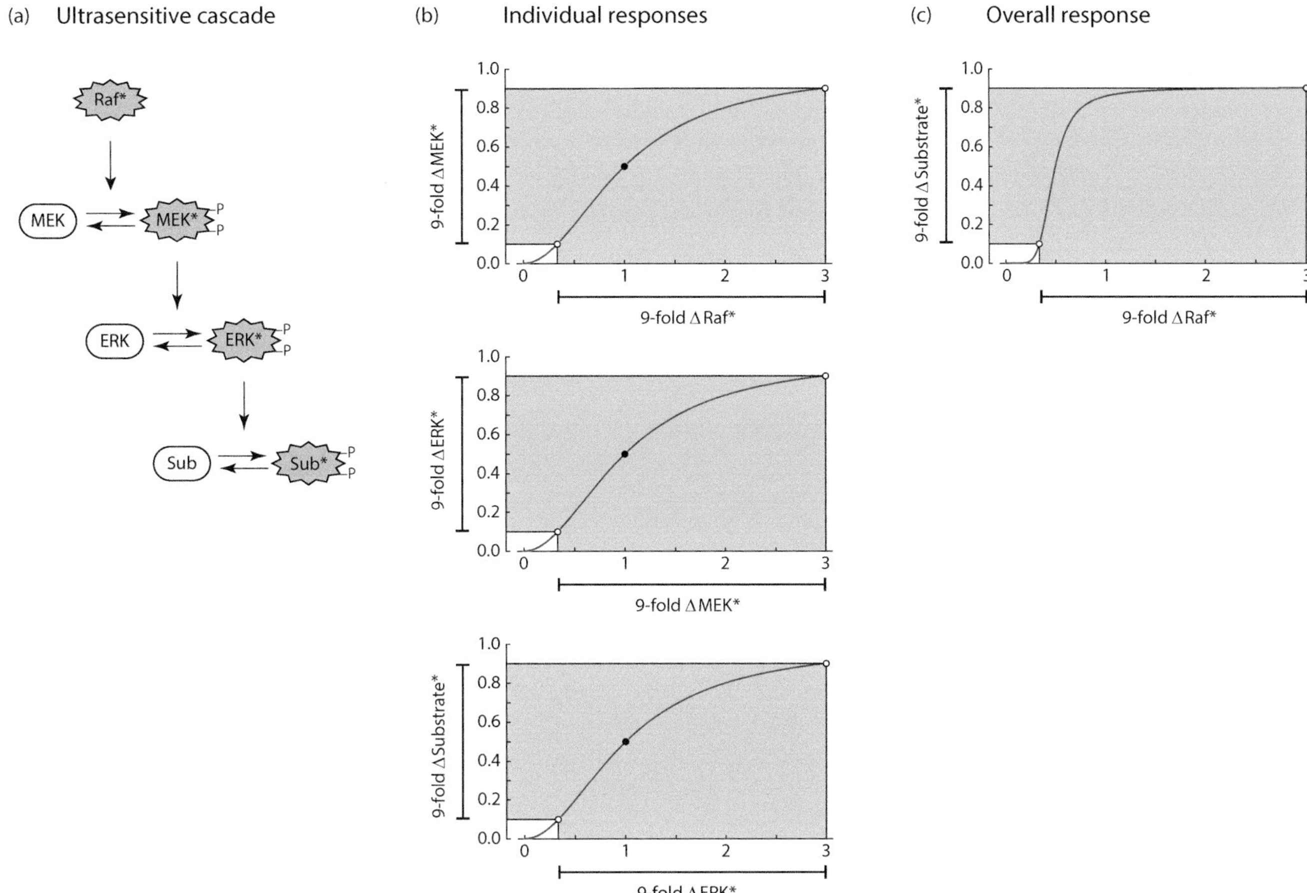

Figure 7.5 The fold-change amplitude of the response is preserved by an ultrasensitive protein kinase cascade. (a) Schematic view of the cascade. (b) Responses of each level to its immediate upstream activator. Each level's steady-state response is assumed to be given by a Hill function with a Hill exponent of 2. In all cases the input stimulus, on the *x*-axis, is expressed in multiples of the *K* value, and the range of input stimuli is geometrically centered on the *K* value to maximize the change in output. For example, *Raf** goes from 3-fold below the *K* value to 3-fold above it, and MEK activity increases from 0.1 to 0.9 (a change of 0.8 units and a fold-change of 9-fold). (c) Overall the fold-change in the output is equal to the fold-change in the input. The effective Hill exponent for the overall response is $n = \sim 5$.

$$S = \frac{d}{dInput}\left(\frac{Input^2}{K^2 + Input^2}\right) \cdot \frac{Input}{\left(\frac{Input^2}{K^2 + Input^2}\right)} \tag{7.18}$$

$$S = \frac{2K^2}{K^2 + Input^2}. \tag{7.19}$$

Equation 7.19 is an inhibitory Hill function multiplied by a factor of 2, and it is plotted in **Figure 7.7b** (the right-most blue curve). The sensitivity S starts at a maximum of 2 and then approaches 0 as the input increases. S turns out to be greater than 1 as long as the input is lower than K, i.e., over the lower half of the response. So if we confine ourselves to the lower half of the response range, the individual sensitivities are all greater than 1, and the overall sensitivity, which is the product of the individual sensitivities (Eq. 7.14), will be greater than any of the individual sensitivities. The multiplicative nature of the sensitivity function made it so that responses lost their oomph on a Michaelian cascade, but now the same multiplicative nature makes it so that the slightly ultrasensitive responses become more and more switch-like as the cascade is descended. This phenomenon is termed sensitivity amplification.

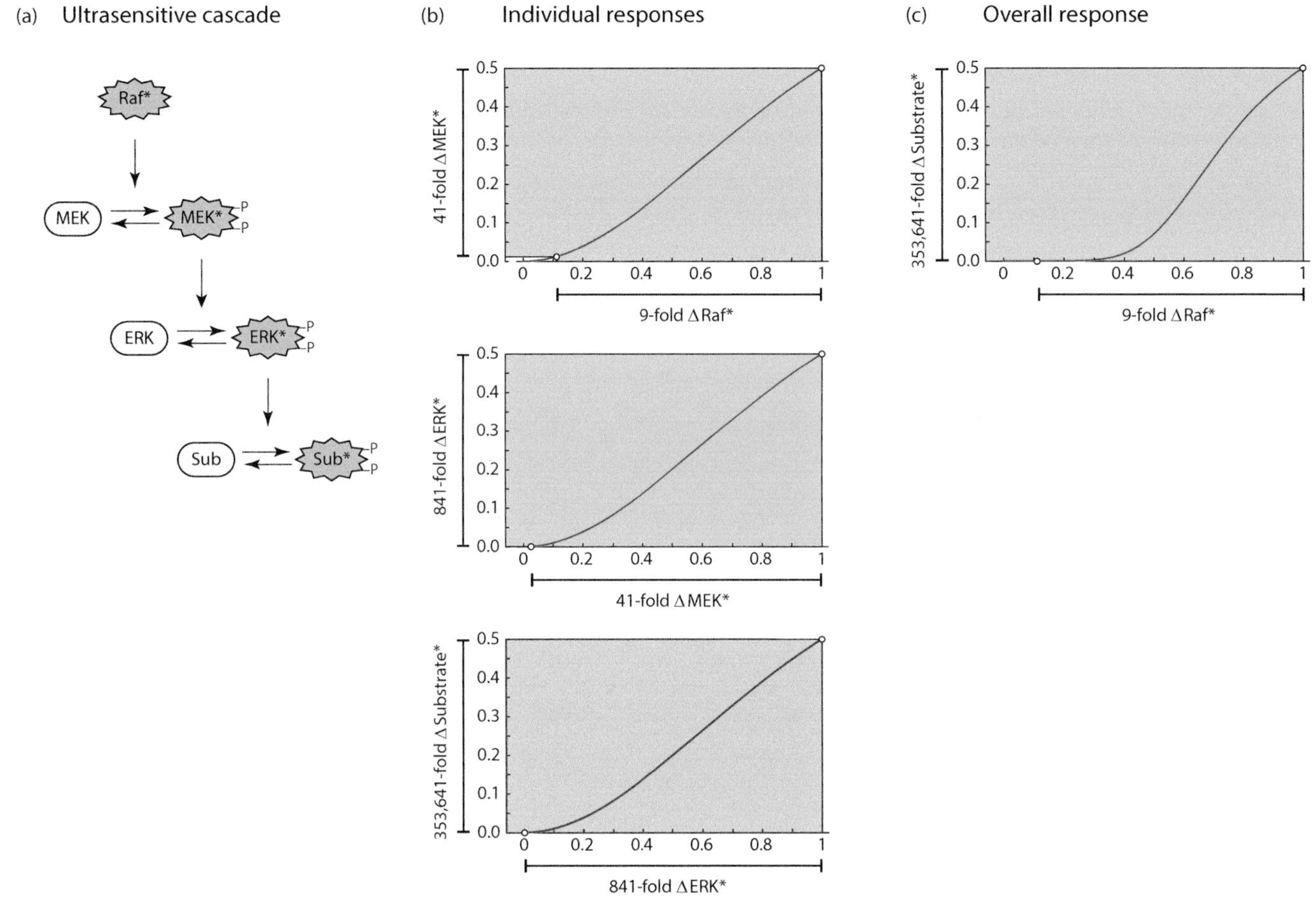

(a) Ultrasensitive cascade (b) Individual responses (c) Overall response

Figure 7.6 Sensitivity amplification by an ultrasensitive protein kinase cascade. (a) Schematic view of the cascade. (b) Responses of each level to its immediate upstream activator. Each level's steady-state response is assumed to be given by a Hill function with a Hill exponent of 2. In all cases the input stimulus, on the x-axis, is expressed in multiples of the corresponding K value. The input is assumed to never exceed the K value. Thus, *Raf** goes from 9-fold below the K value to the K value and *MEK** increases from 1/81 to 1/2—a 41-fold increase. (c) Overall the 9-fold change in input results in a huge 353,641-fold change in the cascade's output, and the effective Hill exponent for the response is n=~3.7.

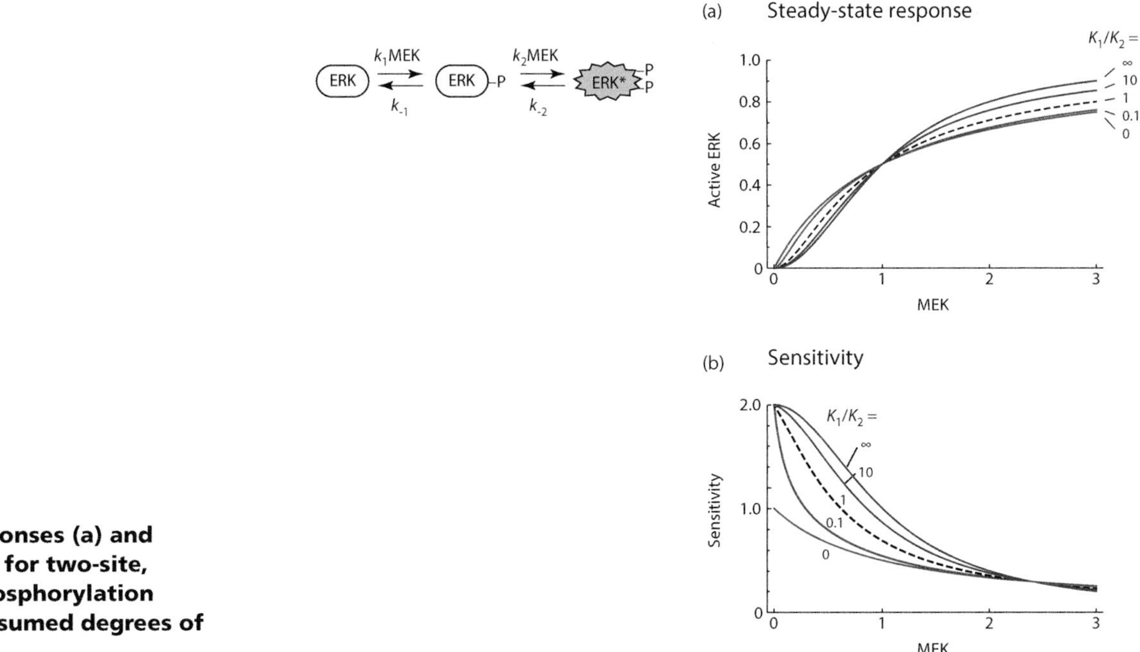

Figure 7.7 Responses (a) and sensitivities (b) for two-site, distributive phosphorylation with various assumed degrees of cooperativity.

So far we have examined two specific cases: a cascade composed of Michaelian responses and one composed of $n=2$ Hill equation responses. From the model developed in Chapter 5 for activation by distributive phosphorylation of two sites, we know that the response can be intermediate between these two extremes. From **Eq. 5.24**, the steady-state response for activation by the ordered, distributive phosphorylation of two sites is:

$$\left(\frac{y^*}{y_{tot}}\right)_{ss} = \frac{x^2}{K_1K_2 + K_2x + x^2}, \tag{7.20}$$

where K_1 is the ratio of the rate constants for the first phosphorylation–dephosphorylation reaction and K_2 is the ratio for the second, so that if $K_1>K_2$ there is positive cooperativity in the dual phosphorylation, and if $K_1<K_2$ there is negative cooperativity. It follows that the sensitivity function is:

$$S = \frac{2K_1K_2 + K_2Input}{K_1K_2 + K_2Input + Input^2}. \tag{7.21}$$

When the input is very small, this function approaches 2; when the input is large it approaches 0.

Figure 7-7 shows responses and sensitivities for two-site phosphorylation–dephosphorylation systems with different degrees of cooperativity (K_1/K_2 ratios). In each case we have fixed the K_1/K_2 ratio and then chosen individual values of K_1 and K_2 so that the $EC50=1$. In the extreme case where the ratio approaches infinity—i.e., very high positive cooperativity—the sensitivity is greater than one until the input exceeds the $EC50$—just as we saw for an $n=2$ Hill function. For the noncooperative case, the sensitivity is greater than 1 until the input exceeds about $0.62*EC50$ (the $EC50$ divided by the Golden Ratio). In the presence of very strong negative cooperativity, the range of inputs that yield sensitivities greater than 1 gets smaller and smaller. Thus, two-site phosphorylation can yield local sensitivities greater than 1, and the more cooperative the two-site phosphorylation, the greater the range of inputs that produce these sensitivity-amplifying $S>1$ sensitivities.

At this point we have only considered a cascade composed of dual phosphorylation activation mechanisms, like the MAPK cascade, where the individual levels will have steady-state responses that are somewhere between Michaelian responses and $n=2$ Hill equation responses. If the individual levels generate even higher degrees of ultrasensitivity, either through additional phosphorylations or some other mechanism, such as zero-order ultrasensitivity or stoichiometric inhibitors, the cascade could generate even more switch-like outputs.

7.5 IN *XENOPUS* OOCYTE EXTRACTS, RESPONSES GET MORE ULTRASENSITIVE AS THE MAPK CASCADE IS DESCENDED

So does sensitivity amplification actually occur in signaling cascades? As usual, there is little experimental evidence on this point, but in at least one system—*Xenopus laevis* oocyte extracts, a good system for quantitative biology studies—sensitivity amplification is known to occur. We will discuss oocyte maturation and oocyte signaling in more detail in Chapter 8. It suffices for now to note that oocytes

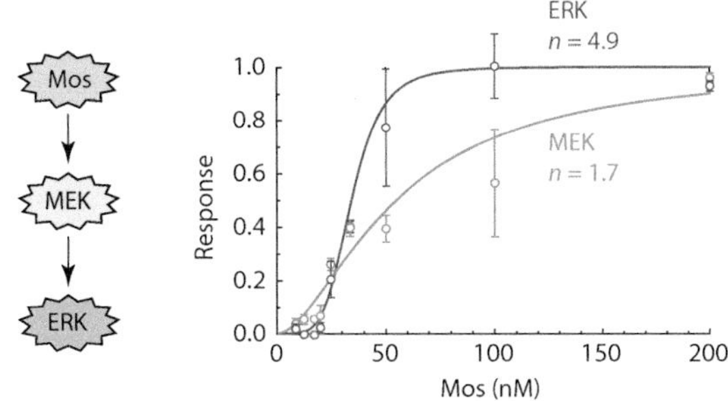

Figure 7.8 Responses get more switch-like as the MAPK cascade is descended. These experimental data are for the steady-state responses of MEK and ERK to different concentrations of purified Mos in *Xenopus* oocyte extracts. (Adapted from Ferrell and Ha, *Trends Biochem Sci.* 2014, and Huang and Ferrell, *Proc Natl Acad Sci USA*. 1996.)

possess a MAP kinase cascade consisting of the Mos oncoprotein (a MAP kinase kinase kinase), a MAP kinase kinase (mainly MEK1), and a MAP kinase (mainly ERK2). Activation of the cascade is driven by synthesis of Mos, which is absent from immature oocytes.

Cell-free extracts from immature oocytes, like the oocytes themselves, are devoid of Mos and have inactive MEK1 and ERK2. Various concentrations of purified recombinant Mos can be added to these extracts, and the steady-state responses of MEK1 and ERK2 can be assessed. The results are shown in **Figure 7.8**. The response of MEK1 to Mos is slightly ultrasensitive, and the response of ERK2 to Mos is quite markedly ultrasensitive, with an apparent Hill exponent of about 5. Thus, responses become more switch-like as the cascade is descended, as the theory we developed in Section 7.4 says it might.

SUMMARY

In Chapters 2–6 we examined the steady-state and dynamical behaviors of the simplest signaling systems. Here we have turned our attention to something a bit more complicated—a linear cascade of signaling proteins. Signaling cascades are reasonably common in eukaryotes, and so we have asked what a cascade might be able to do that a single signaling protein could not.

We showed that signaling cascades can act as two types of amplifiers. First, they can be magnitude amplifiers, meaning that a small number of active upstream signaling molecules are converted into a larger downstream response than a single-tiered system would be able to generate in the same amount of time. Magnitude amplification allows the retina to detect small numbers of photons and the immune system to detect small numbers of peptide antigens.

The second phenomenon is termed sensitivity amplification. If each level in a cascade generates a slightly ultrasensitive response, the overall response can be highly ultrasensitive. This occurs because sensitivities combine multiplicatively. But just as ultrasensitive responses become more ultrasensitive as a cascade is descended, graded responses become more graded, a potential problem in signal transduction.

FURTHER READING

MAGNITUDE AMPLIFICATION

Koshland DE Jr, Goldbeter A, Stock JB. Amplification and adaptation in regulatory and sensory systems. Science. 1982 Jul 16;217(4556):220–5.

Stryer L. Transducin and the cyclic GMP phosphodiesterase: amplifier proteins in vision. Cold Spring Harb Symp Quant Biol. 1983;48 Pt 2:841–52.

Stryer L. Cyclic GMP cascade of vision. Annu Rev Neurosci. 1986;9:87–119.

SENSITIVITY AMPLIFICATION

Brown GC, Hoek JB, Kholodenko BN. Why do protein kinase cascades have more than one level? Trends Biochem Sci. 1997 Aug;22(8):288.

Ferrell JE Jr. How responses get more switch-like as you move down a protein kinase cascade. Trends Biochem Sci. 1997 Aug;22(8):288–9.

Ferrell JE Jr, Ha SH. Ultrasensitivity part III: cascades, bistable switches, and oscillators. Trends Biochem Sci. 2014 Dec;39(12):612–8.

Hooshangi S, Thiberge S, Weiss R. Ultrasensitivity and noise propagation in a synthetic transcriptional cascade. Proc Natl Acad Sci U S A. 2005 Mar 8;102(10):3581–6.

Huang CY, Ferrell JE Jr. Ultrasensitivity in the mitogen-activated protein kinase cascade. Proc Natl Acad Sci USA. 1996 Sep 17;93(19):10078–83.

Koshland DE Jr, Goldbeter A, Stock JB. Amplification and adaptation in regulatory and sensory systems. Science. 1982 Jul 16;217(4556):220–5.

BISTABILITY 1

Systems with One Time-Dependent Variable

8

IN THIS CHAPTER . . .

DOI: 10.1201/9781003124269-8

In the last two chapters we showed how a simple signaling system can generate an ultrasensitive response and how a cascade can amplify the ultrasensitivity. The end result could approach a step function, with the system behaving like a doorbell buzzer (**Figure 8.1a**). Press the doorbell a tiny bit and nothing happens; press it a little harder and the buzzer buzzes at maximal volume; release the button and the buzzing stops. A doorbell buzzer is an all-or-none, reversible, monostable switch.

Another kind of switch, which is common in everyday life and in cell signaling, is the toggle switch (**Figure 8.1b**). Like a doorbell, a toggle switch turns on in an all-or-none fashion, but unlike a doorbell, the response is irreversible; once you have flipped the switch, it stays on indefinitely. A toggle switch is an all-or-none, irreversible, bistable system.

Here we will examine how bistability, with either irreversibility or **hysteresis** (a less-extreme variation on irreversibility), can be generated by a cell signaling system. Bistable switches are found in a variety of biological contexts: the lysis-lysogeny decision in phage-infected bacteria, the induction of the metabolic enzyme β-galactosidase in *E. coli*, and the G1/S and G2/M cell cycle transitions in diverse eukaryotic cells. Bistability also probably underpins apoptotic cell death and memory in the brain. But the archetypal example of biological bistability is cell fate induction, and that is what we will start with here.

8.1 CELL FATE INDUCTION IS TYPICALLY ALL-OR-NONE AND IRREVERSIBLE IN CHARACTER

If you treat a dish of pluripotent stem cells with a cocktail of myogenesis factors and wait long enough, many of the cells will differentiate into myocytes (**Figure 8.2**). This differentiation is all-or-none in character—a cell either differentiates or it does not. Once a cell commits to differentiating, it will continue to differentiate and remain differentiated even if the myogenesis factors are washed away. The same is true for fat cell differentiation, for neuronal differentiation, and for many other examples of cell fate determination in culture, and it is true for differentiation in a developing animal. Thus the biology of differentiation is characteristically all-or-none and irreversible.[1]

Cell differentiation is typically driven by the activation of some cell surface receptor. As we saw in Chapter 2, the activity of a monomeric receptor is often a Michaelian function of the concentration of the ligand, and if the receptor is oligomeric the response may be ultrasensitive. But unless the ultrasensitivity is extreme, it should be possible to end up with one receptor molecule activated, or two, or three, and so one might expect an almost continuously graded array of outcomes to be possible. Yet the biology driven by these activated receptors is generally all-or-none. Likewise, receptor activation should be reversible;

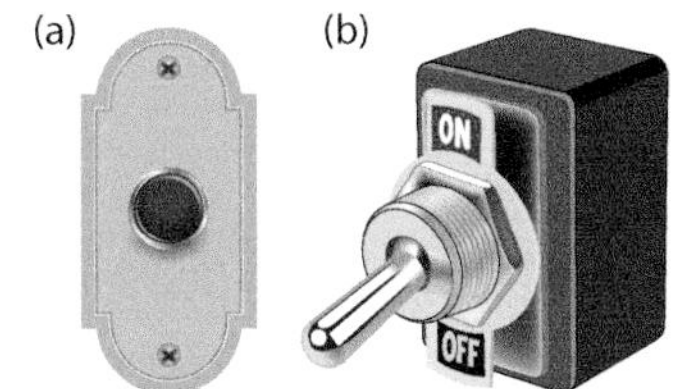

Figure 8.1 Two types of switches. (a) A doorbell switch. (b) A toggle switch.

1 Note that although differentiation is normally irreversible, differentiated cells can be engineered to revert to an undifferentiated, pluripotent state. Shinya Yamanaka was awarded the 2012 Nobel Prize in Physiology or Medicine for his work on the production of pluripotent stem cells from differentiated cells in culture. Likewise, if the nucleus of a differentiated frog cell is transferred into an unfertilized, enucleated egg, it is possible to obtain a viable tadpole with a full complement of cell types. For this dramatic demonstration of de-differentiation, John Gurdon shared the 2012 Nobel Prize with Yamanaka.

wash away the ligand and the receptor quickly becomes inactivated. Yet the biology driven by receptor activation is typically irreversible. How does graded, reversible receptor activation get transformed into an all-or-none, irreversible response?

The answer, we think, is that in these cases, the system of signaling proteins and transcription factors downstream of the receptor includes positive feedback loops, and these loops function as a bistable toggle switch. In this chapter we will explore how positive feedback can generate bistability.

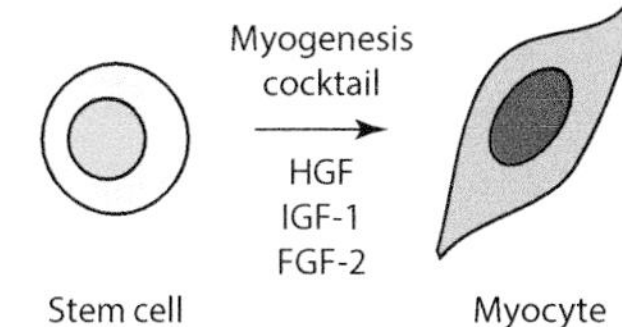

Figure 8.2 Differentiation of a stem cell into a myocyte. A variety of differentiation protocols have been devised, typically including growth factors and signaling inhibitors, that can convert embryonic stem cells or induced pluripotent stem cells into myocytes and ultimately mature myofibrils. Key steps in the cell fate induction process are all-or-none and irreversible in character.

XENOPUS OOCYTE MATURATION

8.2 *XENOPUS* OOCYTE MATURATION IS AN ALL-OR-NONE, IRREVERSIBLE PROCESS

One particularly well-studied cell fate switch, which we will examine here in detail, is *Xenopus laevis* **oocyte maturation** (**Figure 8.3a**). Immature *Xenopus* oocytes are gargantuan cells, about 1.2 mm in diameter, and they are arrested in the equivalent of G2 phase of the cell cycle. This can be regarded as the default fate of the oocyte, and the oocyte can remain in this state for months in the frog's ovary.

In response to environmental cues, the frog's pituitary gland releases gonadotropins that induce the secretion of the steroid hormone progesterone by epithelial cells that surround each oocyte. Progesterone acts on the oocyte through nontraditional, plasma membrane-associated progesterone receptors, initiating the signal transduction process that allows the oocyte to be released from its G2-phase arrest and undergo maturation.

This signal transduction begins with a drop in cAMP levels, activation of the protein kinase Aurora A, and then an increase in Mos translation and decrease in its degradation. Mos sits at the top of a MAP kinase cascade—it is a MAP kinase kinase kinase—and its accumulation brings about the phosphorylation and activation of MEK1, which phosphorylates and activates ERK2, the MAP kinase. ERK2 activation is critical for oocyte maturation. Blocking ERK2 activation blocks progesterone-induced maturation, and activating ERK2 artificially induces maturation in the absence of progesterone (**Figure 8.4**).

Activation of the MAPK cascade ultimately results in activation of cyclin B-Cdk1. Cyclin B-Cdk1 is the universal trigger of M-phase and, when active, it phosphorylates hundreds of substrate proteins. These phosphorylations result in the migration of the oocyte's huge nucleus (traditionally called the germinal vesicle—its relationship to the normal-sized nuclei present in somatic cells was not initially clear) to the animal pole of the cell, which results in the displacement of pigment granules and the appearance of a characteristic

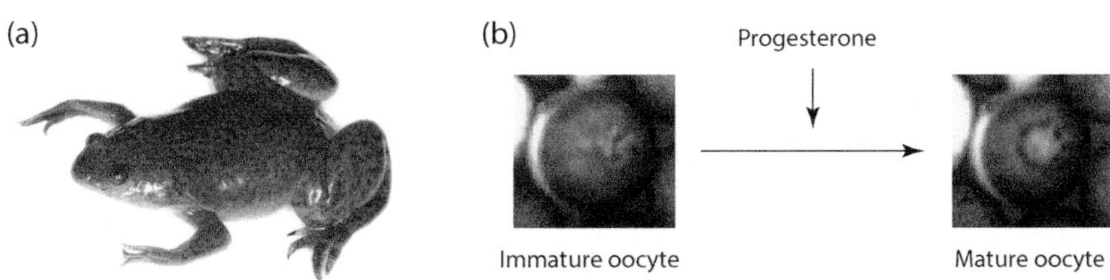

Figure 8.3 *Xenopus* oocyte maturation. (a) A female *Xenopus laevis* frog. The photograph is by Brian Gratwicke, and was downloaded from Wikipedia. (b) The appearance of an oocyte before and after progesterone-induced maturation. Photos by Tony Tsai.

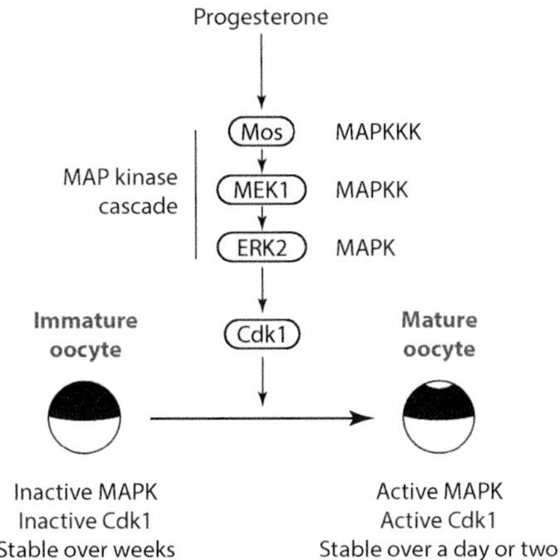

Figure 8.4 The MAPK cascade and its role in activating Cdk1 during *Xenopus* oocyte maturation.

white dot (**Figure 8.3b**). At about the same time, the nuclear envelope of the germinal vesicle breaks down. Next there is a dip in Cdk1 activity, which triggers the first meiotic division and the expulsion of the first polar body, which contains half of the homologous chromosomes. Cdk1 activity promptly rises back to M-phase levels, and the second meiotic spindle is organized. The oocyte then arrests in metaphase of meiosis II with high Cdk1 activity. At this point the oocyte is said to be mature. This whole process typically takes around 10 h.

The metaphase II-arrested state can be thought of as the induced fate of the oocyte, and, like the default state, it is stable. The mature oocyte can stay arrested in meiosis II for at least a day or so, whereupon it will either be ovulated and fertilized, and proceed with the rapid embryonic cell cycles, or it will die an apoptotic death.

The details of oocyte maturation vary from species to species—for example, mature oocytes are not always arrested in metaphase of meiosis II, and ERK2 activation is not always required for maturation. But there are recurring themes, including the all-or-none, irreversible character of the process, and *Xenopus* oocyte maturation is probably the best-understood example of oocyte maturation. Thus *Xenopus* oocytes have become the standard to which others, including human oocytes, are compared.

8.3 THE RESPONSE OF ERK2 IS ALL-OR-NONE IN CHARACTER

At what point along the pathway from the progesterone receptor to Cdk1 activation does the signal become all-or-none and irreversible? When pools of oocytes are examined, the ERK2 response to graded doses of progesterone is graded (**Figure 8.5a**). But this could mean either of two things: it could be that each individual oocyte has a graded response, or, alternatively, the individual oocytes could be exhibiting all-or-none responses, with some heterogeneity in the concentration of progesterone required to flip the switch (**Figure 8.5b**). These two possibilities can be distinguished by single-cell biochemical analysis, which, because oocytes are large enough to be picked up with a pipette and individually lysed, and contain enough protein for standard immunoblots and kinase assays, is not particularly difficult. As shown in **Figure 8.5c**, when a dish of oocytes is treated with a low dose of progesterone, each individual oocyte phosphorylates its ERK2 either fully, shifting up to a higher apparent molecular weight on an immunoblot, or not at all. Even though an oocyte contains about 180,000,000,000 ERK2 molecules, so that there is the potential for an almost infinitely-graded range of ERK2 activity, somehow the ERK2 molecules act not just individually, but also collectively, like a switch.

The all-or-none response to progesterone could, in principle, be generated upstream of the MAPK cascade and simply relayed by the cascade or it could be generated by the MAPK cascade itself. If the former is true, then an injection of recombinant Mos in the absence of progesterone should yield a graded ERK2 response; if the latter, the response should still be all-or-none. Experiments showed that, indeed, the response is still all-or-none (**Figure 8.5c**). Evidently in oocytes the Mos-MEK-ERK cascade can turn a graded stimulus into an all-or-none response.

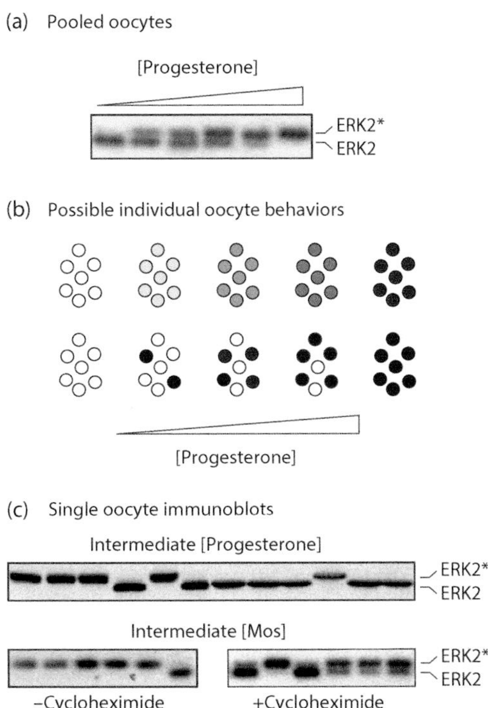

Figure 8.5 All-or-none, irreversible activation of ERK2 during oocyte maturation. (a) When oocytes are pooled, ERK2 activation, which is indicated by a shift to a higher apparent molecular weight on the immunoblot, is a graded function of the progesterone concentration. (b) A graded response at the population level could be due to either a graded or an all-or-none response at the level of individual cells. (c) The single-cell ERK2 response to progesterone (top) or microinjected Mos (bottom) is all-or-none, and blocking protein synthesis makes the response to Mos more graded. (Adapted from Ferrell and Machleder, *Science*. 1998.)

8.4 THERE IS POSITIVE FEEDBACK IN THE OOCYTE'S MAPK CASCADE

The depiction of the MAPK cascade in **Figure 8.4** is actually not complete. It turns out that Mos does activate ERK2 (via MEK), but also ERK2 stimulates Mos production. This constitutes a positive feedback loop. Moreover, compromising the feedback from ERK2 to Mos, either by blocking protein synthesis altogether or by selectively blocking Mos translation with an antisense morpholino oligonucleotide, which binds to the Mos mRNA and inhibits its translation, converts the response from all-or-none to more graded in character (**Figure 8.5c**). Thus, positive feedback is critical for the all-or-none response.

8.5 THE RESPONSE OF ERK2 TO PROGESTERONE IS NORMALLY IRREVERSIBLE

After a frog ovulates and releases an egg (which is a mature oocyte plus a coat of jelly that it acquires during ovulation) into the pond, one would expect that the progesterone that induced the maturation process would become diluted to almost zero. But since mature oocytes do not de-mature when they are laid as eggs, it seems likely that the signaling system that regulates maturation must be able to convert a transitory stimulus into an irreversible response. This is in fact the case; as shown in **Figure 8.6**, once ERK2 is phosphorylated, it stays phosphorylated for many hours after progesterone is washed away. This irreversibility is not due to slow dephosphorylation kinetics; actually, the phosphates that activate ERK2 turn over on a time scale of 5–10 min, so the phosphorylation state of ERK2 must be actively maintained. The irreversibility of ERK2 activation can be compromised by blocking Mos synthesis, which means that positive feedback is critical for both the all-or-none character of the response and for the irreversibility.

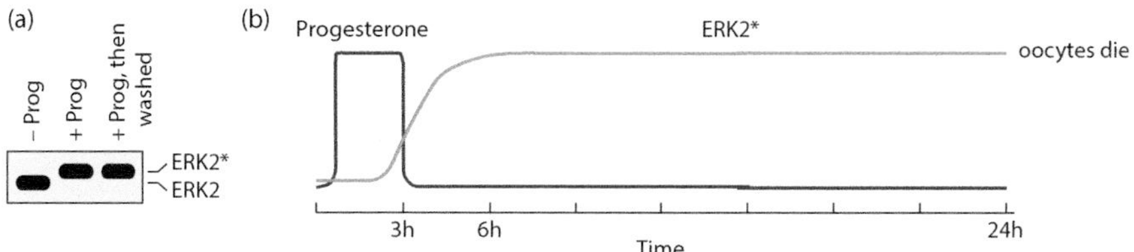

Figure 8.6 ERK2 activation becomes irreversible during oocyte maturation. (a) ERK2 stays activated after progesterone is washed away, and (b) the activation is maintained for many hours. (Adapted from Xiong and Ferrell, *Nature*. 2003.)

These are the experimental results: incubating or injecting oocytes with graded, reversible stimuli (progesterone or Mos) gives rise to an all-or-none, irreversible ERK2 response. To understand how such a response is generated, we will next build a computational model and analyze the model with methods from non-linear dynamics.

8.6 THE MOS/ERK2 SYSTEM CAN BE REDUCED TO A MODEL WITH A SINGLE TIME-DEPENDENT VARIABLE BECAUSE OF A SEPARATION OF TIME SCALES

Let us focus our model on the protein that sits atop the MAPK cascade, Mos. Its regulation is shown schematically in **Figure 8.7**, and we can convert this scheme into a rate equation.

Mos levels are determined by Mos synthesis and degradation; first let us examine Mos degradation. We will make the simplest-case assumption that the rate of Mos degradation is directly proportional to the Mos concentration (mass action kinetics):

$$Degradation\ rate = k_{deg} Mos. \tag{8.1}$$

It is actually not known whether Mos degradation is first order or saturable, but for simplicity we will assume that it is first order.

Mos synthesis is more complicated. There should be a feedback-dependent synthesis term where, in the simplest possible world, the synthesis rate would be directly proportional to the concentration of active ERK2. In addition, there should be some basal synthesis term that gets things started, where the rate could be directly proportional to the progesterone concentration (*prog*). Thus two terms contribute to the overall rate of synthesis:

$$Synthesis\ rate = k_{basal} prog + k_{feedback} ERK2^{*}. \tag{8.2}$$

The time-dependent variable *ERK2** represents the concentration of active ERK2.

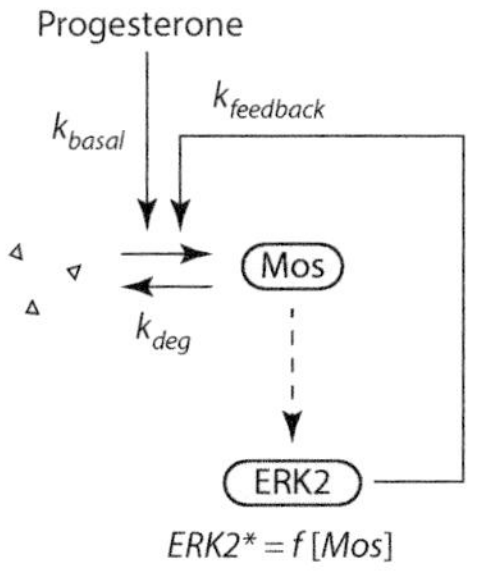

Figure 8.7 Simplified view of the Mos-ERK2 system in oocyte maturation.

The regulation of ERK2 by MEK and Mos occurs via phosphorylation, which takes place on a time scale of a few minutes, whereas the synthesis of Mos occurs over a few hours. We can therefore assume that at any given instant in time, ERK2 will be at its steady-state level of activity for whatever the Mos concentration is. This is referred to as a separation of time scales, and it allows us to replace the rate equations for the fast processes (MEK regulation by Mos and ERK2 regulation by MEK) by a single algebraic expression describing the steady-state response of ERK2 to Mos. For the moment we will not specify the exact functional form for the steady-state response; instead, we say that at any instant in time,

$$ERK2^* = f[Mos], \quad (8.3)$$

which means that

$$Synthesis\ rate = k_{basal} prog + k_{feedback} f[Mos]. \quad (8.4)$$

With equations for the synthesis rate (Eq. 8.4) and degradation rate (Eq. 8.1) in hand, we can explore the steady states of this system and learn something about the dynamics as well, through rate-balance analysis.

8.7 RATE-BALANCE ANALYSIS SHOWS WHAT IS REQUIRED FOR A BISTABLE RESPONSE

For the rate-balance analysis, we plot the rates of Mos synthesis and Mos degradation as functions of the concentration of Mos. Where the rate curves intersect, synthesis and degradation are balanced and Mos is in steady state.

We start by plotting the degradation rate as a function of the Mos concentration. From Eq. 8.1 we know this will be a straight line, with the slope of the line being determined by the rate constant k_{deg}. This is shown as in blue in **Figure 8.8**, where we have arbitrarily chosen the value of k_{deg} to be equal to 1.

Next we plot the synthesis rate curve. First let us ignore the feedback term and just plot the basal component of the synthesis rate, which we have assumed to be directly proportional to the progesterone concentration and independent of Mos. The result is the flat green line in **Figure 8.8**, where we have chosen $k_{basal}=1$ and $prog=25$. The two rate curves intersect at one point; the system has a single steady state, with Mos=25 units and the rate of Mos synthesis and degradation balanced at 25 units per sec. And the steady state is stable: if we raise the Mos concentration, the degradation rate will exceed the synthesis rate and the Mos concentration returns toward the steady state, and if we lower the Mos concentration, the synthesis rate will exceed the degradation rate and the system again returns toward steady state. Thus, if there is no feedback, we will have a monostable system.

If we were to double the progesterone concentration, we would double the height of the green curve and shift the steady state from 25 to 50. On the other hand, if we were to double k_{deg}, we would double the slope of the blue line and shift the steady state from 25 to 12.5. Regardless of what we choose for parameters, there will be a single steady state and it will be stable.

Now let us look at the feedback contribution to Mos synthesis, and for simplicity let us take the progesterone concentration to be zero so that there is no basal synthesis. At this point we need to make some assumption about how the steady-state level of ERK2 activity depends on Mos activity. One simple possibility would be a Michaelian relationship, with the fractional activation of ERK2 given by:

$$ERK2^*_{ss} = \frac{Mos}{K + Mos}. \quad (8.5)$$

For any choice of K and $k_{feedback}$ (and here we have chosen $K=10$ and $k_{feedback}=40$), the rate curve will be a hyperbola, and if the hyperbola is tall enough it will intersect the degradation line in two places: at the

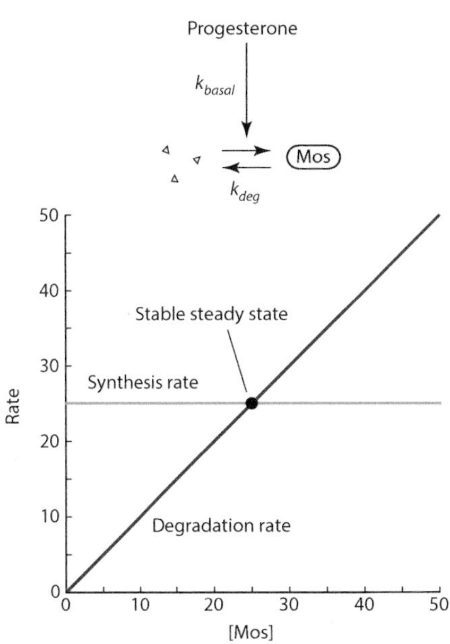

Figure 8.8 Rate-balance analysis of Mos synthesis and degradation assuming no feedback.

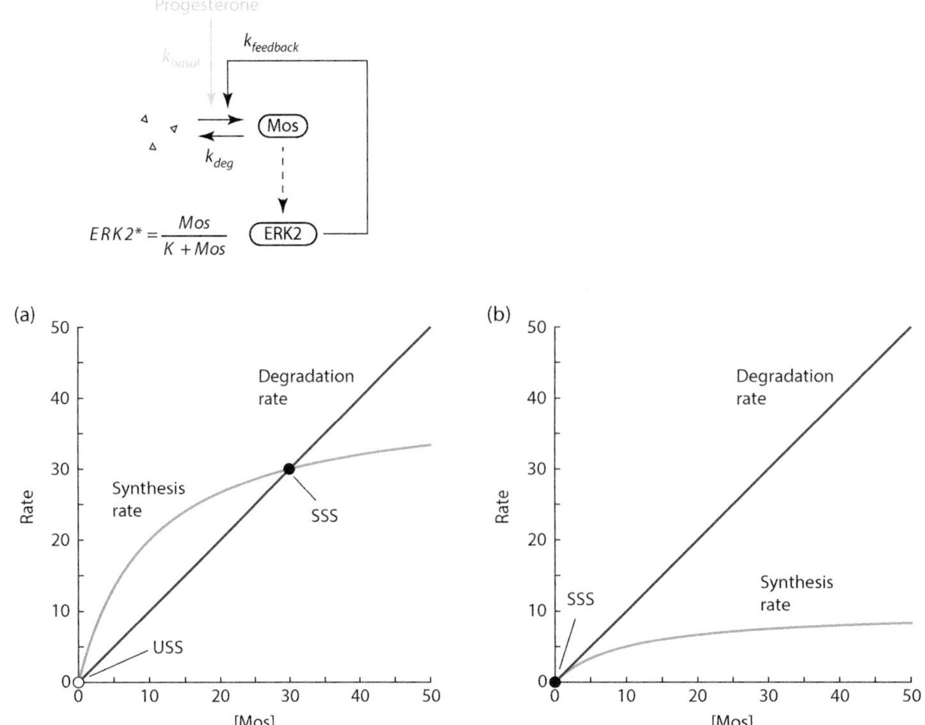

Figure 8.9 Michaelian feedback does not yield a bistable response for either high (a) or low (b) feedback strength. USS stands for unstable steady state, and SSS stands for stable steady state.

origin (Mos=0) and at one other place where Mos>0 (**Figure 8.9a**). The system has two steady states, an off-state and an on-state, rather than the usual one. So is it bistable?

The answer is no. The on-state is stable, but the off-state is not. To signify this difference, we have represented the stable on-state with a solid circle and the unstable off-state with a hollow circle. If the system begins in the off-state, the Mos synthesis and degradation rates would be balanced (both would be zero), but if we perturb the system by adding even a single molecule of Mos, the synthesis rate would increase more than the degradation rate, and so the system would move away from the off-state, not back toward it. Ultimately the perturbed system would settle into the on-state, the only stable steady state the system has.

The problem is that Mos synthesis increases faster than Mos degradation when you nudge the system to the right of the off-state. So what happens if we decrease the initial slope of the green curve, say by decreasing $k_{feedback}$? Surely if the feedback is weak enough, the Mos=0 state will become stable. And that is in fact the case (**Figure 8.9b**); once we decrease $k_{feedback}$ to 10 or lower, the Mos=0 state does become stable. But note that the system no longer has two steady states; as $k_{feedback}$ gets smaller, the on-state moves leftward, and only after it has moved all the way to Mos=0 and the system changes from having two steady states to one does the Mos=0 state become stable. No matter what the choice of parameters, the system will have a single stable steady state. We have still not succeeded in producing a bistable system.

What is needed is for the green curve to snake around the blue curve, starting out increasing more gradually with Mos than the blue curve does but then catching up. That is what we would get with a sigmoidal curve, and, as luck would have it, we already know that the experimentally determined steady-state relationship between Mos and ERK2 activity is, in fact, sigmoidal, well-approximated by the Hill equation with a Hill exponent of about 5 (see **Figure 7.8**):

$$ERK2^*_{ss} = \frac{Mos^5}{K^5 + Mos^5}. \tag{8.6}$$

If $k_{feedback}$ is sufficiently large, as it is in **Figure 8.10**, then there will be three places where the synthesis and degradation curves intersect: a stable off-state at the origin (designated by a filled circle); a stable on-state with $Mos \approx 39$ units (again a filled circle); and an unstable steady state between the two, with $Mos = 20$ units (hollow circle). If one starts with *Mos* anywhere above 20 units, the system will move to the on-state. Below 20 units, the system will move to the off-state. And right at $Mos = 20$, the system is balanced, but precariously balanced, so that the slightest nudge will push it down to the off-state or up to the on-state.

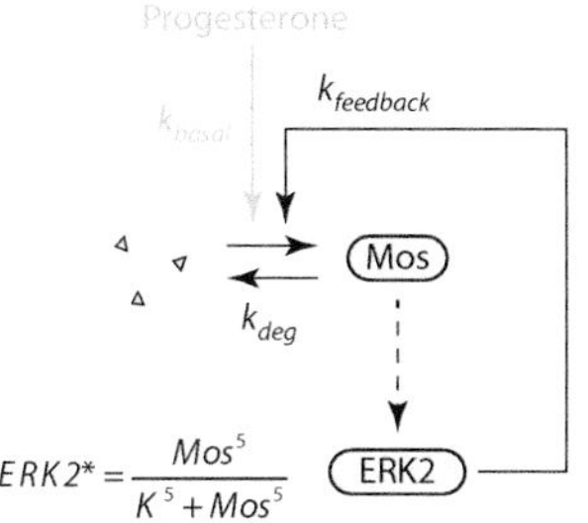

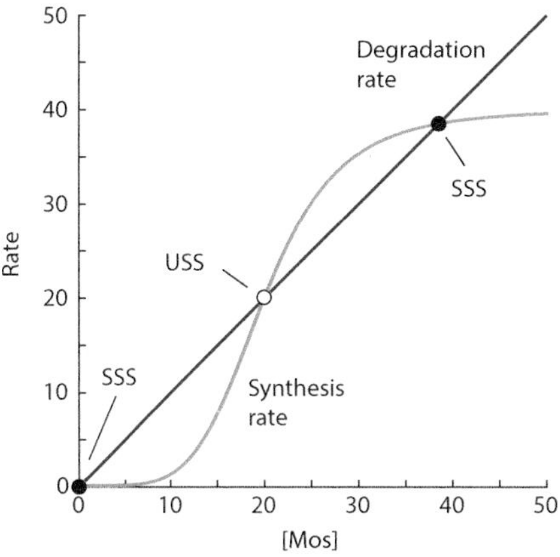

Figure 8.10 Ultrasensitive feedback can yield a bistable response. USS stands for unstable steady state, and SSS stands for stable steady state.

We now have a bistable system, with two stable steady states and one unstable one. And we can see how various features of the model contributed to the bistability. The positive feedback made it so that the rate of Mos synthesis increases with the Mos concentration, which means that the green curve has a positive slope, which is essential if we want it to intersect the blue curve more than once; in fact, it has been shown that for a deterministic biochemical reaction system to exhibit bistability, the system must include positive feedback. The ultrasensitivity within the positive feedback loop enabled the green curve to snake around the blue curve, allowing the system to be bistable. Any Hill exponent greater than one would permit a bistable response, but the bigger the exponent, the easier it is to make the two curves intersect at three points.

8.8 INCREASING THE PROGESTERONE CONCENTRATION PUSHES THE SYSTEM THROUGH A SADDLE-NODE BIFURCATION

The rate curves in **Figure 8.10** assumed that the progesterone concentration was zero. What happens if we raise the concentration to some constant non-zero level?

Adding progesterone does nothing to the degradation rate curve, but it adds a constant level of basal Mos synthesis to the synthesis rate curve, which means that the green curve shifts upward but otherwise stays identical in shape. This is shown in **Figure 8.11a** for 0, 2, 4, 6, 8, and 10 units of added progesterone. As the rate curve shifts upward, the positions of the three steady states change, and, in particular, the off-state moves toward higher Mos concentrations and the unstable steady state moves lower. Eventually (at about 9 units of progesterone) the off-state and the unstable steady state collide with and annihilate each other—boom! The system then transitions from being bistable to monostable, with a single stable steady state, the on-state. Once this transition from bistability to monostability takes place, no matter what the starting Mos concentration is, the system will end up in the on-state.

On the other hand, if we start with a high progesterone concentration and the system in the on-state, and then wash the progesterone away, the steady-state concentration of Mos will drop a bit, down to ~39 units. But it will not drop all the way to zero; the on-state does not disappear, and it is still stable, so there is no reason for the system to leave it. Thus the transition from the off-state to the on-state is irreversible, even though the binding of progesterone to the progesterone

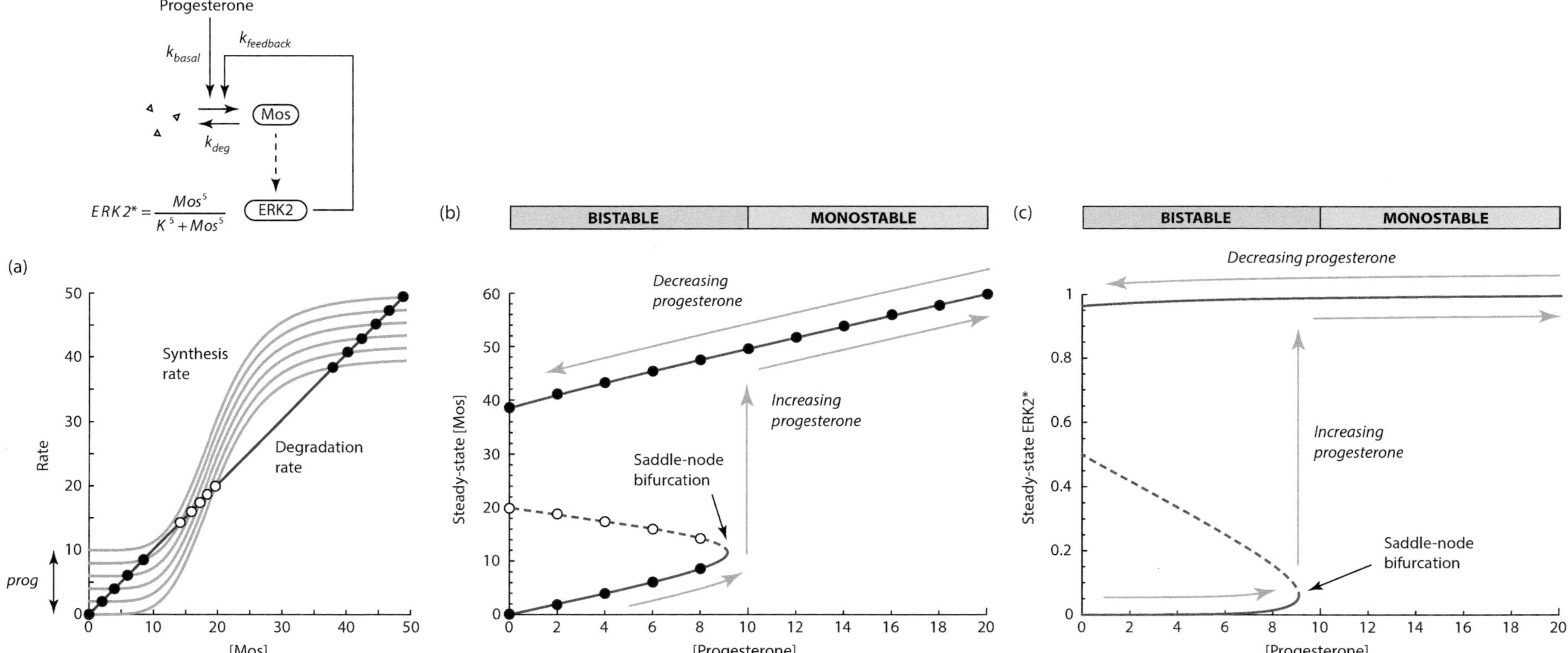

Figure 8.11 Increasing the progesterone concentration (*prog*) makes the off-state disappear. (a) Rate-balance analysis assuming ultrasensitive positive feedback and various concentrations of *prog*. (b) The steady-state concentration of Mos as a function of *prog*. At about 9 units of progesterone, the off-state and unstable steady state annihilate each other at a saddle-node bifurcation. Note that once the system is in the on-state, it stays there even after the progesterone concentration is decreased back to zero. (c) Modeled ERK2 activation assuming that ERK2 activity is an ultrasensitive function of the Mos concentration.

receptor, the synthesis of Mos, and the phosphorylation of ERK2 are all intrinsically reversible.

We can construct a stimulus–response plot based on the intersection points in the rate–balance plot. On the *x*-axis we plot the progesterone concentration; on the *y*-axis, the Mos concentrations at each of the steady states, including the unstable one. The result is an S-shaped curve that can be divided into three sections: a bottom section made up of stable off-states; a top section that includes the stable on-states; and a middle section, designated by the dashed curve, where the unstable steady states reside (**Figure 8.11b**). The middle and bottom sections merge and then disappear at the progesterone concentration where the two rate curves (**Figure 8.11a**) go from having three intersections to one. Thus if one were to start in the off-state and dial up the progesterone concentration, there would be a discontinuity in the steady-state response. The modeled Mos response is not quite all-or-none in character; it is more like a big-or-small response. But since *ERK2** is a steeply sigmoidal function of *Mos*, the modeled *ERK2** response is pretty close to all-or-none (**Figure 8.11c**), which accounts for the experimental results shown in **Figure 8.5c**. Positive feedback plus nonlinearities has yielded a bistable response, and the bistable response converts a graded, reversible stimulus into a nearly all-or-none, irreversible output.

The transition where the system goes from being bistable to monostable is called a saddle-node bifurcation, and because we are varying one parameter of the model (*prog*), a plot like **Figure 8.11b** or **c** is sometimes called a one-parameter bifurcation diagram. Some explanation of this terminology is probably in order. A **bifurcation** is a splitting of one thing into two, and, if you look at the plot from right to left, the lower steady state does bifurcate or split immediately after it appears (splitting into a stable and an unstable steady state). **Node** is another term for a stable steady state; the term and **saddle** is borrowed from the two-variable version of this type of bifurcation, which we will explore in Chapter 9.

What determines the position of the progesterone threshold for the system? That is, what determines the concentration of progesterone at which the off-state and the unstable steady state meet and annihilate each other (**Figure 8.11a**)? For one thing, it depends on the slope of the degradation rate curve; the steeper it is, the higher the threshold. It also depends on the shape of the degradation rate curve. We assumed that mass action kinetics applies, making the blue curve a straight line (**Figure 8.11a**), but we could have assumed a saturable process, and if the blue curve levels off as Mos increases, it would make it easier for the green curve to "catch up" to the blue curve. The height, shape, and *EC50* of the synthesis rate curve all contribute to the position of the threshold as well. For example, the higher the ultrasensitivity, the flatter the curve will initially be, which will shift the threshold to a higher progesterone concentration. The relationship between the progesterone concentration and the basal rate of Mos synthesis also figures into the position of the threshold. For mass action kinetics, the position of the threshold scales with k_{basal}. But if we had assumed something other than a linear relationship between progesterone and the basal synthesis rate, the additional parameters that define that relationship would figure into the threshold as well. Thus the position of the threshold—the saddle-node bifurcation for flipping from the off-state to the on-state—is a systems-level property, influenced by all of the factors that bear on Mos synthesis, Mos degradation, and the balance between the two.

Saddle-node bifurcations show up very commonly in systems that transition in and out of bistability; it is the typical (but not the only) way bistability arises and disappears. All saddle-node bifurcations share certain properties, so what you learn about what happens at the saddle-node bifurcation in a model of oocyte maturation can help you to understand other phenomena involving totally unrelated proteins organized in different circuits.

8.9 TWEAKING THE MODEL CAN CHANGE AN IRREVERSIBLE RESPONSE TO A HYSTERETIC ONE

Bistable systems can be irreversible, but irreversibility is not inevitable. We can demonstrate this by slightly modifying our model.

Suppose we were to start with a synthesis rate curve like that shown in **Figure 8.10**, but made the feedback a little weaker—made $k_{feedback}$ smaller—so that in the absence of progesterone, there was only a single intersection between the synthesis and degradation rate curves (**Figure 8.12a**). The system therefore has a single steady state, an off-state, and it is stable. As the progesterone concentration increases, the green curve moves upward, and at just above 2 units of progesterone, the upper knee of the curve touches the blue degradation rate curve (**Figure 8.12a**). Thus a new steady state appears and then splits into a stable on-state and an unstable steady state through a saddle-node bifurcation (**Figure 8.12a,b**). As the progesterone concentration increases further, the green curve shifts up. The unstable steady state and the off-state approach each other and then disappear (boom!) through a second saddle-node bifurcation. The result is a hysteretic stimulus–response relationship (**Figure 8.12b**). The system requires more stimulus (progesterone) to move from the off-state to

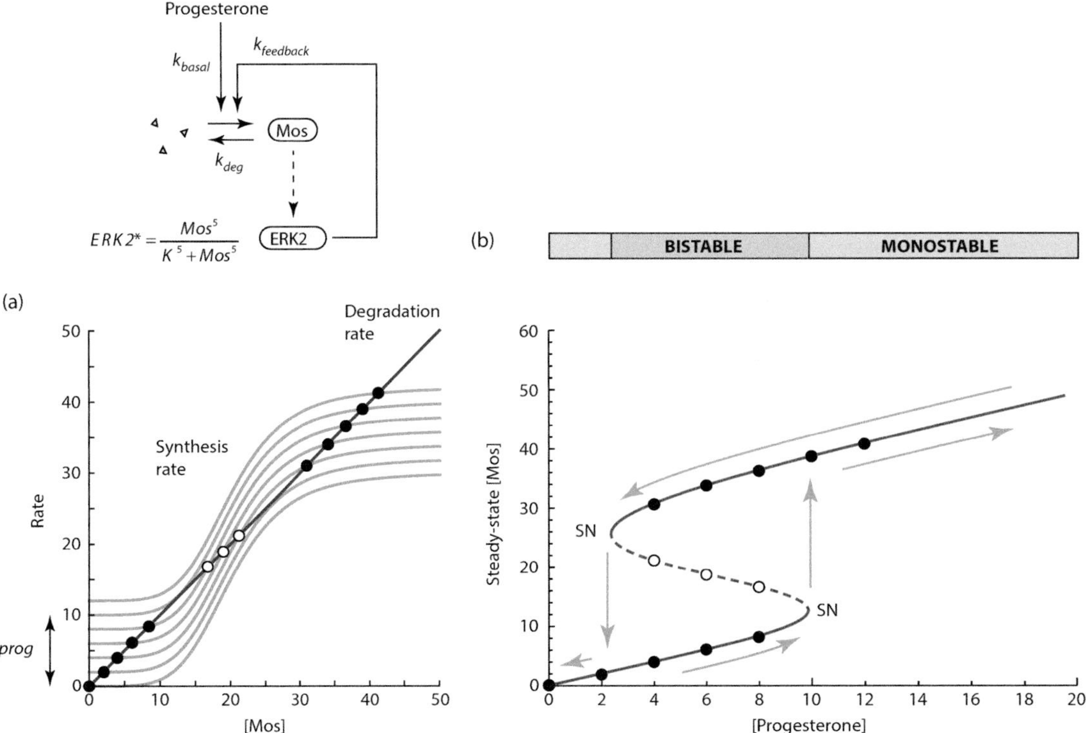

Figure 8.12 A modeled hysteretic, rather than irreversible, response. Panel a shows the rate-balance analysis and panel b shows the corresponding steady-state responses. The feedback strength here was decreased from 40 to 30. SN denotes saddle-node bifurcation.

the on-state than it does to maintain itself in the on-state once it has gotten there, but the response is not irreversible.

Both hysteretic and irreversible responses are important in biological regulation. Oocyte maturation is an example of biological irreversibility that arises out of a bistable control system, and probably many other cell fate induction processes are as well. On the other hand, the activation of the master regulator Cdk1 is an example of a hysteretic process, which we will discuss in more detail when we examine relaxation oscillators in Chapter 15.

The fact that an irreversible bistable response can be converted into a hysteretic bistable response fits nicely with the observation that differentiated cells can be engineered to de-differentiate into pluripotent stem cells. Presumably the bistable circuits that maintain the differentiated state have been nudged into a hysteretic regime and then switched off by, say, the expression of Yamanaka factors or the transfer of a differentiated nucleus into an egg.

8.10 THE DYNAMICS OF THE SYSTEM CAN BE INFERRED FROM THE RATE–BALANCE PLOT

To understand the dynamics of our bistability model, we begin by writing down the ODE for the system by combining Eqs. 8.1, 8.2, and 8.6:

$$\frac{dMos}{dt} = k_{basal}prog + k_{feedback}\frac{Mos^5}{K^5 + Mos^5} - k_{deg}Mos. \tag{8.7}$$

Ideally we would next solve the ODE and explore how the solution varies with different assumed values of the parameters and initial conditions, but this ODE is too complicated to solve analytically. Nevertheless, we can solve the ODE numerically and explore the solutions.

First let us parameterize the system to yield an irreversible response (as it was in **Figure 8.10**), and start with the system in its off-state, with a progesterone concentration of 0 and a Mos concentration of 0. Then let us step the progesterone up to 2, 4, 6, 8, 10, or 12 units and calculate the resulting time courses. As shown in **Figure 8.13a**, for progesterone concentrations of up to 8 units, the system smoothly approaches a low steady state Mos concentration. The curves look qualitatively like the typical exponential approach to steady state, and indeed for very low progesterone concentrations this is exactly the case.

Once the progesterone concentration reaches 10 units—just beyond the saddle-node bifurcation where the off-state disappears (**Figure 8.12**), something strange happens. The system initially looks like it is approaching a steady state with ~10–15 units of Mos, but it sort of does a rolling stop and then accelerates onward toward the on-state (**Figure 8.13a**). We can see why this happens by looking at the corresponding rate–balance plot (**Figure 8.13b**) or the vector field of reaction velocities (**Figure 8.13c**). Even though there is no longer an intersection point between the green and the blue curves in the vicinity of 10–15 units of Mos, the green curve does come close to the blue curve, and so the net velocity of the reaction—the synthesis rate minus the degradation rate—becomes very small (**Figure 8.13b,c**). This is a phenomenon called critical slowing; with progesterone concentrations just past the saddle-node bifurcation, which can be

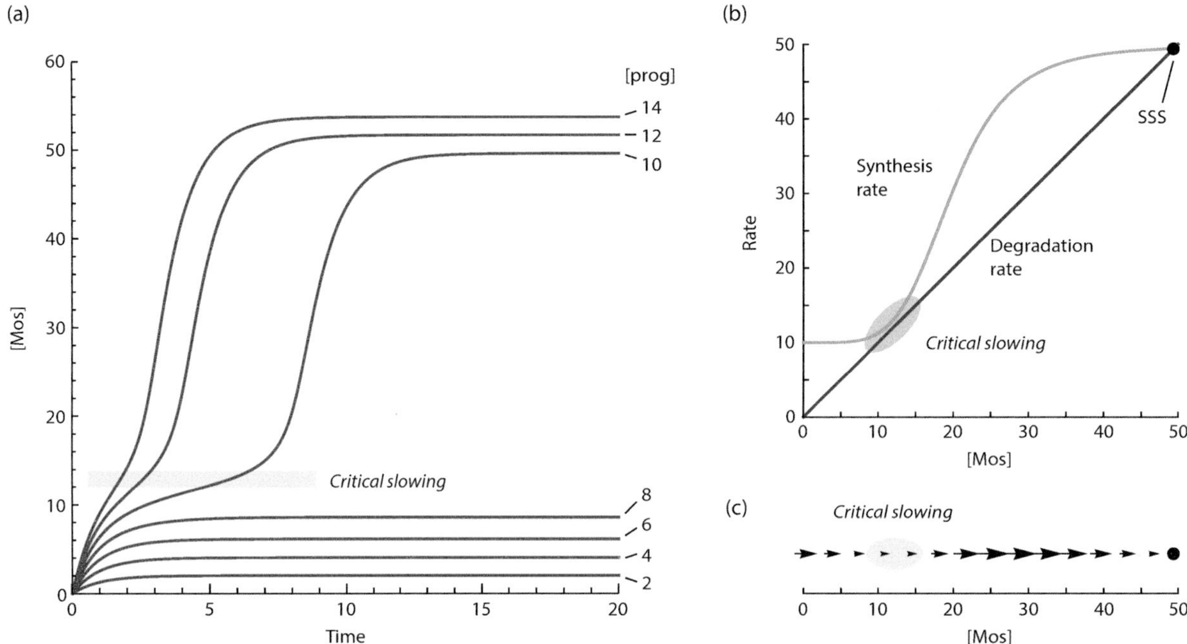

Figure 8.13 Critical slowing. (a) Calculated time courses for Mos accumulation starting from [Mos]=0, assuming various concentrations of progesterone. (b) Rate-balance analysis for a system where the progesterone concentration is a little higher (10 units) than that which corresponds to the saddle-node bifurcation (~9.1 units). (c) The velocity vector field for the system with 10 units of progesterone.

viewed as a critical point, the annihilated steady state exerts a sort of ghostly influence on the reaction dynamics. Critical slowing is one of those general properties of bistable systems with saddle-node bifurcations we alluded to in Section 8.8.

8.11 THE VELOCITY VECTOR FIELD CAN BE REPRESENTED AS A POTENTIAL LANDSCAPE

The energetics of protein folding are often depicted as a potential energy landscape, with the minimum-energy species residing at the bottom of a potential energy well. This landscape is a scalar field—a bunch of energies—and we can intuitively infer the forces on the species—a vector field—from the slope of the landscape. If the landscape is flat, there is no net force; if it slopes upward, there is a force to the left; and if it slopes downward there is a force to the right. The reason for constructing the landscape is that it conveys the dynamics of the system in a particularly easy-to-understand fashion.

We can construct an analogous potential landscape for any one-ODE model, including our Mos model. The vector field of relevance in this case is not a force field, like we have with an energy landscape, but rather a velocity field. Thus we want to construct a scalar field—in other words, a function—which, when differentiated, gives us a vector field like that shown in **Figure 8.13c**. To do this we start with the rate equation (Eq. 8.7) and call the right-hand side $f[Mos]$, the instantaneous net rate of *Mos* production as a function of the *Mos* concentration. Next we integrate f with respect to *Mos*, put a minus sign in front of the integral, and obtain a potential function Φ:

$$\Phi = -\int f[Mos]dMos, \tag{8.8}$$

$$\Phi = -\int \left(k_{basal} prog + k_{feedback} \frac{Mos^5}{K^5 + Mos^5} - k_{deg} Mos \right) dMos. \tag{8.9}$$

The minus sign in front of the integral comes from the fact that we want a positive slope of the potential to drive the Mos concentration toward a smaller Mos concentration, and a negative slope to drive the Mos concentration higher. This makes zero-velocity stable steady states end up at the bottoms of valleys, just as the zero-acceleration states are in energy landscapes.

The first and third terms of the integrand are easy to integrate. The second one is not so easy, but it can be done, and the overall expression for the potential function is:

$$\Phi = -\frac{1}{20}\left(\begin{array}{l} 20k_{feedback} Mos - 10k_{deg} Mos^2 + 20k_{basal} Mos \cdot prog \\ \quad - 2\sqrt{10+2\sqrt{5}}K \cdot k_{feedback}\, \mathrm{ArcTan}\left[\dfrac{\left(-1+\sqrt{5}\right)K + 4Mos}{\sqrt{10+2\sqrt{5}}K} \right] \\ \quad - 2\sqrt{10-2\sqrt{5}}K \cdot k_{feedback}\, \mathrm{ArcTan}\left[\dfrac{\left(-1-\sqrt{5}\right)K + 4Mos}{\sqrt{10-2\sqrt{5}}K} \right] \\ \quad - 4K \cdot k_{feedback}\, \mathrm{Ln}\left[K + Mos\right] \\ \quad - \left(-1+\sqrt{5}\right)K \cdot k_{feedback}\, \mathrm{Ln}\left[K^2 + \tfrac{1}{2}\left(-1+\sqrt{5}\right)K \cdot Mos + Mos^2\right] \\ \quad + \left(1+\sqrt{5}\right)K \cdot k_{feedback}\, \mathrm{Ln}\left[K^2 + \tfrac{1}{2}\left(1+\sqrt{5}\right)K \cdot Mos + Mos^2\right] \end{array} \right). \tag{8.10}$$

Eq. 8.10 is too complicated to gain much insight from—sorry—but still we can plot it and see what the potential curve looks like. **Figure 8.14a** shows the potential for the $prog=0$ case. There is a half-valley whose bottom represents the off-state, a full-valley whose bottom is the on-state, and a local maximum corresponding to the unstable steady state between them.

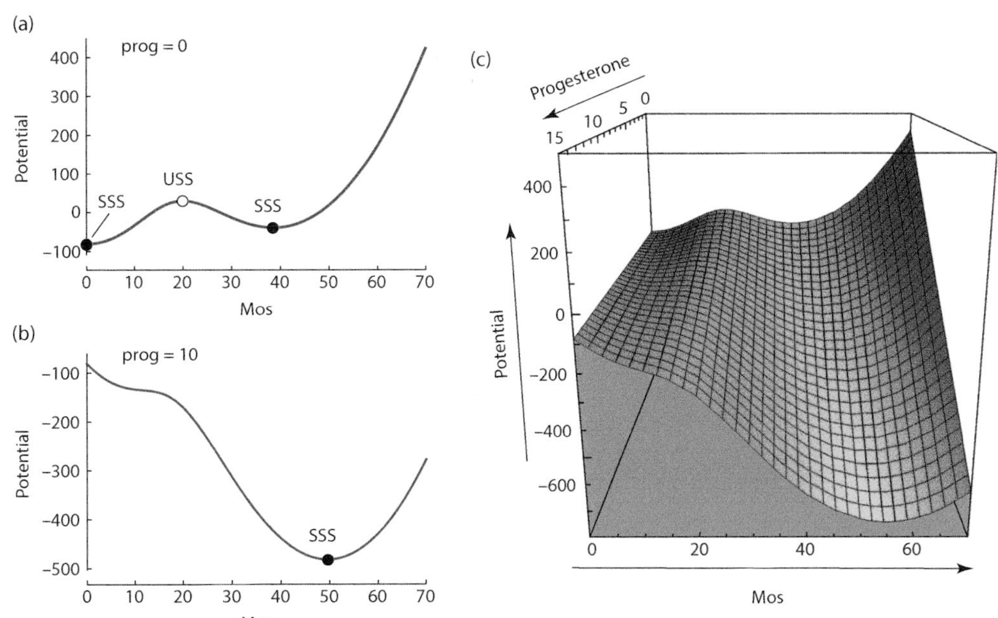

Figure 8.14 Potential curves (a, b) and the potential surface (c) for the Mos system at various concentrations of progesterone.

Dialing the progesterone concentration by up to 10 units skews the potential curve so that height of the unstable steady state decreases to zero and the depth of the on-state's potential well increases. This means that there is a monotonic path from Mos=0 to the on-state, albeit with some slowing (the critical slowing) near where the unstable steady state used to be. Moreover, one can see why the potential varies with the progesterone concentration as it does. Even though Eq. 8.10 is complicated, there is only one term that depends on the progesterone concentration—$-k_{basal}Mos \cdot prog$. As the progesterone concentration increases, this term contributes a steeper and steeper negatively sloped line to the potential function, skewing the curve downward, and this accounts for the disappearance of the off-state and unstable steady state and the deepening of the on-state's potential well.

We can also plot the potential as a two-dimensional landscape, a surface that shows how the potential varies with both the Mos and progesterone concentrations (**Figure 8.14c**). When the progesterone concentration is small (at the back of the plot), the potential surface has two valleys with a small hill between them. As the progesterone concentration increases (moving toward the viewer), the hill gets smaller and the on-state's valley gets deeper. Eventually the off-state and unstable steady state disappear at the saddle-node bifurcation, and from there on the on-state dominates and the system is monostable.

Thus, a potential function is another way, like the rate–balance plot, of conveying the character and dynamics of a bistable system and of understanding what changes about a system at a saddle-node bifurcation.

SUMMARY

Here we have shown how positive feedback can give rise to bistability. Bistability is a systems-level property, and it depends on the feedback, some nonlinearity in the feedback loop, and the proper balance between the positive and negative reactions within the loop. In a bistable system, transitions between the alternative stable steady states typically occur when something changes about the system—for example, the basal rate of activation increases—causing one stable steady state to collide with an unstable steady state or saddle and disappear at a saddle-node bifurcation.

Bistable systems are generally at least hysteretic—it is harder to switch into a new state than it is to maintain the new state—and are sometimes irreversible. An irreversible bistable system behaves much like a toggle switch, flipping on in response to a stimulus and then staying on even after the stimulus is removed. Progesterone-induced *Xenopus* oocyte maturation is one well-studied example of how a bistable signaling system can give rise to an irreversible switch between cell fates; Cdk1 activation during mitotic entry and inactivation during mitotic exit represents an example of a hysteretic bistable switch. Bistability provides nature with a way of building discrete biological states out of continuously variable components and of building irreversible responses out of reversible components.

For one-variable systems, one can make inferences about the global stability of a steady state through rate-balance analysis, and one can visualize the dynamics of the system from either the rate–balance plot or by constructing a potential function whose slope tells the direction and speed at which the system will go toward or away from a steady state.

FURTHER READING

XENOPUS OOCYTE MATURATION

Abrieu A, Dorée M, Fisher D. The interplay between cyclin-B-Cdc2 kinase (MPF) and MAP kinase during maturation of oocytes. *J Cell Sci*. 2001 Jan;114(Pt 2):257–67.

Ferrell JE Jr. Xenopus oocyte maturation: new lessons from a good egg. *Bioessays*. 1999 Oct;21(10):833–42.

Ferrell JE Jr, Machleder EM. The biochemical basis of an all-or-none cell fate switch in Xenopus oocytes. *Science*. 1998 May 8;280(5365):895–8.

Sohaskey ML, Ferrell JE Jr. Distinct, constitutively active MAPK phosphatases function in Xenopus oocytes: implications for p42 MAPK regulation in vivo. *Mol Biol Cell*. 1999 Nov;10(11):3729–43.

Xiong W, Ferrell JE Jr. A positive-feedback-based bistable 'memory module' that governs a cell fate decision. *Nature*. 2003 Nov 27;426(6965):460–5. Erratum in: *Nature*. 2007 Aug 30;448(7157):1076.

BISTABILITY

Delbrück, M. Enzyme systems with alternative steady states. In Unités Biologiques Douées de Continuité Genetique (international Symposium CNRS No. 8). Editions du CNRS, Paris, 1949, pp. 33–34.

Ferrell JE, Xiong W. Bistability in cell signaling: How to make continuous processes discontinuous, and reversible processes irreversible. *Chaos*. 2001 Mar;11(1):227–36.

Monod J, Jacob F. General conclusions: teleonomic mechanisms in cellular metabolism, growth, and differentiation. *Cold Spring Harbor Symp. Quant. Biol*. 1961;26:389–401.

Kauffman S. Homeostasis and differentiation in random genetic control networks. *Nature*. 1969;224:177–8.

Novick A, Weiner M. Enzyme induction as an all-or-none phenomenon. *Proc Natl Acad Sci USA*. 1957 Jul 15;43(7):553–66.

Thomas R. On the relation between the logical structure of systems and their ability to generate multiple steady states or sustained oscillations. *Springer Ser. Synergetics*. 1981;9:180–93.

POTENTIAL SURFACES

Ferrell JE Jr. Bistability, bifurcations, and Waddington's epigenetic landscape. *Curr Biol*. 2012 Jun 5;22(11):R458–66.

Wang J, Xu L, Wang E. Potential landscape and flux framework of nonequilibrium networks: robustness, dissipation, and coherence of biochemical oscillations. *Proc Natl Acad Sci USA*. 2008;105:12271–6.

SADDLE-NODE BIFURCATIONS

Strogatz SH. *Nonlinear Dynamics and Chaos: With Applications to Physics, Biology, Chemistry, and Engineering*. Westview Press, Cambridge, MA, 1994, Chapters 3 and 8.

BISTABILITY 2

Systems with Two Time-Dependent Variables

9

IN THIS CHAPTER . . .

Both the rate–balance plot and the potential landscape, which we used to analyze the bistability of the Mos/ERK2 system in Chapter 8, rely on the fact that our model had a single time-dependent variable, which in turn depended on our identification of a single slowly varying species (Mos), with the activities of all of the other components of the positive feedback loop responding quickly enough that we can invoke a separation of time scales. But sometimes it is not possible to make such a simplification. For example, in cell cycle regulation, the master M-phase regulator Cdk1 inactivates Wee1 through phosphorylation and Wee1 feeds back to inactivate Cdk1 through phosphorylation—a

DOI: 10.1201/9781003124269-9

double-negative feedback loop—and the time scales for both phosphorylation reactions are probably similar. Likewise, in fat cell differentiation, the master transcriptional regulator PPARγ and C/EBPβ stimulate each other's transcription in a positive feedback loop, and again the time scales for both processes are probably similar. Thus, it is useful to have ways of analyzing and understanding two-variable positive feedback systems.

Here we present one such way: a graphical way of looking for steady states in the two-dimensional **phase plane**, coupled with **linear stability analysis** to characterize the stability of each steady state.

9.1 TWO-VARIABLE POSITIVE FEEDBACK AND DOUBLE-NEGATIVE FEEDBACK LOOPS CAN FUNCTION AS BISTABLE SWITCHES

Figure 9.1 depicts two simple two-variable systems that can, under the right circumstances, function as bistable switches: a positive feedback loop, with x activating y and y activating x (**Figure 9.1a**), and a double-negative feedback loop, with each species inhibiting the other (**Figure 9.1b**). Either or both of the species could accept regulatory inputs to flip the bistable switch, and either or both could regulate downstream targets.

We can characterize the properties of these two-variable systems through what is termed phase plane analysis. The phase plane in this case is one quadrant of the x–y plane, with one time-dependent variable (x) plotted on the x-axis and the other variable (y) on the y-axis. Phase plane analysis is not as simple or as directly intuitive as rate-balance analysis, but it is nevertheless a powerful method for graphically understanding the dynamics and steady states of two-variable models.

Let us start with the positive feedback loop (**Figure 9.1a**) and begin by writing down rate equations for x and y. In principle the positive arrows could represent stoichiometric activation, enzymatic activation, or stimulated production; here we will work through an example where both arrows represent stimulated production. Moreover, since ultrasensitivity in the feedback proved to be important in generating bistability in the Mos model, let us assume that rate of production of x is a Hill function (here we chose $n=2$) of the concentration of y, and vice versa. Let us also assume, as we did in the Mos model, that the degradation processes are described by mass action kinetics. And finally, let us assume that there is some tunable basal rate of production of x, which we will use to flip the system between an off-state, with x and y inactive, and an on-state, with x and y active. We then have:

$$\frac{dx}{dt} = k_{basal} + k_1 \frac{y^2}{K_1^2 + y^2} - k_{-1}x \tag{9.1}$$

$$\frac{dy}{dt} = k_2 \frac{x^2}{K_2^2 + x^2} - k_{-2}y. \tag{9.2}$$

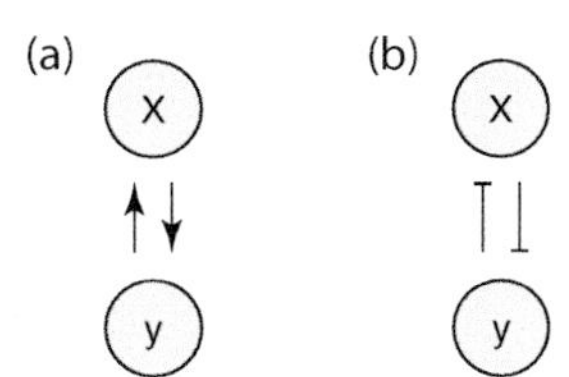

Figure 9.1 Two-variable systems with positive (a) and double-negative (b) feedback.

Next let us choose values for the parameters; for a start, let k_{basal} equal 0, let each of the other rate constants equal 1, and let $K_1=K_2=0.4$.

With the model specified and parameterized, let us see how many steady states the system has. For the system to be in steady state, neither x nor y can be changing with respect to time. This means that:

$$0 = \frac{y^2}{0.4^2 + y^2} - x \tag{9.3}$$

$$0 = \frac{x^2}{0.4^2 + x^2} - y. \tag{9.4}$$

Equations 9.3 and 9.4 define curves in the phase plane, which are plotted in **Figure 9.2**. We call the blue curve, which comes from setting $\frac{dx}{dt} = 0$, the x-nullcline and the green curve the y-nullcline. The two curves intersect at three points, {0, 0}, {0.2, 0.2}, and {0.8, 0.8}, and each intersection point represents a steady state; that is, a point where both $\frac{dx}{dt}$ and $\frac{dy}{dt} = 0$. The ability of the nullclines to intersect three times arises out of the positive feedback in the model—positive feedback means that x increases with y (from the x-nullcline) and y increases with x (from the y-nullcline), and that keeps the nullclines pointed in the same general direction—and the ultrasensitivity, which allows the two nullclines to snake around each other and intersect multiple times.

By analogy to our one-variable Mos model, you probably suspect that two of the steady states are stable and one is unstable (and in fact that is how we have colored the points in **Figure 9.2**). But how do we know for sure? With the one-variable model, we could use the rate–balance plot to quickly determine the stability of each of the steady states, but not so for the two-variable model.

One way to approach the stability question is to plot some sample trajectories and see if they go toward or away from each of the steady states. As shown in **Figure 9.3**, the on-state at {0.8, 0.8} is stable; all of the nearby trajectories converge to it. In fact, any trajectory that begins in the pink region of the phase plane will ultimately approach the on-state. For this reason, a stable steady state like this on-state is sometimes called an **attractor**, and the pink region is called the stable manifold of this attractor, or the basin of attraction for the on-state. The trajectories do not necessarily approach the steady state monotonically, unlike the trajectories in our one-variable model; for example, if we start in the bottom right-hand corner of the phase plot, x will initially decrease and then subsequently increase on its way to the stable on-state. In fact, the trajectories appear to initially be attracted to the diagonal, and then once they get close to it, veer off toward the on-state. Overall the trajectories look like a collection of side streets that all empty into a main thoroughfare running along the diagonal. Or like the veins of a banana leaf connecting to the midrib.

The off-state {0, 0} is also stable, and all of the trajectories that start in the yellow region ultimately approach the off-state, although again they tend to initially head toward the diagonal and then turn toward the off-state. So far so good.

The behavior of the third steady state, at {0.2, 0.2}, is a bit more complicated. The trajectories right on the boundary between the stable manifolds of the off- and on-states—the **separatrix**, designated by the dashed black line—are attracted by this steady state. However, trajectories close to the separatrix, but not on it, appear to be initially attracted to the steady state, but then veer away toward either the on-state or the off-state. A steady state like this one is termed a **saddle** or a **saddle point**, because if you imagine a potential surface that might produce trajectories like this, it would be shaped like a saddle. A saddle point is conventionally denoted by an open circle just like

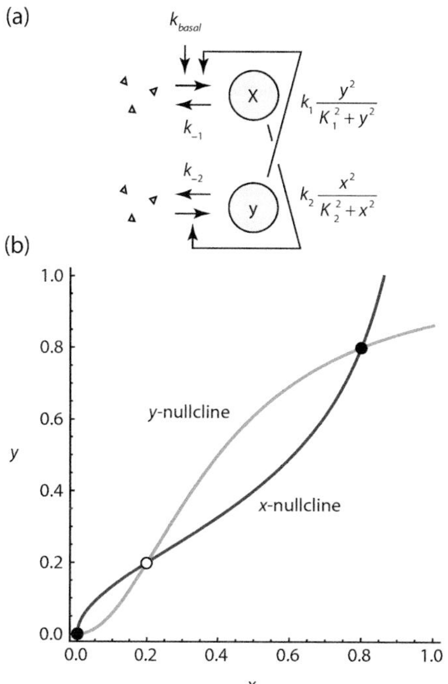

Figure 9.2 Nullclines and steady states in the phase plane for the two-variable positive feedback system. (a) Detailed schematic of the positive feedback system, with x stimulating the production of y and vice versa. (b) Nullclines and steady states. The filled circles will prove to be stable steady states and the open circle a saddle.

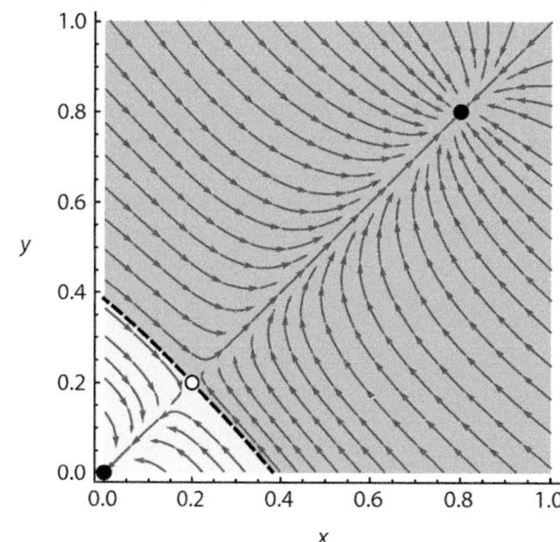

Figure 9.3 Trajectories in the phase plane for the two-variable positive feedback system. The pink region is the basin of attraction for the on-state; the yellow region is the basin of attraction for the off-state; and the dashed line is the separatrix.

the unstable steady state in the Mos model was. From one direction a saddle behaves like an attractor and from another direction—here the +45° angle—it behaves like a repeller. The dashed black line represents the stable manifold of the saddle, and it separates the basins of attraction of the two stable steady states. The diagonal is the unstable manifold of the saddle, and the saddle has the peculiar property of attracting trajectories and then deflecting them toward one of the stable steady states.

9.2 LINEAR STABILITY ANALYSIS EXPLAINS THE DYNAMICS OF THE SYSTEM NEAR EACH OF THE STEADY STATES

To understand why the off-state and on-state are stable and to get a better understanding of what the saddle represents, we turn to what is called linear stability analysis, which tells us whether trajectories that start out very close to a steady-state will converge to it or not. The basic idea behind linear stability analysis is to (1) choose a steady state; (2) perturb the system away from that steady state by an infinitesimal amount; (3) calculate the rate at which the system returns to or is repelled from the steady state; and (4) repeat for the rest of the steady states. Linear stability analysis is easier to do for a one-variable system than a two-variable system, so let us start by returning to our one-variable Mos model from Chapter 8.

We start by re-plotting **Figure 8.10**, which showed the rate–balance plot for the system with *prog*=0, with three intersection points corresponding to two stable and one unstable steady state (**Figure 9.4a**). Before, we considered and plotted the synthesis and degradation rates individually; here we will combine them into an expression for

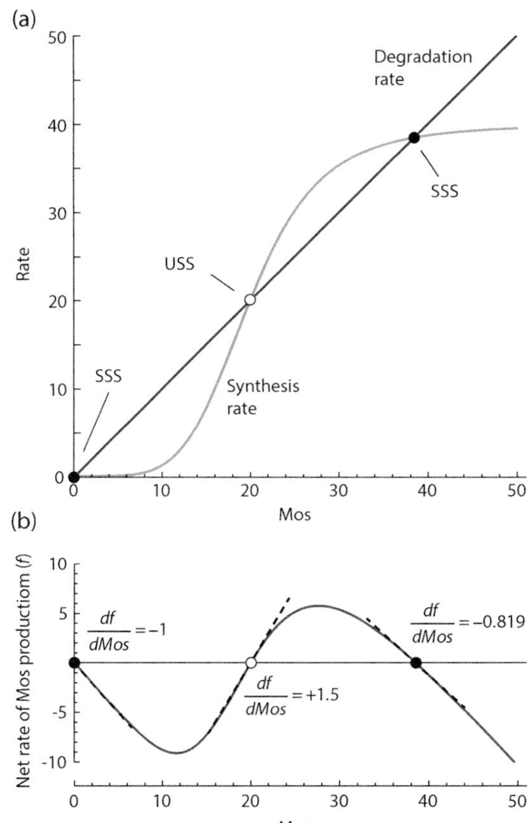

Figure 9.4 Linear stability analysis for the one-variable Mos model. (a) Rate–balance curves. The stable steady states are denoted by filled circles and the unstable steady state by an open circle. (b) The net rate of Mos synthesis as a function of the Mos concentration is shown in red. The tangents to the curve at the three steady states are illustrated by the dashed black lines, and the corresponding slopes are shown as well.

the instantaneous net rate of Mos synthesis as a function of the Mos concentration. From Eq. 8.7, this rate, which we will call *f*, is given by:

$$f = k_{feedback} \frac{Mos^5}{K^5 + Mos^5} - k_{deg} Mos. \tag{9.5}$$

This function equals zero when Mos is zero, goes negative immediately thereafter, crosses zero at the unstable steady state and becomes positive, then dives back to zero at the on-state and stays negative thereafter (**Figure 9.4b**). At Mos concentrations where *f* is negative, Mos will decrease in abundance, and where *f* is positive it will increase.

Now let us look at what happens if you are initially sitting on a steady state—let us start with the off-state—and perturb the system by increasing the Mos concentration by a small amount, δMos. The net rate of Mos production will fall below zero, and the function that defines how negative the *f* will be is Eq. 9.5, a complicated nonlinear equation. But for small values of δMos, we can approximate *f* (the red curve in **Figure 9.4b**) by the dashed black line, the tangent to the red curve at the off-state, as long as the function *f* is smooth and continuous at the steady state (which it is). The slope of the tangent is $\left.\left(\frac{df}{dMos}\right)\right|_{SS}$, and so the net rate of Mos production at $Mos = Mos_{SS} + \delta Mos$ is given by:

$$f\left[Mos_{ss} + \delta Mos\right] = \left.\left(\frac{df}{dMos}\right)\right|_{SS} \cdot \delta Mos. \tag{9.6}$$

This is a linear equation—how fast you come back toward or go away from the steady-state is linearly proportional to δMos—which is why this way of assessing the dynamics of the system near the steady states is called linear stability analysis.

Note that Eq. 9.6 is the same in form as the equation for the exponential decay or explosion of some variable z:

$$\frac{dz}{dt} = k_{apparent} z, \tag{9.7}$$

where the sign of $k_{apparent}$ determines whether the system decays or explodes, and the magnitude of $k_{apparent}$ tells you how fast. Thus, the slope of *f* at the steady state is a gauge of whether the steady state is stable or unstable and of how stable or unstable it is. If the slope of *f* is a negative number, the steady state is an attractor and is stable. If it is positive, the steady state is a repeller and is unstable. The larger the magnitude of the slope, the more stable (if the slope is negative) or unstable (if the slope is positive) the steady state is. It is possible for the slope at a steady state to be zero, although we will not see this in the examples we examine.

We can evaluate the slope of the rate function by differentiating the right-hand side of Eq. 9.5:

$$\frac{df}{dMos} = \left.\frac{5k_{feedback} Mos^4}{K^5 + Mos^5} - \frac{5k_{feedback} Mos^9}{\left(K^5 + Mos^5\right)^2} - k_{deg}\right|_{Mos_{ss}}. \tag{9.8}$$

For the values of the parameters we chose for **Figure 9.4** ($k_{feedback} = 40$; $k_{deg} = 1$; $K = 20$), when we evaluate this expression at $Mos_{SS} = 0$, we get:

$$\frac{df}{dMos} = -1. \tag{9.9}$$

Because $\frac{df}{dMos}$ is a negative number, this steady state is stable.

For the steady state at Mos=20, the slope evaluates to:

$$\frac{df}{dMos} = +1.5. \tag{9.10}$$

This positive number means the distance between the perturbed system and the steady state initially grows exponentially, and so the steady state is unstable. For the on-state at Mos=~38.6, the slope is:

$$\frac{df}{dMos} = -0.819 \tag{9.11}$$

Thus the on-state is stable, but not quite as stable as the off-state; the magnitude of the negative slope is smaller. Note that this fits nicely with the potential surface picture in **Figure 8.14a**; the sides of the potential well for the on-state are slightly less steep than they are for the off-state.

This is the essence of linear stability analysis. One calculates the slope of the net rate function at the steady state, and from its sign and magnitude, makes an inference about the local stability of the steady state.

9.3 TO APPLY LINEAR STABILITY ANALYSIS TO A TWO-VARIABLE SYSTEM, WE CALCULATE EIGENVECTORS AND EIGENVALUES

Let us return now to the two-variable positive feedback system given by Eqs. 9.1 and 9.2 and apply linear stability analysis to it. Let us call the right-hand side of Eq. 9.1 *f*, and the right-hand side of Eq. 9.2 *g*:

$$f = k_{basal} + k_1 \frac{y^2}{K_1^2 + y^2} - k_{-1}x \tag{9.12}$$

$$g = k_2 \frac{x^2}{K_2^2 + x^2} - k_{-2}y. \tag{9.13}$$

Suppose we begin at the off-state {0, 0} and perturb the system by some small amount δx in the *x*-direction. The perturbation will produce a response in the *x*-direction, but also potentially one in the *y*-direction. We can calculate each of the components of the velocity by evaluating the partial derivatives of *f* (for the *x*-component of the velocity) and *g* (for the *y*-component of the velocity) with respect to *x* at the steady state {0,0}:

$$\left.\frac{\partial f}{\partial x}\right|_{\{0,0\}} = -k_1 = -1 \tag{9.14}$$

$$\left.\frac{\partial g}{\partial x}\right|_{\{0,0\}} = \left.\left(\frac{2k_1x}{K^2 + x^2} - \frac{2k_1x^3}{\left(K^2 + x^2\right)^2}\right)\right|_{\{0,0\}} = 0. \tag{9.15}$$

As luck would have it, the velocity in the *y*-direction (Eq. 5.23) is zero. Thus, if you perturb the system by some small δx, it will return straight back to the steady state and the proportionality constant is –1. Likewise, if you perturb the system along the *y*-axis, the two components of the velocity are:

$$\left.\frac{\partial f}{\partial y}\right|_{\{0,0\}} = \frac{2k_1 y}{K^2+y^2} - \left.\frac{2k_1 y^3}{\left(K^2+y^2\right)^2}\right|_{\{0,0\}} = 0 \tag{9.16}$$

$$\left.\frac{dg}{dy}\right|_{\{0,0\}} = -k_{-1}\big|_{\{0,0\}} = -1. \tag{9.17}$$

So again you come straight back toward the steady state, and again the proportionality constant is –1 (which you probably already guessed because of the symmetry of the equations). And if you perturb the system in any other direction, the response is still a vector pointing straight back toward the steady state. Thus, the steady state is locally stable, with a $k_{apparent}$ of –1 for perturbations in any direction. So far so good.

Let us now examine the steady state at {0.2, 0.2}, where some trajectories come toward the steady state and some go away from it. We will need all four partial derivatives here; for compactness, we will lay them out in matrix form:

$$\begin{bmatrix} \frac{\partial f}{\partial x} & \frac{\partial f}{\partial y} \\ \frac{\partial g}{\partial x} & \frac{\partial g}{\partial y} \end{bmatrix}. \tag{9.18}$$

This is referred to as the **Jacobian matrix** of the rate equations {*f*, *g*}. Pulling together the results from Eqs. 9.14–9.17, we have:

$$\begin{bmatrix} -k_1 & \frac{2k_1 y}{K^2+y^2} - \frac{2k_1 y^3}{\left(K^2+y^2\right)^2} \\ \frac{2k_1 x}{K^2+x^2} - \frac{2k_1 x^3}{\left(K^2+x^2\right)^2} & -k_1 \end{bmatrix}, \tag{9.19}$$

and evaluating the partial derivatives at {0.2, 0.2} yields:

$$\begin{bmatrix} -1 & 1.6 \\ 1.6 & -1 \end{bmatrix}. \tag{9.20}$$

These results show that if you perturb the system slightly in the positive *x*-direction, the response is to come toward the steady state in the *x*-direction (since $\frac{\partial f}{\partial x} = -1$) but also to go away from the steady state in the *y*-direction (because $\frac{\partial g}{\partial x} = 1.6$). The response to the perturbation is not in the same direction as the perturbation. This is shown schematically in **Figure 9.5**. Likewise, if you perturb the system in the positive *y*-direction; it comes back in toward the steady state in the *y*-direction but away from it in the *x*-direction, and one can see this behavior from the sample trajectories in **Figure 9.3**. The velocity vector $\{v_x, v_y\}$ that results from these perturbations is not equal to $k_{apparent} \bullet \{\delta x, \delta y\}$ the way it was for the {0, 0} steady state.

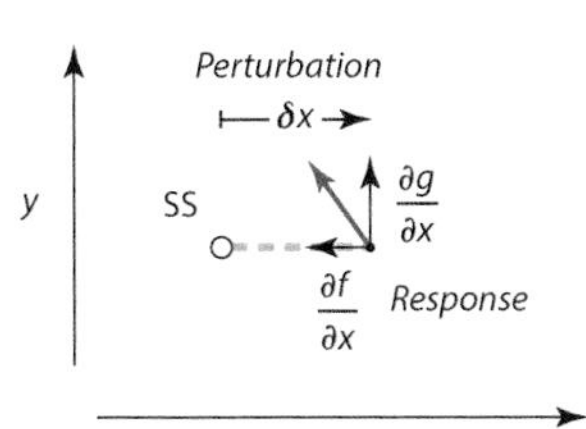

Figure 9.5 Perturbation of a two-variable system from a steady state (SS) can generate responses that are not in the same direction (or in the opposite direction) as the perturbation.

However, just by looking at **Figure 9.3**, one can see that there are two special directions—the ±45° diagonals going through the {0.2, 0.2} steady state—where the trajectories do go straight back (for the 45° diagonal with negative slope) or straight away (for the 45° diagonal with positive slope) from the steady state. We could do a change of variables using a rotation matrix to recast the problem in this coordinate system, but we can take a shortcut: from matrix algebra, the special directions seen here correspond to the **eigenvectors** of the Jacobian matrix, and the proportionality constants for trajectories in these special directions are the **eigenvalues** of the Jacobian. The eigenvectors and eigenvalues can be calculated by hand, but for now we can simply use software like *Mathematica* to evaluate them for us. The eigenvectors turn out to be $\left\{\frac{\sqrt{2}}{2},-\frac{\sqrt{2}}{2}\right\}$ and $\left\{\frac{\sqrt{2}}{2},\frac{\sqrt{2}}{2}\right\}$, which are vectors of unit length at ±45°, and the corresponding eigenvalues are –2.6 and 0.6. Thus, along the –45° direction, the steady state is an attractor (with its eigenvalue λ_1=–2.6), and along the +45° direction the steady state is a repeller (with its eigenvalue λ_2=0.6). This is one definition of a saddle: a steady state of a two-variable system with one positive and one negative eigenvalue. In addition, the magnitude of the two eigenvalues indicates that it attracts faster than it repels, and by plotting the time courses of a few trajectories (**Figure 9.6**), we can verify that this is in fact the case.

If we apply the same procedure to the steady state at {0, 0}, we get eigenvectors of {0, 1} and {1, 0}—that is, the unit vectors along the *x*- and *y*-axes—and eigenvalues of λ_1=–1 and λ_2=–1. Since the eigenvalues are equal, any combination of the eigenvectors would qualify as an eigenvector with the same eigenvalue; for this system and this steady state, all directions are equally special. For the on-state at {0.8, 0.8}, we get eigenvectors of $\left\{\frac{\sqrt{2}}{2},-\frac{\sqrt{2}}{2}\right\}$ and $\left\{\frac{\sqrt{2}}{2},\frac{\sqrt{2}}{2}\right\}$ and eigenvalues of λ_1=–1.4 and λ_2=–0.6—two negative eigenvalues, confirming that the steady state is locally stable.

Thus we have analyzed the stabilities of the three steady states for our two-variable model. And we have a general plan for how to carry out linear stability analysis:

1. Write down the Jacobian matrix for the system.
2. Choose a steady state.
3. Evaluate the elements of the Jacobian matrix at the steady state.
4. Calculate the eigenvectors and eigenvalues of the matrix at the steady state.

The eigenvectors show you the special directions for trajectories close to the steady state in the phase plane. The eigenvalues tell you

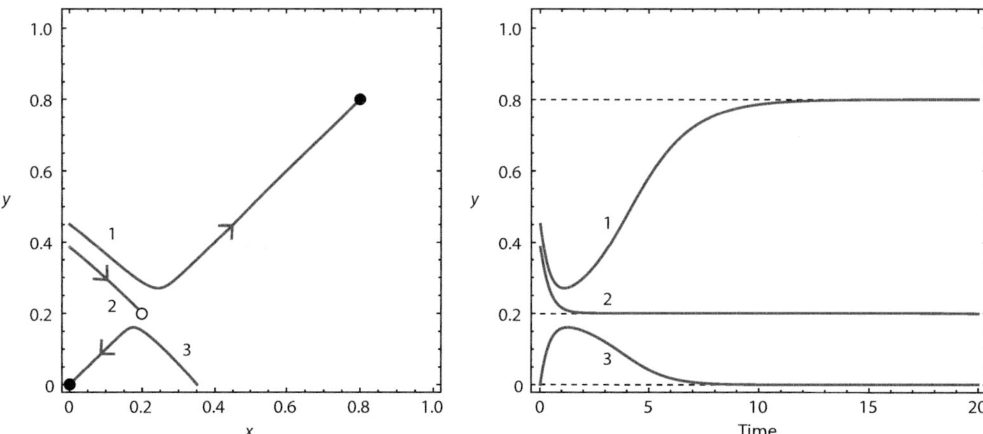

Figure 9.6 Trajectories in the phase plane (left) and in a time course plot (right) for the two-variable positive feedback system. In the time course plot, the dashed black lines represent the three steady states.

whether the steady state is locally stable (two negative eigenvalues), a saddle (one positive and one negative eigenvalue), or unstable (two positive eigenvalues). Eigenvalues of zero are possible too, but we will not encounter them in any of the systems we will analyze.

5. Repeat steps 2–4 for the other steady states.

This procedure works not only for two-variable systems but for any number of variables—the Jacobian matrix gets bigger, and the eigenvalue calculation gets more complicated, but the basic recipe is the same.

9.4 THE SYSTEM CAN CHANGE BETWEEN STATES VIA A SADDLE-NODE BIFURCATION

In our two-variable positive feedback model, the input into the system was the term k_{basal} in the x rate equation (Eq. 9.1). How do the nullclines and the steady states change as k_{basal} is varied?

Each increment of k_{basal} shifts the x-nullcline to the right, without affecting the y-nullcline (**Figure 9.7a**). The result is that the off-state (a node) and the saddle move toward each other and annihilate each other through a saddle-node bifurcation (and the terminology now makes sense!), beyond which the system has no alternative but to settle into the on-state. The resulting transition is irreversible; once the system makes it into the on-state, it remains there even if k_{basal} decreases to zero. (**Figure 9.7b**). Although it is not obvious from looking at the phase plane picture, which shows the positions of the trajectories but not the speeds, the same critical slowing we saw with the Mos model occurs when trajectories come near to where the off-state and saddle have annihilated each other.

Note that if we were to weaken the positive feedback—say by decreasing the value of k_1 and/or k_2 (Eqs. 9.1 and 9.2)—we could make the system be hysteretic rather than irreversible, with the transitions from the off-state to the on-state and from the on-state to the off-state both taking place through saddle-node bifurcations.

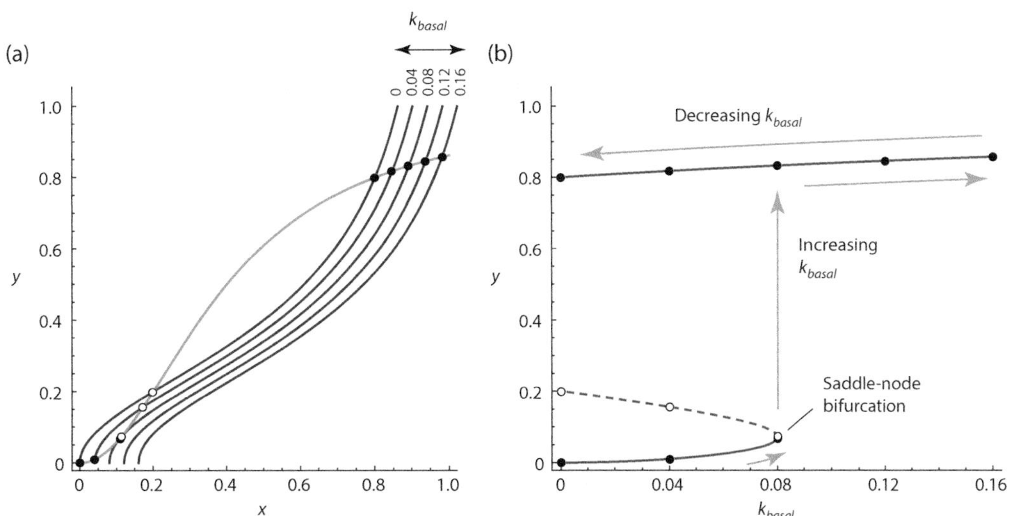

Figure 9.7 The two-variable positive feedback system changes between steady states via a saddle-node bifurcation. (a) Phase plot showing the y-nullcline in green and the x-nullclines in blue for 5 values of k_{basal}. (b) Steady-state concentration of y as a function of k_{basal}. At a k_{basal} value of just over 0.08, the off-state and saddle collide and annihilate each other at a saddle-node bifurcation. The result is an irreversible switch to the on-state.

9.5 DOUBLE-NEGATIVE FEEDBACK PLUS ULTRASENSITIVITY CAN YIELD BISTABILITY

As a final example, let us analyze the double-negative feedback circuit shown in **Figure 9.1b** and, in more detail, in **Figure 9.8a**. We begin by writing down the rate equations, with x stimulating the degradation of y and vice versa. As was the case with our other bistable models, some source of ultrasensitivity will prove to be important; here we acknowledge this by making the degradation rate terms be described by Hill functions. We assume there is some basal rate of degradation of x that does not depend on y (and vice versa), and that degradation is opposed by constitutive production processes. Finally, we assume there is some basal level of degradation. The resulting equations are:

$$\frac{dx}{dt} = k_1 - k_{-1}x - k_{feedback1}x\frac{y^2}{K_1^2 + y^2}, \tag{9.21}$$

$$\frac{dy}{dt} = k_2 - k_{-2}y - k_{feedback2}y\frac{x^2}{K_2^2 + x^2}. \tag{9.22}$$

We calculate the nullclines by setting each rate equation equal to zero:

$$k_1 - k_{-1}x - k_{feedback1}x\frac{y^2}{K_1^2 + y^2} = 0 \tag{9.23}$$

$$k_2 - k_{-2}y - k_{feedback2}y\frac{x^2}{K_2^2 + x^2} = 0. \tag{9.24}$$

For our example we have chosen $k_1 = k_2 = k_{-1} = k_{-2} = 1$; $k_{feedback1} = k_{feedback2} = 20$; and $K_1 = K_2 = 0.5$; the resulting nullclines are shown plotted in the phase plane in **Figure 9.8b**.

Next we solve Eqs. 9.23 and 9.24 simultaneously and obtain the coordinates of the steady states, which are (to 3 significant figures) {0.0683, 0.732}, {0.227, 0.227}, and {0.732, 0.0683}. We can also plot some sample trajectories in the phase plane (**Figure 9.8a**) and show that the first and the third steady states appear to be stable—a high y, low x state and a high x, low-y state—and the middle steady state appears to be a saddle, and that is the way the steady states are denoted in **Figure 9.8b**. The +45° diagonal is the separatrix between the stable manifolds of the two attractors or stable steady states.

Finally, we can carry out linear stability analysis to establish what we have provisionally concluded about the stabilities of the three steady states. We define the right-hand side of Eq. 9.21 to be f, and the right-hand side of Eq. 9.22 to be g, and write the Jacobian matrix of partial derivatives:

$$\begin{bmatrix} \frac{\partial f}{\partial x} & \frac{\partial f}{\partial y} \\ \frac{\partial g}{\partial x} & \frac{\partial g}{\partial y} \end{bmatrix} = \begin{bmatrix} -k_{-1} - k_{feedback1}\frac{y^2}{K_1^2 + y^2} & -\frac{2k_{feedback1}x \cdot y}{K_1^2 + y^2} + \frac{2k_{feedback1}x \cdot y^3}{\left(K_1^2 + y^2\right)^2} \\ -\frac{2k_{feedback2}x \cdot y}{K_2^2 + x^2} + \frac{2k_{feedback2}x^3 \cdot y}{\left(K_2^2 + x^2\right)^2} & -k_{-2} - k_{feedback2}\frac{x^2}{K_2^2 + x^2} \end{bmatrix}. \tag{9.25}$$

At the first steady state, this matrix evaluates to:

$$\begin{bmatrix} -14.6 & -0.811 \\ -7.71 & -1.37 \end{bmatrix}. \tag{9.26}$$

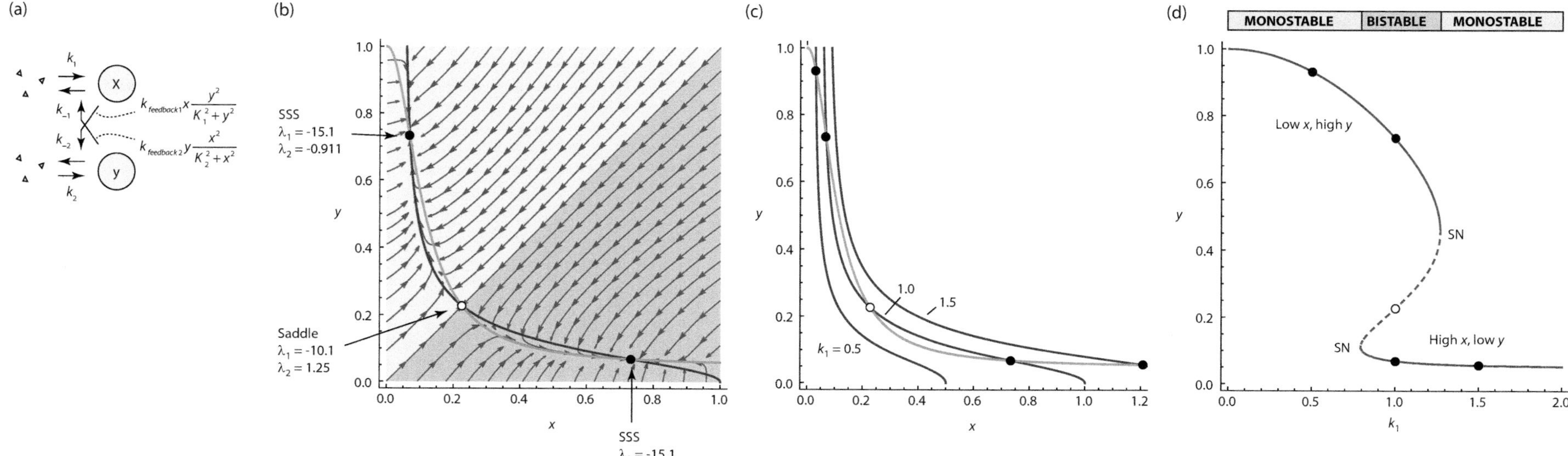

Figure 9.8 Bistability from double-negative feedback. (a) Schematic of a double-negative feedback system where *x* stimulates the degradation of *y* and vice versa. (b) Phase plane analysis showing the *x*-nullcline (blue), the *y*-nullcline (green), the three steady states (two stable steady states, designated SSS, and one saddle, with their eigenvalues), some sample trajectories (arrows), and the basins of attraction for the two stable steady state (yellow for the high-*y* state and pink for the high-*x* state). (c) Switching between states. Decreasing the value of k_1 makes the high-*x* state collide with the saddle and disappear. Increasing k_1 makes the high-*y* state collide with the saddle and disappear. (d) The steady states of the system as a function of k_1. The solid red curves are the stable states and the dashed curve is the saddle. Note that the overall response is hysteretic and that steady states appear and disappear through saddle-node bifurcations (SH). The positions of the steady states for $k_1 = 0.5$, 1, and 1.5 are depicted by filled or hollow circles.

The eigenvalues of this matrix are λ_1=–15.1 and λ_2=–0.911—it is indeed a stable steady state, since both eigenvalues are negative—and the corresponding eigenvectors are {–0.872, –0.490}, which is a unit-length vector at an angle of 209.3°, and {0.0590, –0.998}, which corresponds to –86.6°.

At the second steady state, the Jacobian matrix is:

$$\begin{bmatrix} -4.41 & -5.66 \\ -5.66 & -4.41 \end{bmatrix}. \tag{9.27}$$

The eigenvalues are λ_1=–10.1 and λ_2=1.25; the steady state is a saddle, and it attracts faster than it repels. The corresponding eigenvectors are $\left\{\frac{\sqrt{2}}{2}, \frac{\sqrt{2}}{2}\right\}$ and $\left\{-\frac{\sqrt{2}}{2}, \frac{\sqrt{2}}{2}\right\}$, so the attracting direction is the +45° diagonal and the repelling direction is the +135° diagonal. At the third steady state, λ_1=–15.1 and λ_2=–0.911, just as they were for the first steady state, so it is stable as well. The corresponding eigenvectors are {0.490, 0.872} or +60.7°, and {0.998, –0.0590} or –3.4°. Thus, we have a bistable system with two stable steady states and a saddle.

To make a transition from the high-*x* state to the high-*y* state, or vice versa, we manipulate the input to the system, which we will take to be k_1, the rate of production of *x*. If we make k_1 large enough, the high-*y* state collides with the saddle and disappears, making the system monostable with only a high-*x* state (**Figure 9.8c**). Conversely, if we make k_1 small enough, the saddle moves toward and then annihilates the high-*x* state, making the system monostable with only a high-*y* state. The result is the hysteretic input–output relationship shown in **Figure 9.8d**. Note that if we were simply to limit the minimum or maximum possible values of k_1, we could make the system irreversible rather than hysteretic.

9.6 PERFECT SYMMETRY CAN PRODUCE A PITCHFORK BIFURCATION

To push our double-negative system from one state to the other, we varied the relative values of k_1 and k_2. When k_1 (the rate of *x* synthesis) was large enough, the high-*x* state predominated; when the two were equal, both the high *x* and the high *y* states co-existed; and when k_1 was small enough, the high *y*-state predominated. This makes intuitive sense.

Let us consider another way that the parameters of the system might be regulated by some input—let us suppose that the input makes *x* better at inhibiting the production of *y and y* better at inhibiting the production of *x*. For example, *x* and *y* could be transcriptional inhibitors, and the input could be something that promotes the translocation of both *x* and *y* to the nucleus. This is modeled in **Figure 9.9**; we start with $k_1 = k_2 = k_{-1} = k_{-2} = 1$ and $K_1 = K_2 = 0.5$, set $k_{feedback1} = k_{feedback2}$, and vary the feedback strength. The resulting nullclines and steady states are shown in **Figure 9.9a**. When the feedback strength is zero, neither species affects the expression of the other. The result is a blue vertical line for the *x*-nullcline, and the green horizontal line for the *y*-nullcline, which yields a single stable steady state at {1, 1}. When the feedback strength increases to 1, the steady state moves down the diagonal to about {0.622, 0.622}—still a single stable steady state. But then at $k_{feedback}$ ~10, something remarkable happens; the steady state splits in three, yielding two alternative stable steady states, one with high *x* and one with high *y*, and a saddle on the diagonal.

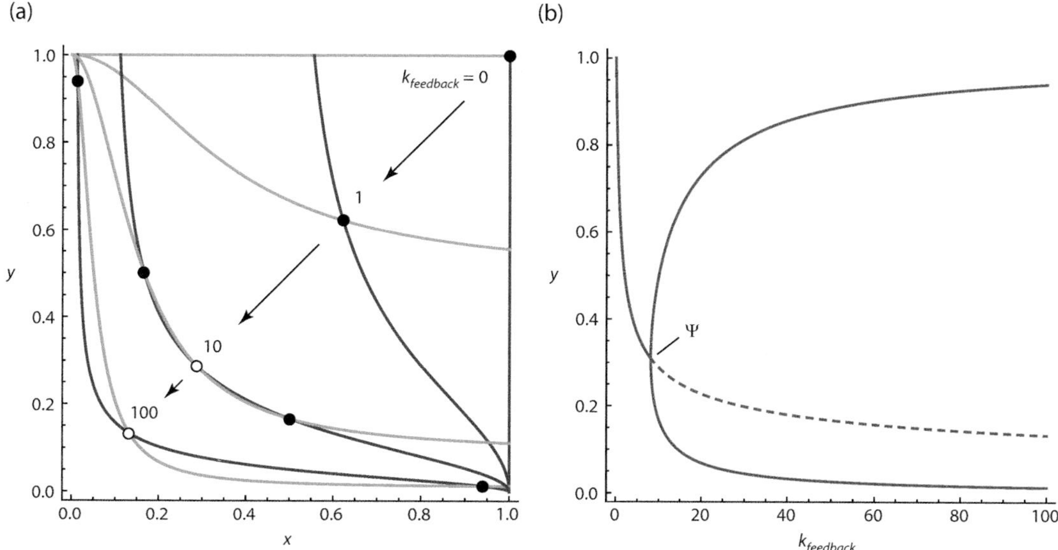

Figure 9.9 Pitchfork bifurcation. (a) Nullclines and steady states for a double-negative feedback system where the strength of x's inhibition of y *and* y's inhibition of x ($k_{feedback}$) is varied together. (b) One-parameter bifurcation plot showing a transition between one stable steady state and three steady states at $k_{feedback} \sim 10$ units. At this pitchfork bifurcation, the middle state (with $x=y$) becomes a saddle and the stable high-x and high-y states appear.

As $k_{feedback}$ increases further, the high x and high y states move farther apart. This behavior can be seen in the phase plane plot (**Figure 9.9a**) and in the input–output plot (or, if you prefer, one-parameter bifurcation diagram) shown in **Figure 9.9b**.

The transition that takes place at $k_{feedback}$ ~10 is called a **pitchfork bifurcation**, because the curves shown in **Figure 9.9b** sort of look like a pitchfork, and it is designated by the Greek letter Ψ, because Ψ resembles a pitchfork. (In cardiology, the splitting of a blood vessel into three is called a trifurcation, which makes etymological sense, but in nonlinear dynamics the term bifurcation is used). A familiar example of a pitchfork bifurcation from everyday life would be a vertical ruler being pushed on from above; above some critical strength of push, the straight ruler becomes unstable and two alternative bowed states become stable (**Figure 9.10a**). Related processes in biology include the buckling of epithelial sheets (**Figure 9.10b**) as well as certain signaling processes. For example, the Notch protein and its ligand Delta are cell surface proteins involved in a phenomenon termed lateral inhibition. Delta on one cell can repress, via its binding to Notch, the expression of Delta in its neighbors (**Figure 9.10c**). At low expression levels, the antagonistic Delta proteins on adjacent cells may be able to co-exist, but once the mutual inhibition becomes strong enough, in theory the system will traverse a pitchfork bifurcation and one or another of the two cells will win out.

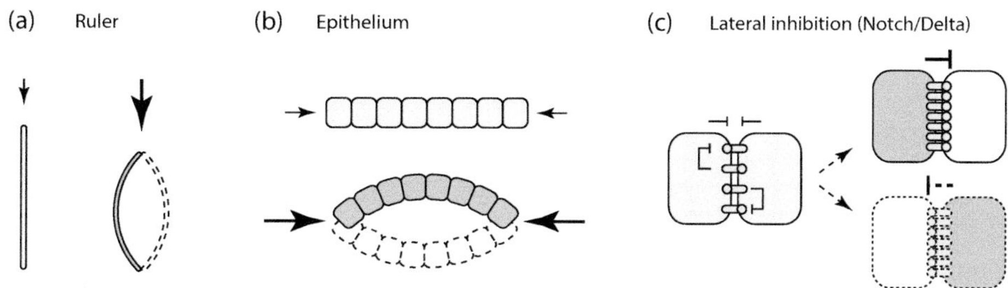

Figure 9.10 Pitchfork bifurcations in mechanics and signaling. (a) The bending of a ruler. (b) The buckling of an epithelial sheet. (c) Lateral inhibition through the Notch/Delta system.

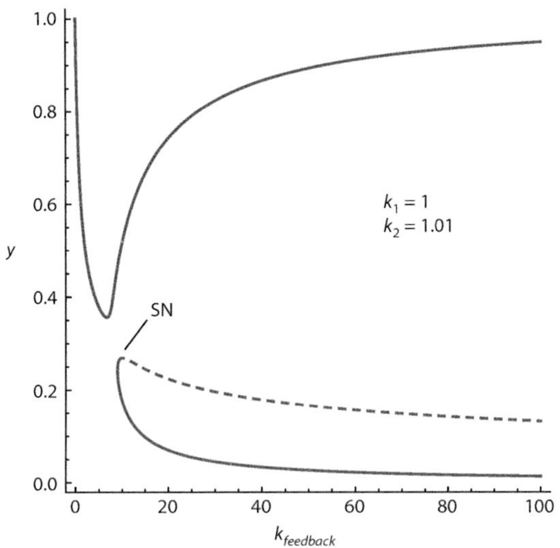

Figure 9.11 Imperfect symmetry changes a pitchfork bifurcation to a saddle-node bifurcation. Here we assumed $k_1 = 1$ and $k_2 = 1.01$. The other parameters for the x and y rate equations are identical to those used in Figure 9.9.

9.7 IN THE ABSENCE OF PERFECT SYMMETRY, A PITCHFORK BIFURCATION MORPHS INTO A SADDLE-NODE BIFURCATION

The pitchfork bifurcation seen in **Figure 9.9** arose because we assumed that the system was perfectly symmetrical. What would happen if the symmetry was not quite perfect? This is shown in **Figure 9.11**, where we have assumed that the synthesis rates of x and y are close but not exactly the same ($k_1 = 1$ vs. $k_2 = 1.01$). The answer is that the pitchfork bifurcation changes into a saddle-node bifurcation. As the feedback strength is increased, the steady state value of y falls, as it did in **Figure 9.7**, but when the critical feedback strength is reached, a new steady state appears and splits into a saddle and a stable low-y state, and the high-y state moves back upward (**Figure 9.11**). There is no chance for the system to switch from its initial state to the low-y state; it was destined for the high-y state right from the start thanks to the slightly higher rate of synthesis of y. Of course if there is any noise in the system—for example, fluctuations in the concentrations of x and/or y—it might be possible for the system to jump from one stable steady state over the intervening saddle and into the alternative stable steady state. The smaller the difference is between the k_1 and k_2, the easier it is for the system to jump between states and the more the saddle-node bifurcation behaves like a pitchfork bifurcation.

SUMMARY

Here we have shown how positive feedback and double-negative feedback can give rise to bistability in models with two time-dependent variables, where rate-balance analysis is not applicable. We identify the steady states by plotting nullclines in the phase plane and looking for their intersection points. And we analyze the stability of the steady states through linear stability analysis. This involves setting up a Jacobian matrix of partial derivatives, evaluating the derivatives at the steady states, and then calculating the eigenvalues and eigenfunctions of the matrix. The same approach can be taken for systems with more than two variables, although it is harder to visualize a higher dimensional phase plot.

Linear stability analysis will continue to be an important tool when we analyze oscillators in Chapters 14 and 15 and excitable systems in Chapter 16.

FURTHER READING

SADDLE-NODE AND PITCHFORK BIFURCATIONS

Oster G, Alberch P. Evolution and bifurcation of developmental programs. *Evolution*. 1982;36:444–59.

Strogatz SH. *Nonlinear Dynamics and Chaos: With Applications to Physics, Biology, Chemistry, and Engineering*. Westview Press, Cambridge, MA, 1994, Chapters 3 and 8.

TRANSCRITICAL BIFURCATIONS IN PHASE SEPARATION AND INFECTIOUS DISEASE

10

IN THIS CHAPTER . . .

DOI: 10.1201/9781003124269-10

INTRODUCTION

Along the way to formulating our bistable model of Mos and ERK2 regulation during *Xenopus* oocyte maturation (Chapter 8), we examined a model with Michaelian rather than ultrasensitive responses in the positive feedback loop. We discarded this model because it did not yield bistability and did not account for the system's all-or-none, irreversible responses. But positive feedback does not always involve ultrasensitivity, and bistability is not the only useful behavior that can emerge out of a positive feedback system.

Here we will explore two examples of this, both taken from biology, although not cell signaling biology. The first is liquid–liquid phase separation, a process that can concentrate proteins and other cell components in functional compartments that lack membrane boundaries. The second is the spread of an infectious disease through a population. In both cases, positive feedback yields a system with a critical point, above which you get one kind of behavior (phase separation in the one case and pandemic disease spread in the other) and below which you get another (no phase separation in the one case and sporadic disease in the other).

LIQUID–LIQUID PHASE SEPARATION

10.1 LIQUID–LIQUID PHASE SEPARATION CAN PRODUCE DISCRETE FUNCTIONAL DOMAINS THAT LACK MEMBRANES

The current interest liquid–liquid phase separation arose in part out of the realization that P granules, which are protein- and RNA-rich species that localize to one pole of a fertilized *C. elegans* egg, behave more like liquid droplets than solid granules. They are spherical, they flow and fuse, and their contents are more dynamic than would be expected in a solid.

To explain why these liquid droplets maintain a discrete identity even though they are not separated from the bulk cytoplasm by a membrane, it was proposed that they represent a distinct liquid phase, just as oil droplets dispersed in water are. The significance of being a phase-separated compartment is that it potentially promotes binding processes and reactions among components of the compartment while minimizing off-target interactions with components of the bulk cytoplasm. A number of other membraneless cell compartments have also been proposed to be separated liquid phases, including nucleoli, centrosomes, and heterochromatin.

Here we aim to understand the mechanism of liquid–liquid phase separation better by modeling it. Phase separation is often modeled by starting with an expression for the free energy or the chemical potential of the system, and then working out what is needed for the system to be in equilibrium. However, there is a simple rate equation approach, similar to what we have been doing for cell signaling processes, which allows both the equilibria and dynamics of the process to be explored. This theory is a good way for biologists to begin to explore these fascinating processes.

10.2 PHASE SEPARATION CAN BE MODELED BY A SINGLE RATE EQUATION WITH POSITIVE FEEDBACK AND A TRANSCRITICAL BIFURCATION

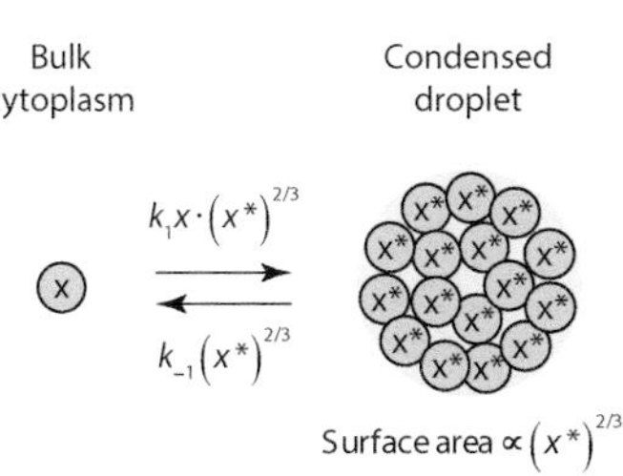

Figure 10.1 Schematic view of the equilibration of a macromolecule between the cytoplasm and a condensed droplet.

Figure 10.1 shows a simple scheme for a phase separation process. We suppose that there is a species that can slowly equilibrate between the well-mixed bulk cytoplasm and a single phase-separated liquid droplet, and we assume that this species is the main macromolecular component of the droplet. We initially consider a single but will extend the model to multiple droplets later.

We call the equilibrating species x when it is in the cytoplasm and x^* when it is in the droplet. And, finally, we will assume that each compartment is spatially homogeneous. With these assumptions, we can model the two-compartment system with an ODE that describes the exchange of the species between the compartments.

The rate of association of x with the droplet should be proportional to x and to the surface area of the droplet (since the molecule x would have to interact with the surface of the droplet to enter it). If the volume of the droplet is proportional to x^*, then the surface area of a spherical droplet will be proportional to $(x^*)^{2/3}$. Thus it follows that:

$$\textit{Association Rate} = k_1 x \cdot (x^*)^{2/3} \tag{10.1}$$

The rate at which x^* is lost back to the cytoplasm is assumed to be proportional to the surface area of the droplet. It follows that:

$$\textit{Dissociation Rate} = k_{-1}(x^*)^{2/3} \tag{10.2}$$

The net rate of production of x^* is therefore:

$$\frac{dx^*}{dt} k_1 x \cdot (x^*)^{2/3} - k_{-1} \cdot (x^*)^{2/3}. \tag{10.3}$$

We can eliminate the variable x using the conservation equation $x_{tot} = x + x^*$:

$$\frac{dx^*}{dt} = k_1(x_{tot} - x^*)(x^*)^{2/3} - k_{-1}(x^*)^{2/3}. \tag{10.4}$$

This is our one-ODE model of the equilibration of a species x between the bulk cytoplasm and phase-separated liquid droplet. Note that there is positive feedback in the production of x^*; the bigger the droplet is, the faster it grows.

At equilibrium, the time derivative must be zero:

$$0 = K_1(x_{tot} - x^*) \cdot (x^*)^{2/3} - k_{-1} \cdot (x)^{2/3} \tag{10.5}$$

There are two solutions for the equilibrium concentration of x^*. Either

$$x^*_{eq} = 0 \tag{10.6}$$

or

$$x^*_{eq} = x_{tot} - \frac{k_{-1}}{k_1}. \tag{10.7}$$

Note that the rate constants enter into the equilibrium solutions only as their ratio, and the ratio is the equilibrium constant for the process.

The first solution (Eq. 10.6) is valid for any value of x_{tot}. However, the second solution (Eq. 10.7) only applies when $x_{tot} \geq k_{-1} / k_1$, because it yields negative values for x^* when $x_{tot} < k_{-1} / k_1$. Thus, as shown in **Figure 10.2a**, the system has a single equilibrium, with $x^*=0$, when x_{tot} is less than a critical concentration given by $x_{crit}=k_{-1}/k_1$, and two equilibria when $x_{tot}>k_{-1}/k_1$.

Likewise, we can plot the equilibrium concentration of x as a function of x_{tot} (since $x=x_{tot}-x^*$) (**Figure 10.2b**). The equilibrium concentration of x grows with x_{tot} until x_{tot} reaches k_{-1}/k_1, and it then splits to either plateau at that value or continue rising with x_{tot}.

To determine which of the equilibria are stable, we can carry out rate-balance analysis, since it is a one-variable system. To this end we plot Eqs. 10.1 and 10.2 and look for where the two rate curves intersect each other. **Figure 10.2c,d** show rate–balance plots for two values of x_{tot}, one below the critical concentration ($x_{tot}=0.5\ k_{-1}/k_1$) and one above it ($x_{tot}=1.5\ k_{-1}/k_1$). In the first case, there is one intersection point between the two rate curves, at $x^*=0$, and it represents a globally stable equilibrium; perturb the system to the right and the dissociation rate will exceed the association rate, and so the system will return back toward that equilibrium. In the second case, there

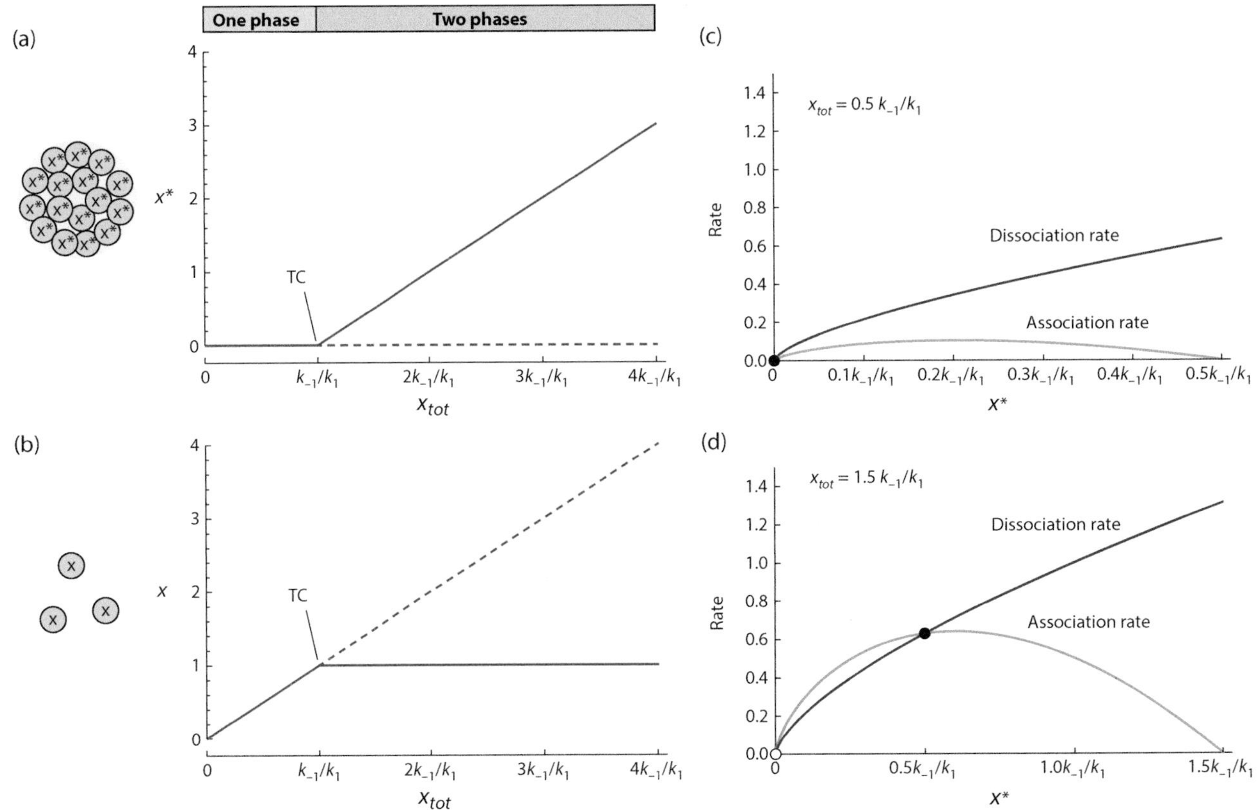

Figure 10.2 The behavior of a one-ODE model of equilibration of a species between the bulk cytoplasm and a condensed droplet. (a, b) The condensed phase (a) and the dispersed phase (b) as a function of x_{tot}. The solid red lines represent stable equilibria and the dashed red lines unstable equilibria. TC designates the transcritical bifurcation where the single equilibrium splits into two, and where the condensed droplet appears. (c, d) Rate-balance analysis, showing the association and dissociation rates as a function of x^*, the amount of x in the condensed droplet, for one concentration of x_{tot} below the transcritical bifurcation (c) and one above it (d).

are two intersection points, at $x^*=0$ and 0.5. The $x^*=0$ equilibrium is now unstable; perturb the system to the right of $x^*=0$ and the association rate exceeds the dissociation rate, so that the system will move further to the right toward the $x^*=0.5$ equilibrium, which is stable.

Thus, when the total concentration of x is less than a critical value of $x_{crit} = \frac{k_{-1}}{k_1}$, the system has one equilibrium, with $x^*_{eq} = 0$ and $x_{eq} = x_{tot}$, and it is a stable equilibrium. The system also has only one phase: all of the x is in the bulk cytoplasmic phase and none of it is in a condensed droplet.

But when the total concentration of x is greater than $x_{crit} = \frac{k_{-1}}{k_1}$, the system has two equilibria: an unstable equilibrium with $x^*_{eq} = 0$ and $x_{eq} = x_{tot}$, and a stable equilibrium with $x^*_{eq} = x_{tot} - \frac{k_{-1}}{k_1}$, and $x_{eq} = \frac{k_{-1}}{k_1}$. This means that when x_{tot} is above the critical concentration there will be a constant concentration of x in the bulk phase, irrespective of the total concentration of x, as long as the dispersed phase is not in the supersaturated unstable equilibrium state. Momentary changes in x will be buffered by opposing movements of x into or out of the condensed phase, a potential mechanism for keeping free protein concentrations constant in the face of fluctuations in their production rate. This has been proposed as a mechanism for suppressing noise in biological processes mediated by x in the bulk cytoplasm.

Note that $x_{crit} = \frac{k_{-1}}{k_1}$ represents a critical point in two senses. It is where the single stable equilibrium bifurcates into an unstable equilibrium and stable one, a type of bifurcation referred to as a **transcritical bifurcation**. It is where the system goes from having one (dispersed) phase, with no condensed droplet, to two phases (dispersed and condensed), with x at its maximum-possible concentration of $x = \frac{k_{-1}}{k_1}$ and the condensed droplet taking up all of the excess. These are the characteristics of the equilibria in our simple model of liquid–liquid phase separation.

What if we were to assume that x was equilibrating not with just one droplet, but several, or many? In this case we could write a rate equation for each droplet:

$$\frac{dx_1^*}{dt} = k_1\left(x_{tot} - x_1^* - x_2^* - \cdots - x_n^*\right)\left(x_1^*\right)^{2/3} - k_{-1}\left(x_1^*\right)^{2/3} \tag{10.8}$$

$$\frac{dx_2^*}{dt} = k_1\left(x_{tot} - x_1^* - x_2^* - \cdots - x_n^*\right)\left(x_2^*\right)^{2/3} - k_{-1}\left(x_2^*\right)^{2/3} \tag{10.9}$$

...

$$\frac{dx_n^*}{dt} = k_1\left(x_{tot} - x_1^* - x_2^* - \cdots - x_n^*\right)\left(x_n^*\right)^{2/3} - k_{-1}\left(x_n^*\right)^{2/3}, \tag{10.10}$$

where the variables of the form x_n^* represent the amount of the x species in the nth droplet. The system will still have an equilibrium when all of the x_n^* variables are zero in concentration, and the system will have another equilibrium when the total concentration of x^* in all of the droplets is equal to $\left(x^*_{tot}\right)_{eq} = x_{tot} - \frac{k_{-1}}{k_1}$. There is still a critical point $x_{crit} = \frac{k_{-1}}{k_1}$ where the system transitions from having a single dispersed phase to having coexisting droplets and dispersed x. Thus, the model is completely agnostic about how the x molecules are distributed among the droplets.

10.3 THE TIME COURSE OF DROPLET FORMATION IS SIGMOIDAL

So far we have considered the system's equilibria; its dynamical properties are interesting as well. Suppose that we start with $x_{tot}=0$ and $x^*=0$, and then step x_{tot} up above the critical point. In principle the system can stay in the unstable $x^*=0$ state indefinitely, but it is supersaturated; any fluctuation that results in the production of a little x^* will result in a transition to the stable equilibrium, with $x^*_{eq} = x_{tot} - \frac{k_{-1}}{k_1}$ and $x_{eq} = \frac{k_{-1}}{k_1}$. Because the two rate curves are close together when x^* is close to zero, and then get farther apart as x^* grows (**Figure 10.2d**), the time course for such a transition will be sigmoidal (**Figure 10.3**); there will be a time lag in the production of the condensed phase. The time course makes it appear that the process consists of an initial slow nucleation phase followed by a rapid growth phase and then saturation, but note that our model did not assume that it is more difficult to add the first couple of x molecules to the condensed phase than the 100th or 1000th. Instead the slow initial kinetics emerge from the dynamical properties of a system with positive feedback.

To sum it up, the main features of this simple model of liquid–liquid phase separation are: (1) at equilibrium, there will be a condensed phase only if the concentration of x_{tot} is above a critical value given by $x_{crist} = \frac{k_{-1}}{k_1}$; (2) when the condensed phase does exist, the bulk cytoplasmic concentration of x is a constant maximal value; (3) supersaturation of the cytoplasm is possible, although the supersaturated state is unstable; and (4) there will be a temporal lag in the production of the condensed phase from a dispersed system. All of these predictions are characteristic of a positive feedback system that traverses a transcritical bifurcation.

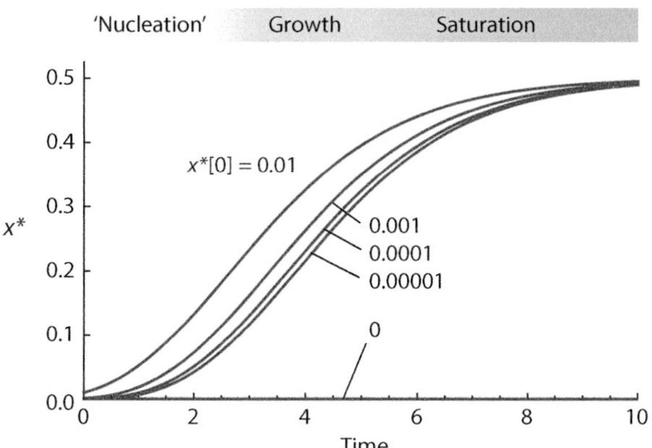

Figure 10.3 The dynamics of droplet growth above the transcritical bifurcation. Here we have assumed that we are above the transcritical bifurcation (at $x_{tot}=1.5\ k_{-1}/k_1$, as in Figure 10.2d), and the system begins with various low concentrations of x^*. The assumed parameters for the simulation were $k_{-1}=k_1=1$.

10.4 THE SAME PRINCIPLES UNDERPIN THE FORMATION OF PHOSPHOLIPID VESICLES

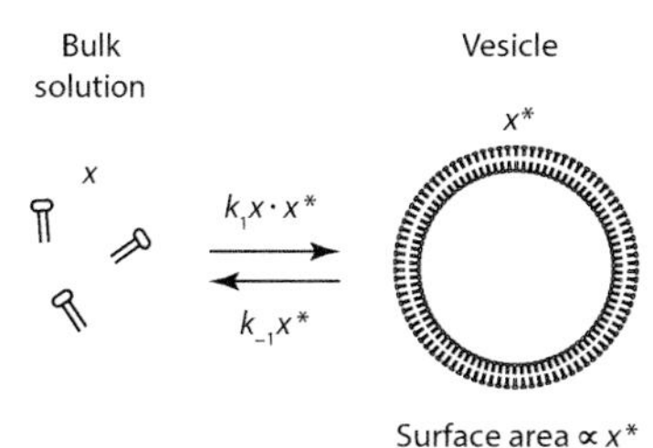

Figure 10.4 Schematic view of the equilibration of a phospholipid between a bulk solution phase and a vesicle.

Decades before it was appreciated that membraneless organelles such as P granules and nucleoli might represent discrete liquid phases, it was known that certain phospholipids could, when dispersed in water, self-organize into micelles or vesicles, and the vesicle membranes resembled natural biological membranes in terms of their structure and permeability properties. A schematic of the process is shown in **Figure 10.4**.

To model this process, note that the situation is very similar to that shown in **Figure 10.1**, except that if the phospholipid vesicle is hollow and there is a single bilayer, its surface area will be proportional to x^* rather than to $(x^*)^{2/3}$. It follows that the rate equation for the system is:

$$\frac{dx^*}{dt} = k_1(x_{tot} - x^*) - k_{-1}x^*, \tag{10.11}$$

and the equilibrium solutions are:

$$x^*_{eq} = 0 \quad \text{and} \quad x^*_{eq} = x_{tot} - \frac{k_{-1}}{k_1}. \tag{10.12}$$

Thus once again we have one equilibrium ($x^*_{eq} = 0$) when x_{tot} is below a critical value of k_{-1}/k_1, which, in this case, is usually termed the critical micelle concentration; a transcritical bifurcation when $x_{tot}=k_{-1}/k_1$; and then coexisting vesicle and monomer phases when x_{tot} is above the critical micelle concentration. Once again the model makes no predictions about whether one or many vesicles will form, or of the size distribution of the vesicles.

INFECTIOUS DISEASE

10.5 THE SIR (SUSCEPTIBLE-INFECTED-RECOVERED) MODEL EXPLAINS WHY INFECTIOUS DISEASES SOMETIMES SPREAD EXPLOSIVELY

As I write (May 2020) the world is in the grip of the SARS-CoV-2 coronavirus, which is causing the COVID-19 pandemic. Modeling the spread of the virus is of obvious importance for planning mitigation strategies and allocating resources, and several different types of models have been proposed. One category is statistical models, which use data from countries that experienced early outbreaks to try to predict the course of later outbreaks. A second is agent-based modeling, where individual members of the population are treated individually. The third is ordinary differential equation modeling, very similar in flavor to the modeling we have been carrying out for signal transduction. As usual, ODE models are particularly useful for understanding the basic principles of the disease, including its dynamics (will the disease explode in my neighborhood, or just fizzle out?) and its steady-state behavior (what proportion of the population will ultimately be infected?), and so we will work through one of these models here, the SIR model.

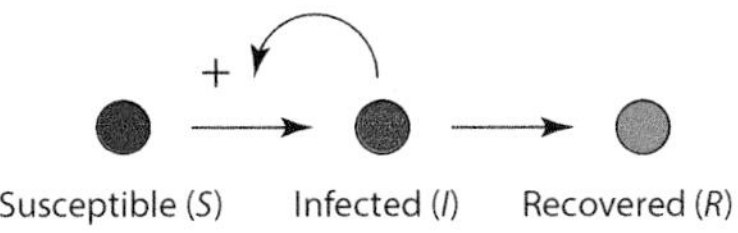

Figure 10.5 The SIR model of infectious disease. Individuals in the population get transferred from susceptible (S) to infected (I) to recovered (R) pools through one-way processes.

The SIR model was developed in the 1920s by the Scottish epidemiologists Anderson Gray McKendrick and William Ogilvy Kermack. In its simplest form, the model divvies up the population (of a city, a county, a country, or the earth) into three compartments or well-mixed pools (**Figure 10.5**). Those who are susceptible to the disease are said to be in the S pool, and in our case that is initially everyone, since for a new disease like COVID-19 the expectation is that none of us are immune. Those who have caught the virus and are infectious are in the I pool. Those who had caught the virus but are no longer infectious, either because they have recovered to health and immunity or have died, are in the R pool. The conversions from S to I and I to R are assumed to be one-way processes.

With these assumptions stated, we can write equations for the rates of infection and recovery. We assume that the rate of infection is directly proportional to the fraction of the population that is susceptible (S) and the fraction that is contagious (I):

$$\textit{rate of infection} = k_1 S \cdot I. \tag{10.13}$$

Note that there is positive feedback built into this expression: the rate of infection goes up as the fraction of the population that is infected (and infectious) goes up.

The rate of recovery is proportional to I:

$$\textit{rate of recovery} = k_2 I. \tag{10.14}$$

We can then combine Eqs. 10.1 and 10.2 to yield ordinary differential equations (rate equations) for the net rate of change of each of the three time-dependent species (S, I, and R):

$$\frac{dS}{dt} = -k_1 S \cdot I \tag{10.15}$$

$$\frac{dI}{dt} = k_1 S \cdot I - k_2 I \tag{10.16}$$

$$\frac{dR}{dt} = k_2 I. \tag{10.17}$$

In addition, we have a conservation equation:

$$S + I + R = 1 \tag{10.18}$$

This constitutes the SIR model of infectious disease spread.

10.6 THE SIR MODEL PREDICTS EXPONENTIAL GROWTH FOLLOWED BY EXPONENTIAL DECAY

Figure 10.6 shows the time course for a modeled infection, assuming rate constants and initial conditions that are probably reasonably close to what was initially true for the COVID-19 pandemic: $k_1 = 2.4\,\text{week}^{-1}$ and $k_2 = 1.2\,\text{week}^{-1}$. The model shows an initial exponential rise in infections, which peak and then fall, with the fall eventually approaching a negative exponential. The rise is quicker than the fall. Ultimately about 80% of the population becomes infected, and then recovers (or dies), whereas 20% never become infected and remain susceptible.

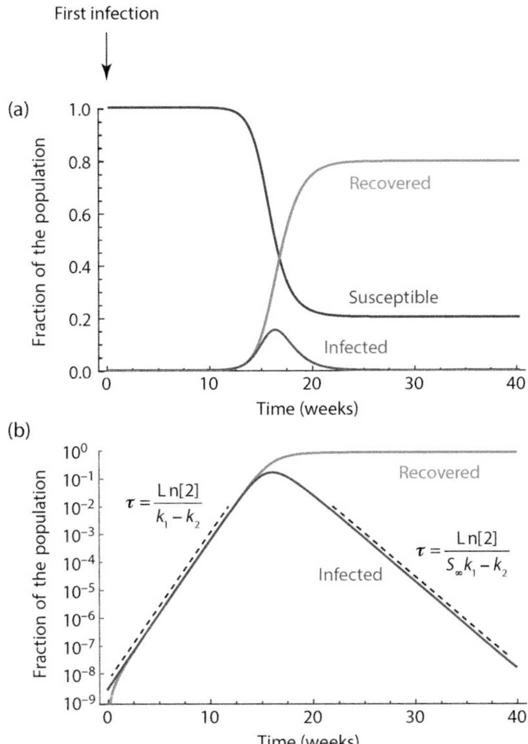

Figure 10.6 Simulated infection dynamics for the COVID-19 pandemic according to the SIR model. We have assumed an infection rate constant k_1 of 2.4 infections per week and a recovery rate constant k_2 of 1.2 per week, which means that R_0, the basic reproduction number, is 2. We have also assumed that no quarantines or social distancing measures are adopted, so that R_0 remains equal to 2 throughout the course of the infection. Finally, we assumed that at time zero there was one infected individual in a population of 330,000,000—the population of the United States.

To understand why the rise is initially exponential, we focus on Eq. 10.16, which is perhaps easier to understand when rearranged:

$$\frac{dI}{dt} = (k_1 S - k_2)I. \tag{10.19}$$

Both S and I are time-dependent variables, but at least initially the susceptible population will not change much from its initial value $S[0] = 1$, and so we can assume $S=S[0]$. This assumption allows us to solve Eq. 10.19, and the solution is a simple exponential function:

$$I[t] = I[0]e^{(k_1 S[0] - k_2)t}. \tag{10.20}$$

If $k_1 S[0] > k_2$, which for our case means $k_1 > k_2$, we have an increasing exponential function. Infections will grow exponentially with a doubling time $\tau = \frac{Ln[2]}{k_1 - k_2}$, which agrees with what is seen in our modeled time course (**Figure 10.6**). However, if $k_2 > k_1$, we have a decreasing exponential and infections will fall exponentially, with a half-time $\tau = \frac{Ln[2]}{k_2 - k_1}$. Thus, it makes a huge difference whether the ratio k_1/k_2 is greater than or less than 1.

At the end of the outbreak shown in **Figure 10.6**, S becomes nearly constant again, this time with $S = S[\infty]$. From Eq. 10.19 it follows that:

$$I[t] = I[0]e^{(k_1 S[\infty] - k_2)t}. \tag{10.21}$$

Since we assumed k_1 to be twice as big as k_2, and $S[\infty]$ proved to be 0.2, $k_1S[\infty] - k_2$ is negative and we have exponential decay of the infection. Because $S[\infty] < S[0]$, the exponential decay will be slower than the exponential rise was. Again, this agrees with what the simulated time course shows (**Figure 10.6**).

10.7 THE BASIC REPRODUCTION NUMBER R_0 DETERMINES WHETHER AN INFECTION WILL GROW EXPONENTIALLY

The ratio of k_1–k_2 is critical for determining whether the infection will initially grow exponentially or decline exponentially, and it is traditionally termed R_0 (R naught). R_0 specifies how many secondary infections, on average, are going to be produced from each primary infection when 100% of the population is susceptible to infection. It is often referred to as the basic reproduction number for the virus (so *R* here stands for "reproduction number," not "recovered fraction"; it is a bit confusing to have two different capital R's in the equations, but this is standard nomenclature). However, note that its value depends both on the intrinsic properties of the virus and the behavior of the susceptible and infectious populations. For COVID-19, R_0 is estimated to be 2–3; for comparison, $R_0 \approx 1.3$ for influenza, a less highly contagious disease. Note that with an $R_0 = 1.3$ infection, in 10 "generations" one initial infection will yield a total of $1.3^{10} \approx 14$ cases, whereas with $R_0 = 2$ there will be $2^{10} = 1{,}024$ cases and with $R_0 = 3$ there will be 59,049 cases. Thus, small changes in R_0 can yield huge changes in the disease's dynamics.

To highlight the importance of R_0, **Eq. 10.19** is often written as:

$$\frac{dI}{dt} = k_2(R_0S - 1)I. \tag{10.22}$$

If the product R_0S is greater than one, the disease will grow exponentially; if it is smaller than one, the disease will fizzle out.

With COVID-19, $S[0]$ is believed to have been equal to approximately 1 through at least the first half of 2020. Eventually, as more and more of the population becomes infected, recovered, and (we hope) immune, or immunized and immune, the value of S will fall. For an $R_0 = 2$ disease (the optimistic estimate for COVID-19), once S falls below 0.5, a new spark of infection would fizzle out rather than growing exponentially. This is an illustration of the concept of herd immunity; if half of the people that you, as an unwittingly infected and infectious person, interact with are not susceptible to the virus, then a new spark of disease will likely fizzle out rather than explode in an epidemic fashion. The minimum proportion of the population that must be immune in order to get herd immunity is $1 - \frac{1}{R_0}$. So if COVID-19 has an R_0 of 3, the more pessimistic end of the range, then herd immunity will kick in once 2/3 of the population is immune. And for a disease like measles, mumps, or chicken pox, which have R_0 values of at least 12, the immune population must be at least 92% to ensure against a new epidemic. This is why people who opt not to be vaccinated against diseases like measles, or not to have their children be vaccinated, can be a substantial danger to themselves, their children, and their neighbors (especially if their neighbors cannot be vaccinated for one reason or another). They are providing tinder that will allow the infection to spread like a blaze.

10.8 THE PROPORTION OF THE POPULATION THAT WILL ULTIMATELY BECOME INFECTED DEPENDS ON R_0

At first you might think that an epidemic would eventually infect every susceptible member of a population, like a fire that burns every flammable twig, or you might think that for an R_0=2 infection, the epidemic would ultimately affect only half the population, since once S falls to 0.5, the product R_0S falls below one. But this is not what our time course showed—with R_0=2 we ended up with 80% of the population infected and recovered, with 20% remaining susceptible. To see why this is the case, we need to derive an expression for the steady-state fraction of the population that will become infected (R_{SS} or $R[\infty]$) as a function of the model's parameters.

Normally the way we find the conditions required for steady state is to set all of the derivatives in the model equal to zero, since steady state occurs when none of the time-dependent variables are changing with time. Eqs. 10.15–10.18 become:

$$0 = -k_1 S \cdot I \tag{10.23}$$

$$0 = k_1 S \cdot I - k_2 I \tag{10.24}$$

$$0 = k_2 I. \tag{10.25}$$

From Eq. 10.24, we can see that I_{SS}=0, no matter what the choice of k_1 and k_2 was. However, these equations do not yield any information on the steady-state fraction of the population in either the S or the R pool. We need a different approach.

For this we go back to the rate equations. If we divide Eq. 10.15 by Eq. 10.17, we get:

$$\frac{dS}{dt}\Big/\frac{dR}{dt} = \frac{-k_1 S \cdot I}{k_2 I} = -\frac{k_1}{k_2} S = -R_0 S. \tag{10.26}$$

Keep in mind that R_0 is not R; the former is the basic replication number (which is k_1/k_2) and the latter is the fraction of the population in the recovered pool. Rearranging, integrating, and exponentiating both sides, we get:

$$\int \frac{dS}{S} = -R_0 \int dR \tag{10.27}$$

$$\mathrm{Ln} S = -R_0 R + C \tag{10.28}$$

$$S[t] = S[0] e^{-R_0 (R[t] - R[0])}. \tag{10.29}$$

We have chosen the constant of integration such that $S[t]=S[0]$ when t=0. In the case of COVID-19, $S[0]$ was believed to be initially 1—none of us had encountered the virus previously, so none of us had antibodies to it and all of us were susceptible—and $R[0]$ is 0. Even after we move a few individuals in the S pool into the I pool to start the infection, $S[0]$ will still be very close to 1. Thus:

$$S[t] = e^{-R_0 R[t]}. \tag{10.30}$$

This defines the relationship between R and S at any time point. Now since $I[\infty] = 0$ and $S+I+R = 1$, it follows that as $t \to \infty$,

$$1 - R[\infty] = e^{-R_0 R[\infty]}. \tag{10.31}$$

Equation 10.31 is a transcendental equation with a single variable, $R[\infty]$, and a single parameter, R_0. For a given value of R_0, it implicitly defines the value of $R[\infty]$. The equation cannot be solved in closed form, although it can be solved in terms of a particular named function, the Lambert W function or product logarithm. For example, if you ask *Mathematica* to solve Eq. 10.31, this is what you will get:

$$R[\infty] = 1 + \frac{1}{R_0}\text{ProductLog}\left[-R_0 e^{-R_0}\right]. \tag{10.32}$$

However, even without delving into the properties of product logarithms, we can get a good idea of how $R[\infty]$ varies with R_0 through a graphical approach.

We plot the expressions on the left- and right-hand sides of Eq. 10.31 as functions of $R[\infty]$; $R[\infty]$ is on the x-axis and $1 - R[\infty]$ and $e^{-R_0 R[\infty]}$ on the y-axis (**Figure 10.7a**). When the two curves intersect, the left-hand side of Eq. 10.31 equals the right-hand side, and the value of $R[\infty]$ at the intersection is its steady-state value.

Repeating this for various assumed values of R_0, we see that if R_0 is less than or equal to 1, there will be a single intersection point, at $R[\infty] = 0$. Nobody gets infected (at least in the limit where $I[0]$ is infinitessimal), and nobody needs to recover. However, once R_0 exceeds 1, the curves have two intersection points (**Figure 10.7a**), meaning that Eq. 10.31 has two solutions—one with $R[\infty] = 0$ and the other with $R[\infty] > 0$. The higher the value of R_0, the higher this second steady-state value of $R[\infty]$ is (**Figure 10.7b**). The system undergoes a transcritical bifurcation at $R_0 = 1$, where it goes from having a single steady state to two. Below the bifurcation point, the $R[\infty] = 0$ steady state is stable as indicated by the solid line in **Figure 10.7b**; above it, the $R[\infty] = 0$ steady state is unstable (dashed line) and ready to explode to the stable $R[\infty] > 0$ steady state (solid curve) upon the introduction of a single infected individual.

Thus, the basic replication number R_0 determines what proportion of the population will eventually get the disease, and, since some proportion of those who get the disease will die, it determines the fatalities as well. When R_0 is below 1, almost no one will get the disease (exactly no one if $I[0] = 0$ and $S[0] = 1$, and a small but non-zero number if, say, $I[0] = 1$

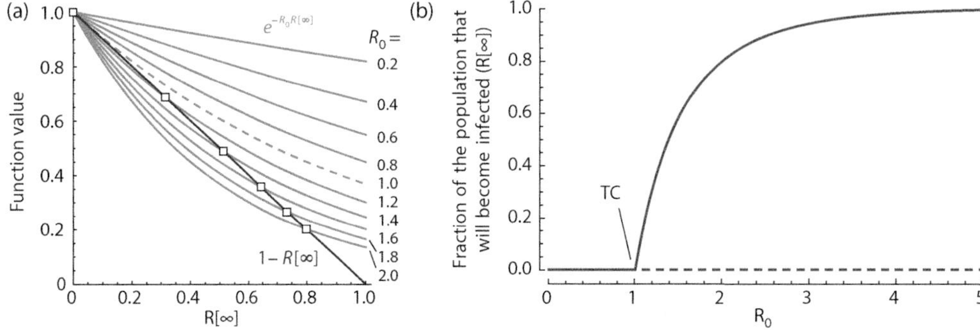

Figure 10.7 The final infection toll as a function of R_0. (a) Solving the transcendental equation $1 - R[\infty] = e^{-R_0 R[\infty]}$ graphically. The blue line shows the value of the left-hand side of the equation as a function of $R[\infty]$. The green curves show the value of the right-hand side for various assumed values of R_0. When R_0 exceeds 1 (where the $R_0 = 1$ curve is the dashed green curve), the green and blue curves go from having two intersection points to one. (b) The fraction of the population that will ultimately be infected, $R[\infty]$, as a function of R_0. There is a transcritical bifurcation (designated TC) at $R_0 = 1$.

person in a population of 330 million, the current US population, and $S[0]$=329,999,999/330,000,000). As R_0 rises above 1, the proportion that will ultimately become infected rises, so that when R_0=2, ultimately 80% of the population will be expected to become infected. Note that this proportion is larger than that required to obtain herd immunity (which was 50%). This is because a rapidly-spreading infection infects a substantial number of people even after the fraction in the susceptible pool has fallen below the threshold required for exponential growth.

10.9 MANIPULATING R_0 CAN DELAY AN EPIDEMIC, DECREASE THE PEAK, AND DIMINISH THE FINAL NUMBER OF INFECTED INDIVIDUALS

Note that even though R_0 is called the basic reproductive number of the virus, its value depends both upon the biology of the virus and on the behavior of the infected population. So if a virus has a value of R_0=2 when a population's physical interactions are at baseline, decreasing encounters within the population by a factor of, say, 3 will decrease R_0 to 2/3, well below the threshold required for exponential growth and epidemic spread. This is why public health officials and epidemiologists put so much emphasis on measures to decrease R_0, like masks and stay-at-home orders.

Figure 10.8 shows how the time course of the disease would be expected to change if a country instituted a social distancing policy

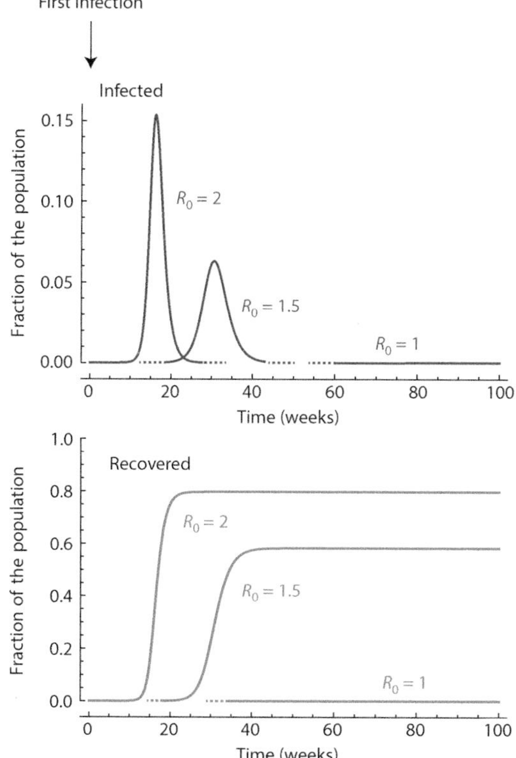

Figure 10.8 Decreasing R_0, for example, through social distancing, both "flattens the curve" (top) and decreases the total number of infections (bottom). Note that the fraction of the population that has recovered is the same as the fraction that has had the infection. We have assumed a country of 330,000,000 people with a single individual infected at time zero, a recovery rate constant $k_2 = 1.2\ \text{week}^{-1}$, and values of k_1 to make R_0=2, 1.5, or 1.

that reduced the number of secondary cases per primary case from its nominal value of $R_0 \approx 2$ to something less. A relatively modest decrease in R_0, from 2 to 1.5, would delay the epidemic and "flatten the curve," shifting the peak of infections from 16 weeks to 31 weeks and decreasing the peak height from 15% to 6% (**Figure 10.8**). This would not only buy the country some time but also make it so that the worst days of the infection were less likely to overwhelm the health-care system and therefore improve the ultimate mortality rate.

Moreover, the proportion of the population that would ultimately be infected (and hence the proportion that would ultimately die of the infection) would drop, from 80% to 58% (**Figure 10.8**). This is an improvement—58 million infections is certainly better than 80 million—but this hypothetical scenario is still catastrophic; it would result in hundreds of thousands of deaths.

But if R_0 can be lowered to 1 or less, something remarkable happens: the peak of infected individuals vanishes, and the proportion of the population that will ultimately become infected drops precipitously. The disease fizzles out instead of growing exponentially. For $R_0=1$, an initial case of one individual in a population of 330 million would ultimately result in ~26,000 cases—only 0.008% of the population (**Figure 10.8**). If $R_0=0.5$, our initial case would, on average, result in only a single additional case.

Of course this is all predicated on the assumption that the SIR model is actually applicable to real diseases like the COVID-19 pandemic. To be sure, some of the assumptions built into the model are suspect. For example, the model assumes that the population of the county, state, country, or whatever, is a well-mixed system, where every person interacts with every other person on the time scale of the epidemic. This is clearly not true—there is spatial structure to the evolving epidemic. But the most basic lessons of the model, that the infectiousness of a virus determines whether the infection will fizzle out or explode and that the infectiousness also determines the fraction of the population that will ultimately become infected, probably are true.

SUMMARY

Here we have examined two different examples of biological processes that involve positive feedback but not bistability: liquid–liquid phase separation and the spread of an infectious disease. In both cases, the modeled system undergoes a dramatic change in behavior at a critical point, a transcritical bifurcation.

In liquid–liquid phase separation, positive feedback is present because the larger the separated phase, the faster it takes up more dispersed molecules. The critical point, above which the separated phase becomes able to persist, corresponds to the maximum solubility of the species x, and it is given by $x_{crit}=k_{-1}/k_1$, where k_{-1} is the rate constant for the dissociation of x from the condensed phase and k_1 is the rate constant for the association of x with condensed phase. When the total concentration of x is below x_{crit}, x will be dispersed throughout the cytoplasm or buffer and the system will have a single phase; above it, the system changes to an inhomogeneous, two-phase system with some of the x dispersed and some condensed. This basic behavior is shared by simple models of other condensation phenomena, including the formation of vesicles from phospholipids.

In the spread of an infectious disease, the SIR model includes positive feedback because the larger the infected pool, the faster new infections occur. The critical point in the model occurs when $R_0S[0]=1$, where R_0 is the number of secondary infections per primary infection and $S[0]$ is fraction of the population that is initially susceptible to the infection. If $R_0S[0]>1$, the infections will increase exponentially, and if $R_0S[0]<1$ they will fizzle out.

Thus, positive feedback can provide a system with a threshold and a critical point, where the traversal of a transcritical bifurcation makes the system switch from one type of behavior (dispersed molecules or a disease that fizzles out) to a second, qualitatively different behavior (separated phases or a disease that grows exponentially).

FURTHER READING

LIQUID–LIQUID PHASE SEPARATION IN BIOLOGY

Alberti S, Gladfelter A, Mittag T. Considerations and challenges in studying liquid-liquid phase separation and biomolecular condensates. Cell. 2019 Jan 24;176(3):419–34.

Brangwynne CP, Eckmann CR, Courson DS, Rybarska A, Hoege C, Gharakhani J, Jülicher F, Hyman AA. Germline P granules are liquid droplets that localize by controlled dissolution/condensation. Science. 2009 Jun 26;324(5935):1729–32.

Hyman AA, Weber CA, Jülicher F. Liquid-liquid phase separation in biology. Annu Rev Cell Dev Biol. 2014;30:39–58.

Klosin A, Oltsch F, Harmon T, et al. Phase separation provides a mechanism to reduce noise in cells. Science. 2020;367(6476):464–8.

MICELLE AND VESICLE FORMATION

Israelachvili JN, Mitchell DJ, Ninham BW. Theory of self-assembly of hydrocarbon amphiphiles into micelles and bilayers. J Chem Soc Faraday Trans 2: Mol Chem Physics 1976;72:1525–68.

Tanford C. *The Hydrophobic Effect: Formation of Micelles and Biological Membranes*. 1st Edition. John Wiley & Sons, New York, 1973.

TRANSCRITICAL BIFURCATIONS

Albeck JG, Burke JM, Spencer SL, Lauffenburger DA, Sorger PK. Modeling a snap-action, variable-delay switch controlling extrinsic cell death. PLoS Biol. 2008;6(12):2831–52.

Strogatz SH. *Nonlinear Dynamics and Chaos: With Applications to Physics, Biology, Chemistry, and Engineering*. Westview Press, Cambridge, MA, 1994, Chapter 3.

INFECTIOUS DISEASE, THE SIR MODEL, AND COVID-19

Anderson RM, Heesterbeek H, Klinkenberg D, Hollingsworth TD. How will country-based mitigation measures influence the course of the COVID-19 epidemic? Lancet. 2020 Mar 21;395(10228):931–4.

Dong E, Du H, Gardner L. An interactive web-based dashboard to track COVID-19 in real time. *Lancet Infect Dis*; published online Feb 19 2020. Accessed March 13, 2020.

Kermack WO, McKendrick AG. A contribution to the mathematical field of epidemics. Proc Royal Soc A. 1927;115:700–721.

Zhang S, Diao M, Yu W, Pei L, Lin Z, Chen D. Estimation of the reproductive number of novel coronavirus (COVID-19) and the probable outbreak size on the Diamond Princess cruise ship: A data-driven analysis. Int J Infect Dis. 2020 Apr;93:201–204.

NEGATIVE FEEDBACK 1

Stability and Speed

11

IN THIS CHAPTER . . .

INTRODUCTION

Negative feedback is ubiquitous in cellular regulation. It is probably most famous for its involvement in adaptation, allowing a system to recover after the introduction of a stimulus, and we will examine perfect adaptation through negative feedback in Chapter 12. Negative feedback can also generate oscillations and is believed to be at the heart of all biological oscillators. We will examine oscillations in Chapters 14 and 15.

Here in this brief chapter, we examine two other things that negative feedback can accomplish: it can make a stable steady state more stable and can allow a system to approach the steady state faster than it otherwise could. Both of these properties can help a system to adjust to an uncertain, fluctuating environment.

11.1 NEGATIVE FEEDBACK CAN INCREASE THE STABILITY OF A STEADY STATE

To construct an adapting system we will need to use a model with at least two time-dependent variables—one-variable systems always approach their steady states monotonically. But even a one-variable system can be stabilized and speeded through negative feedback, so we will start with a simple one-variable, one-ODE model here.

Suppose we have a mass action synthesis-destruction system, like that shown in **Figure 11.1a**. The rate equation for the system is:

DOI: 10.1201/9781003124269-11

$$\frac{dy}{dt} = k_1 x_{tot} - k_{-1} y. \quad (11.1)$$

Species x could be an enzyme that produces a second messenger y from some highly abundant precursor, with y being destroyed by some unspecified, unregulated, unsaturated enzyme, or x could be an mRNA being translated by unsaturated ribosomes, and y could be the resulting protein product.

The rate–balance plot for this system is shown in **Figure 11.1c**, with a horizontal line for the synthesis reaction and a line of positive slope for the destruction reaction; it is just the same as what we had in **Figure 6.1**. Arbitrarily we have chosen $k_1 = k_{-1} = 1$ for the values of the rate constants, and $x = 500$ for the concentration of the enzyme x, so that the steady-state value of y is given by:

$$y_{ss} = \frac{k_1}{k_{-1}} x_{tot}, \quad (11.2)$$

is 500 (**Figure 11.1c**).

Next let us add a negative feedback loop to the system. We will suppose that y can bind to and stoichiometrically inhibit x (**Figure 11.1b**). The rate equation for the production of y becomes:

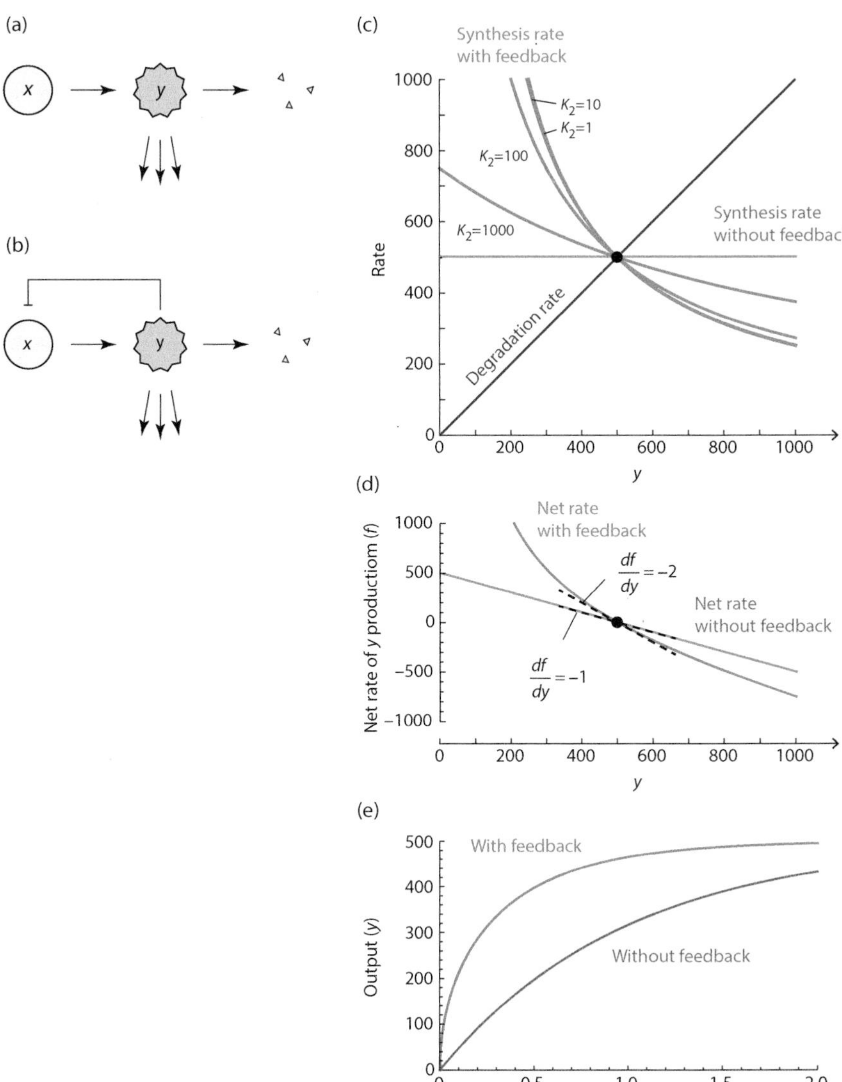

Figure 11.1 Negative feedback can stabilize steady states and speed responses. (a, b) Schematic view of a production/destruction circuit with (b) or without (a) stoichiometric negative feedback. (c) Rate–balance plots. Various values of the parameter K_2, which bears on the strength of the negative feedback (the smaller the value of K_2, the stronger the feedback), were chosen, and the value of k_1 was adjusted to make the steady state level of y, and the flux through the system, equal to 500 in all cases. d) Net production rates as a function of the instantaneous concentration of y. (d) Time course of approach to steady-state for a system with negative feedback (and with K_2 being vanishingly small) vs. no negative feedback.

$$\frac{dy}{dt} = k_1\left(x_{tot} - c_{xy}\right) - k_{-1}y, \tag{11.3}$$

where c_{xy} represents the concentration of the inhibited xy complex. Note that c_{xy} is a time-dependent quantity, but if we assume that the equilibration of x and y with c_{xy} is rapid compared to the synthesis and destruction of y, we can substitute an expression for the equilibrium concentration of c_{xy} into Eq. 11.3. Assuming that the concentration of y is much higher than the concentration of x, so that x is depletable but y is not, the equilibrium concentration of c_{xy} is given by the Langmuir equation:

$$c_{xy} = \frac{y}{K_2 + y} x_{tot}, \tag{11.4}$$

where K_2 is the equilibrium constant for the binding of x to y. It follows that:

$$\frac{dy}{dt} = k_1 \frac{K_2}{K_2 + y} x_{tot} - k_{-1}y. \tag{11.5}$$

This is our one-ODE model for the production of y in the presence of stoichiometric negative feedback from y to x.

We can carry out a rate-balance analysis of this model and use this as the basis for comparing the stability of the two systems at steady state. We have not altered the degradation rate term, and we will choose $k_{-1}=1$ again, so both models will have the same degradation rate curves—the blue diagonal line on **Figure 11.1c**. The difference then is the synthesis rate, which is constant in the no-feedback model and decreases as y increases in the negative feedback model.

Probably the fairest way to compare the stabilities of the steady states in the two models is to arrange that the two systems have the same steady-state level of output y and the same steady-state flux. Since we have assumed that the two models have the same destruction curves, this means that the value of the synthesis rate at the steady state for the feedback model,

$$k_1 \frac{K_2}{K_2 + y_{ss}} x_{tot} \tag{11.6}$$

must equal the value of the synthesis rate for the no-feedback model:

$$k_1' x_{tot}. \tag{11.7}$$

Note that since we have not assumed that the two models have the same synthesis rate constants—it would not be possible to make their synthesis rate curves meet up at the steady state if we did—we are calling the rate constant for the no-feedback model by a new name, k_1'. With a little algebra we can see that the two models will have the same steady state value of y if we choose values for k_1 such that:

$$k_1 = k_1' \frac{K_2 + y_{ss}}{K_2} = k_1' \frac{K_2 + \frac{k_1'}{k_{-1}} x_{tot}}{K_2}. \tag{11.8}$$

So now for our choices of $k_1' = 1$, $k_{-1} = 1$, and $x_{tot} = 500$, we can choose various values for K_2, calculate the appropriate value for k_1, and plot the resulting stimulus rate curves. The results are shown in **Figure 11.1c**. As the value of K_2 decreases, meaning that the inhibition of x by y gets stronger, the negative slope of the stimulus rate curve at the steady state gets larger and larger. The steeper the curve, the faster the system will approach the steady state, and so the more stable the steady state will be.

We can calculate how much more stable the steady state is through the procedure introduced in Section 9.2. We define a function to be the net rate of production of y as a function of y. This yields:

$$f_1 = k_1' x_{tot} - k_{-1} y \tag{11.9}$$

for the no-feedback model, and:

$$f_2 = k_1 \frac{K_2}{K_2 + y} x_{tot} - k_{-1} y \tag{11.10}$$

for the model with negative feedback. These curves are plotted in **Figure 11.1d**; the parameters assumed are $k_1' = 1$ (for the first model), $x_{tot} = 500$ (for both models), $k_{-1} = 1$ (for both models), and for the second model, K_2 is small and k_1 is given by **Eq. 11.8**, because this gives the steepest slope for the net rate curve. We can then calculate the slopes of the net rate curves, $\frac{df_1}{dy}$ and $\frac{df_2}{dy}$>, and evaluate them at $y = y_{ss} = 500$. We obtain $\frac{df_1}{dy} = -1$ and $\frac{df_2}{dy} = -2$. Both are negative numbers, so the steady state is stable for both models, and the slope for the negative feedback model is twice that of the no-feedback model. This means that the negative feedback has doubled the stability of the system.

Recall that in Section 6.2, we compared the stability of a synthesis/destruction model to that of a phosphorylation/dephosphorylation cycle with the same steady-state output and flux and we found that the latter was twice as stable as the former. The reason for this was substrate depletion. Now we have shown that negative feedback can also double the stability of a steady state, and the reason for it can be viewed as enzyme depletion—inhibition of x by its product y. Negative feedback and depletion are conceptually similar, and both phenomena yield a rate equation where the time-dependent variable negatively affects its own rate of production. Perhaps it is not too surprising then that complex systems-level behaviors such as adaptation, which can arise from negative feedback (Chapter 12), can also arise from depletion mechanisms like state-dependent inactivation (Chapter 13). We will see another example of this equivalence in Chapter 15 when we examine relaxation oscillators.

11.2 NEGATIVE FEEDBACK CAN ALLOW A SYSTEM TO RESPOND MORE QUICKLY

In addition, the speed at which the system responds to a stimulus is substantially faster for the negative feedback system than for the no-feedback system, provided the two circuits have the same steady state. This can be seen from the rate–balance plot (**Figure 11.1c**) or the net rate plot (**Figure 11.1d**); the net rate at which the system approaches steady state is always higher, at any value of y, for the negative feedback system. This is because to get the two models to approach the same steady state, the synthesis rate constant for the negative feedback system, k_1, had to be substantially larger than the corresponding rate constant (k_1').

So how much faster is the approach? For the no-feedback system, we have simple exponential approach to steady state with a halftime of $\frac{\text{Ln}2}{2}$ (**Figure 11.1e**); this is a good gauge for how fast the system is. However, for the negative feedback system the approach to steady state is not a simple exponential. Nevertheless, we can calculate numerically how long the system takes to get to various levels of output and compare the results to those of the no-feedback system.

To get 90% of the way to steady state, the negative feedback system is 2.8-fold faster. To get 50% of the way, the negative feedback system is 4.8-fold faster. And to get 10% of the way, the negative feedback system is 21-fold faster. Negative feedback has allowed the system to respond faster to x, especially at the start of the response when the feedback is weak.

It is not an exact analogy, but one way to see why negative feedback allows for a faster response is to think about two cars approaching a stop sign from a dead stop at the bottom of a little hill. The no-feedback car is constrained to not use the brakes, so it has to start at a relatively low speed and let the car gradually come to a stop at the top of the hill. However, the negative feedback car can start fast and wait to apply the brakes until it is almost to the stop sign. This ability to put on the brakes allows the negative feedback car to go faster, especially at first, and to get to the stop sign faster than the no-feedback car can.

Curiously, negative feedback can also destabilize a steady state if you have a long enough negative feedback loop. It is this destabilization that allows negative feedback loops to sometimes function as biochemical oscillators, and we will explore the properties of a famous negative feedback oscillator, the Goodwin oscillator, in Chapter 14.

FURTHER READING

NEGATIVE FEEDBACK: STABILITY AND SPEED

Becskei A, Serrano L. Engineering stability in gene networks by autoregulation. *Nature*. 2000 Jun 1;405(6786):590–3.

Rosenfeld N, Elowitz MB, Alon U. Negative autoregulation speeds the response times of transcription networks. J Mol Biol. 2002 Nov 8;323(5):785–93.

NEGATIVE FEEDBACK 2

Adaptation

12

IN THIS CHAPTER . . .

INTRODUCTION

So far we have focused mainly on how signaling systems initiate a response. But how a system terminates a response is just as important, and cells have evolved numerous strategies to ensure that their responses are not too protracted. Sometimes a cell stops responding by getting rid of the stimulus. This is true in the case of acetylcholine, the neurotransmitter that mediates the contraction of skeletal muscle and the relaxation of smooth muscle in the peripheral nervous system and functions as a neuromodulator in the central nervous system. Cholinergic signals are terminated largely through the hydrolysis of

DOI: 10.1201/9781003124269-12

acetylcholine by the enzyme acetylcholinesterase, and compounds that inhibit acetylcholinesterase are used in the treatment of Alzheimer's disease and for killing insects. The effects of the neurotransmitters dopamine, serotonin, and norepinephrine are limited by degradation too, and drugs that inhibit one of the degrading enzymes, monoamine oxidase, are used to treat depression. These neurotransmitters are also pumped back into the neuron that released them, and this reuptake plays a role in terminating their action. The most commonly prescribed antidepressants inhibit one or more of these reuptake pumps, and so does cocaine.

But often the signal stops before the stimulus is gone; the receptor, or the signaling proteins downstream of the receptor, changes in response to the signal in a way that limits the duration of the response. These changes can be viewed as adaptation, with the system adapting to the presence of the stimulus. If a system initially responds and then returns to exactly its pre-stimulus output, the adaptation is said to be perfect.

There are at least two common motifs for adapting to the presence of stimuli: negative feedback loops and incoherent incoherent systems. Another motif, state-dependent inactivation (which some consider to be a variation of either feedback or incoherent regulation), can yield adaptation as well. In Chapter 13 we will analyze examples of incoherent feedforward regulation and state-dependent inactivation. Here will start with negative feedback, and begin with one example from bacterial chemotaxis and a second from receptor tyrosine kinase signaling and the MAP kinase cascade.

BACTERIAL CHEMOTAXIS

12.1 BACTERIA FIND FOOD SOURCES THROUGH A BIASED RANDOM WALK

One of the best-studied examples of adaptation in biology is chemotaxis in *E. coli*. As mentioned in Chapter 1, *E. coli* can detect and thence swim toward food. The sensing is accomplished by a receptor (e.g. Tsr, the serine receptor, or Tar, the aspartate receptor) that is linked through an adaptor protein (CheW) to a histidine kinase, CheA (with Che standing for chemotaxis) (**Figure 12.1a**). The receptors are multimeric and clustered at the poles of the rod-shaped bacterium, which allows them to function cooperatively and translate small changes in chemoattractant concentration into large changes in receptor output.

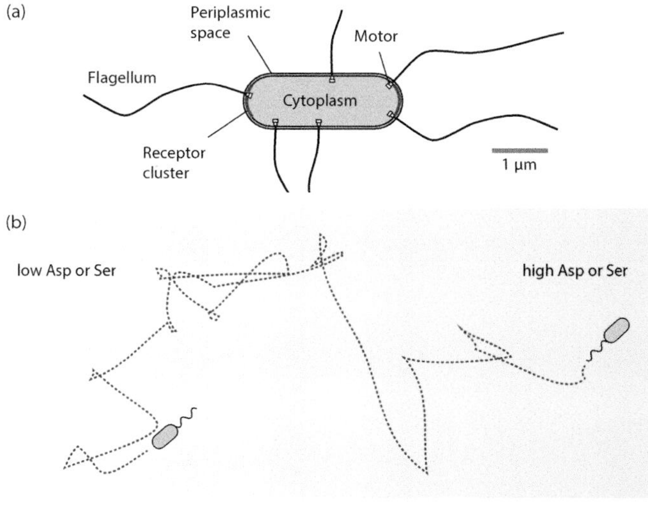

Figure 12.1 Chemotaxis in *E. coli*. (a) Location of the receptors that detect chemoattracts and the motors and flagella that propel the bacterium. The receptors span the membrane between the periplasmic space and the cytoplasm, and are clustered at the two poles. The flagella and the motors that rotate them are located at various positions. (b) Swimming toward food through a biased random walk. The cell swims for a second or two, then tumbles and reorients. If the cell detects that the chemoattractant concentration is increasing, it increases the length of time between tumbles.

The swimming is powered by flagella that propel the bacterium. *E. coli* has peritrichous flagella, meaning that they are present all over the surface of the cell rather than being confined to just one pole as is true of some bacteria. These flagella are rotated by a multiprotein motor, which can make them turn either clockwise or counterclockwise. If the rotation is counterclockwise, the flagella form a neat bundle, and they work together to propel the bacterium in a smooth, more-or-less straight trajectory (**Figure 12.1b**). If the rotation is clockwise, the flagella fly apart and the bacterium tumbles (or "twiddles," the word Berg and Brown originally used in their studies of the phenomenon). The tumbling bacterium changes direction, and then sets off in a new direction. Thus, the cell switches between two discrete modes of locomotion—smooth swimming and tumbling.

If a bacterium is swimming in a medium where food is present at a constant concentration, it will swim for a couple of seconds, covering ~20 μm of distance, then tumble, reorient, and take off in a new direction. The turns are not completely random in direction—small turn angles are more common than large ones—but overall the process resembles a random walk, and once a few tumbles have occurred, on average a bacterium's distance from its original starting point will be proportional to $\sqrt{t}$. This is the default swimming mode of the bacterium.

But if there is a food source at some particular location, and so there is a spatial gradient of chemoattractant concentration (**Figure 12.1b**), the bacteria will bias their random walk. When they are going up the gradient toward the food source, they will maintain their smooth swimming for longer than they do when they are swimming down the gradient. They accomplish this biasing by increasing the time between tumbles.

12.2 BACTERIA SUPPRESS TUMBLING IN RESPONSE TO CHEMOATTRACTANTS AND THEN ADAPT PERFECTLY

Figure 12.2 shows quantitative data on the *E. coli* chemotactic response. After addition of a sufficiently high concentration of the chemoattractant aspartate, the cells rapidly suppress their tumbling, so that the average duration of their smooth swimming runs increases from about 2 s to about 50 s. Then, over the next 15 min, the tumbling rate returns back to basal and, within experimental error, the adaptation is perfect.

In addition, although changing the expression level of various chemotaxis proteins changes the basal tumbling frequency and/or the adaptation time of the system, the adaptation always remains perfect. Thus the precision of the adaptation is robust with respect to perturbations in pathway components, although neither the time scale of adaptation nor the steady-state frequency of tumbling is. This suggests that there was stronger evolutionary pressure for the system to

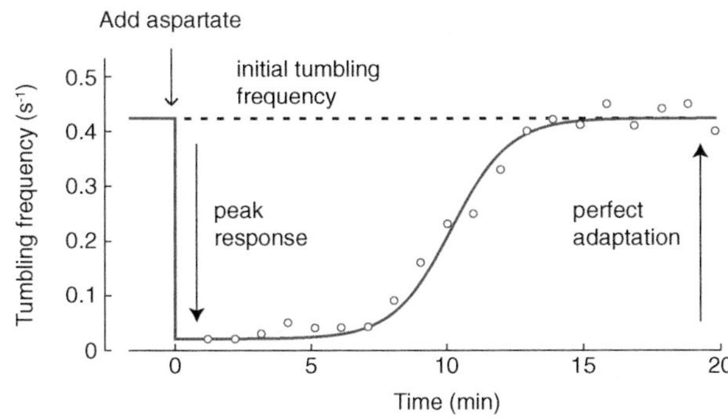

Figure 12.2 Perfect adaptation in the response of *E. coli* to a step change in the concentration of the chemoattractant aspartate. (Based on data in Alon et al., *Nature*. 1999 and used with permission.)

adapt perfectly than to adapt to a particular basal level of tumbling, or to adapt at a particular speed.

So is there a simple way to achieve perfect adaptation in receptor signaling, compatible with what is known about the downstream biochemistry of this two-component signaling system?

12.3 A PLAUSIBLE NEGATIVE FEEDBACK MODEL CAN ACCOUNT FOR PERFECT ADAPTATION

The prevailing model was introduced by Barkai and Leibler and further explored by Alon, Leibler, and colleagues; here we will expand on Alon et al.'s approach. The basic logic of the model is shown in **Figure 12.3a** and in more molecular detail in **Figure 12.3b**.

The model begins with the chemotaxis receptor/histidine kinase complex. For the chemoattractant aspartate, the relevant receptor is Tar, and it is linked through the CheW adaptor protein to the histidine kinase CheA. This complex can be considered a single functional unit. Next, it is assumed that three states are possible for the receptor complex. In the first state, the receptor is methylated and is not bound to aspartate. This is the active form of the receptor. In the second state, the methylated receptor is bound to aspartate. This is an inactive form of the receptor; aspartate turns the kinase off. The third state is also inactive. This state represents receptors inactivated by demethylation, via the enzyme CheB, rather than by ligand binding, and the inactive, demethylated receptors can be viewed as receptors held in reserve. We could also consider a fourth state that is both demethylated and bound to ligand, but, as it turns out, the model works better if it is assumed that CheB only demethylates active receptors. That is, the inactivation of CheA is state-dependent, an aspect of this model that we will see again in Chapters 13 and 15.

Next we add negative feedback to the model by assuming that active receptors activate the demethylase CheB. The demethylation (by CheB) and methylation (by the unregulated CheR protein) reactions eventually come into balance, and the hope is that at this steady state, the activity of the receptor and the tumbling rate of the bacterium will have returned to what they were before the chemoattractant was applied.

At this point we have five time-dependent species—the three states of the receptor complex and the two states of CheB—and we can write rate equations for each of them. First, for the receptor complex, which we call A for CheA, we write a rate equation for the ligand-bound

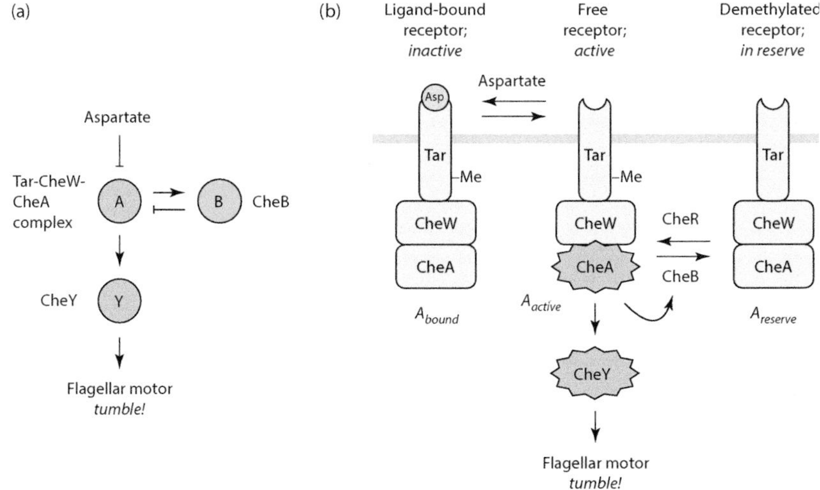

Figure 12.3 Two views of the chemotaxis adaptation circuit. (a) Course-grained view, emphasizing the negative feedback between CheA and CheB. (b) Detailed view, showing the three states of the receptor: inactive (methylated but ligand bound), active (methylated and ligand free), and in reserve (demethylated and ligand free).

receptor A_{bound}, which we assume is produced by the binding of ligand to only the active, methylated receptors—again a form of state-dependent inactivation:

$$\frac{dA_{bound}}{dt} = k_1 Asp \cdot A_{active} - k_{-1} A_{bound}. \tag{12.1}$$

where, as usual, $K_1 = k_{-1}/k_1$.

Next, for the active, non-ligand-bound, methylated state of the receptor:

$$\frac{dA_{active}}{dt} = -k_1 Asp \cdot A_{active} + k_{-1} A_{bound} - k_2 B^* \cdot A_{active} + k_{-2} R^* \cdot A_{reserve}. \tag{12.2}$$

B^* represents the active form of the demethylase CheB. Note that for the moment we are assuming that the methylation and demethylation rates are described by mass action kinetics, with the rates being directly proportional to the concentrations of the enzymes and the substrates. Note also that R^*, the activity of the methylase CheR, is assumed to be constitutive, so we can include it in a redefined k_{-2} for simplicity:

$$\frac{dA_{active}}{dt} = -k_1 Asp \cdot A_{active} + k_{-1} A_{bound} - k_2 B^* \cdot A_{active} + k_{-2} A_{reserve}. \tag{12.3}$$

For the demethylated, reserve form of the receptor, we have:

$$\frac{dA_{reserve}}{dt} = k_2 B^* \cdot A_{active} - k_{-2} A_{reserve}. \tag{12.4}$$

Next we turn to the activation and inactivation of the demethylase CheB. We could write two ODEs, one for active CheB and one for inactive CheB, but it is simpler to write a single ODE for active CheB and eliminate inactive CheB from the equation with the conservation relationship $B_{total}=B^*+B$:

$$\frac{dB^*}{dt} = k_3 A_{active}(B_{tot} - B^*) - k_{-3} B^*. \tag{12.5}$$

Eqs. 12.2–12.5 constitute a four-ODE model of *E. coli* chemotaxis.

To see how well the modeled system responds and adapts, we can choose some arbitrary values for the rate constants and concentrations, and solve the ODEs numerically. **Figure 12.4** shows typical results. In response to a step increase in aspartate, the system responds with a rapid decrease in the proportion of active receptors. This is followed by a return back toward the initial level of receptor activity, but the adaptation is not perfect. We can quantify how good the adaptation is, defining the percent adaptation (here called α) to be how big the adaptation is relative to how big the initial response was. This amounts to:

$$\alpha = \frac{(A_{active})_{peak} - (A_{active})_{final}}{(A_{active})_{peak} - (A_{active})_{initial}} \cdot 100\%. \tag{12.6}$$

For the first increment of stimulus shown in **Figure 12.4**, the adaptation is 66%, and for each subsequent increment it is a little worse. The model does not adapt nearly as well as the actual bacterium does. One can improve the performance of the model with a judicious

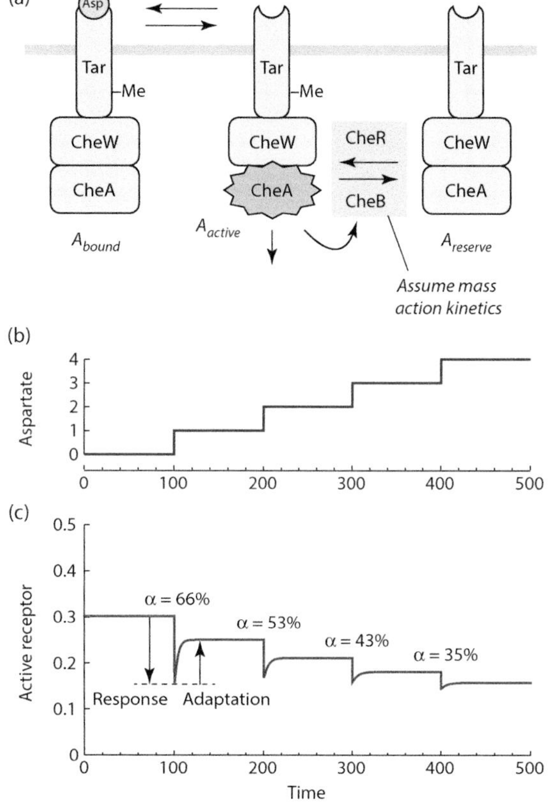

Figure 12.4 Partial adaptation in a model of chemotaxis that assumes mass action kinetics in the negative feedback. (a) Schematic of the adaptation circuit. (b) Step changes in input (aspartate). (c) Changes in receptor activity. The parameters for the model (**Eqs. 12.1** and **12.3–12.5**) were $k_1=k_{-1}=100$; $k_2=1$; $k_3=0.1$; $k_4=k_{-4}=1$; $A_{tot}=1$; and $B_{tot}=1$. The model responds to changes in the chemoattractant, and then adapts, but does not adapt as well as the bacterium actually does.

choice of parameters, but, as Barkai realized, there is a better way forward.

So far we have assumed that the methylation and demethylation reactions are described by mass action kinetics. What if we assume instead that the kinetics of these two processes are described by the Michaelis–Menten equation? This would mean that:

$$\frac{dA_{active}}{dt} = -k_1 Asp \cdot A_{active} + k_{-1}A_{bound} - k_2B^* \frac{A_{active}}{K_{demeth} + A_{active}} + k_{-2}\frac{A_{reserve}}{K_{meth} + A_{reserve}}, \tag{12.7}$$

$$\frac{dA_{reserve}}{dt} = k_2B^* \frac{A_{active}}{K_{demeth} + A_{active}} - k_{-2}\frac{A_{reserve}}{K_{meth} + A_{reserve}}. \tag{12.8}$$

It turns out that, other things being equal, this improves the percent adaptation for the model. And the more saturated the two reactions are, the better the adaptation. If one assumes near saturation—that is, that there is a very high degree of zero-order ultrasensitivity in the methylation/demethylation steady-state response—then the model can produce near-perfect adaptation. In the limiting case where the methylation and demethylation kinetics are of zero order, and so the steady-state response of the methylation/demethylation system is a step function, then the adaptation becomes perfect.

We can implement this by taking:

$$\frac{dA_{active}}{dt} = -k_1 Asp \cdot A_{active} + k_{-1}A_{bound} - k_2B^* + k_{-2}, \tag{12.9}$$

$$\frac{dA_{reserve}}{dt} = k_2B^* - k_{-2}. \tag{12.10}$$

The response of this zero-order model is shown in **Figure 12.5**. The system always returns to exactly the same level of A_{active}, irrespective of the stimulus. The dynamics of the response do vary with the stimulus—the first response has the highest amplitude and the quickest overall time course—but the adaptation is always ultimately perfect.

We can see why the adaptation is perfect by deriving an expression for the steady-state level of A_{active} in the zero-order model defined by Eqs. 12.1, 12.5, 12.7, and 12.8. At steady state, all of the time derivatives must equal zero. This yields three independent algebraic equations:

$$0 = k_1 Asp \cdot A_{active} - k_{-1} A_{bound}, \tag{12.11}$$

$$0 = k_3 A_{active}(B_{tot} - B^*) - k_{-3} B^*, \tag{12.12}$$

$$0 = k_2 B^* - k_{-2}. \tag{12.13}$$

They can be solved simultaneously to derive the steady-state levels of the three time-dependent variables:

$$A_{active} = \frac{K_2 K_3}{K_2 - B_{tot}}, \tag{12.14}$$

$$A_{bound} = \frac{Asp}{K_1} \frac{K_2 K_3}{(K_2 - B_{tot})}, \tag{12.15}$$

$$B^* = K_2, \tag{12.16}$$

where $K_1 = k_{-1} / k_1$, $K_2 = k_{-2} / k_2$, and $K_3 = k_{-3} / k_3$. We can also calculate $A_{reserve}$ from the conservation relationship $A_{tot} = A_{active} + A_{bound} + A_{reserve}$.

Note that the steady-state level of A_{active}, which is the output of the system, as well as of B^*, is independent of the stimulus level of Asp. This explains why the adaptation is perfect. Moreover, the adaptation is perfect, albeit at different steady-state values of A_{active}, irrespective of the affinity of the receptor for aspartate (K_1), the zero-order rate constants (V_{max} values) for receptor methylation and demethylation (which figure into K_2), and the first-order rate constants for the mass action activation and inactivation of the demethylase CheB (which figure into K_3). Thus the adaptation is not only perfect, but the perfection is robust with respect to perturbations in the activities and affinities of all of the proteins that make up the circuit, as is true of adaptation in the actual bacterium.

In summary, the observed adaptation in bacterial chemotaxis can be accounted for by a negative feedback model, and we can make the adaptation perfect and robust by assuming that the reactions of the negative feedback loop are zero-order. The zero-order kinetics postulated in the Barkai model have not been looked for experimentally, but if the methylation and demethylation enzymes really are running very close to saturation it could be far and away the most dramatic example of zero-order ultrasensitivity to be found since Goldbeter and Koshland proposed the idea in the early 1980s. Note though that any of the other mechanisms for ultrasensitivity outlined in Chapters 4 and 5 could possibly work in the place of zero-order ultrasensitivity—for example, some multistep process (not implausible given that there are multiple methylation sites on each receptor, and each receptor functions as part of a multimeric complex) or some as yet unidentified stoichiometric inhibitor.

12.4 THE RESPONSE OF THE ERK MAP KINASES TO MITOGENIC SIGNALS IS TYPICALLY TRANSITORY

Here we will explore a second example of adaptation, the transitory response of the Ras/MAP kinase system to a prolonged dose of a mitogen. The transitory nature of the pathway's activation is biologically important—mutations in the pathway that result in abnormally prolonged MAPK activation, such as the Val 12-KRas point mutation, contribute to malignancy in a substantial fraction of human cancers. As it turns out, we can formulate a satisfactory model of the process that is a bit simpler and easier to analyze than the Barkai model of bacterial chemotaxis.

If a dish of tissue culture cells is treated with serum or EGF and the downstream activity of the main relevant MAP kinase, ERK2, is assessed by immunoblotting or kinase assays, typically one sees the phosphorylation and activity rise over the course of a few minutes and then fall over the course of an hour or so (**Figure 12.6a**). Note that this is qualitatively different from the irreversible, all-or-none ERK2 responses seen in *Xenopus* oocyte maturation (Chapter 8), probably because different feedback loops are present in the two situations. In some cell lines, this population-level ERK2 response hides interesting single-cell behaviors like excitability, and we will see an example of this in Chapter 16. But sometimes the individual cells behave much the way the population does (**Figure 12.6b**). The transitory response is not due to degradation of the mitogenic signal; the cells turn their responses off.

Activated ERK2 translocates to the nucleus, and transcription factors are prominent among the targets of active ERK2 (recall **Figure 1.3**), raising the possibility that ERK2-stimulated transcription might contribute to the termination of ERK2 signaling. One of the early clues

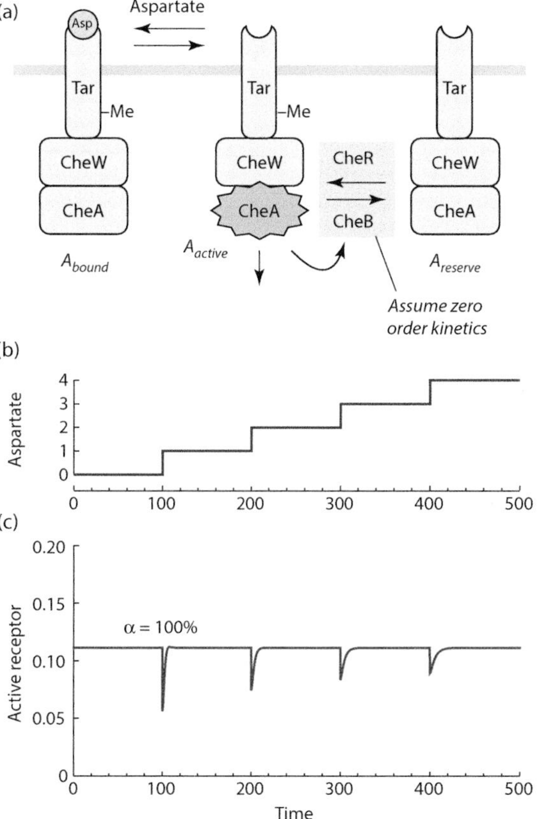

Figure 12.5 Perfect adaptation in a model of chemotaxis that assumes zero-order kinetics in the negative feedback. (a) Schematic of the adaptation circuit. (b) Step changes in input (aspartate). (c) Changes in receptor activity. As in **Figure 12.4**, the parameters for the model (**Eqs.** 12.1, 12.5, 12.9 and 12.10) were $k_1 = k_{-1} = 100$; $k_2 = 1$; $k_3 = 0.1$; $k_4 = k_{-4} = 1$; and $A_{tot} = 1$; $B_{tot} = 1$. Note that although the responses vary in height and duration, they always return exactly to the baseline.

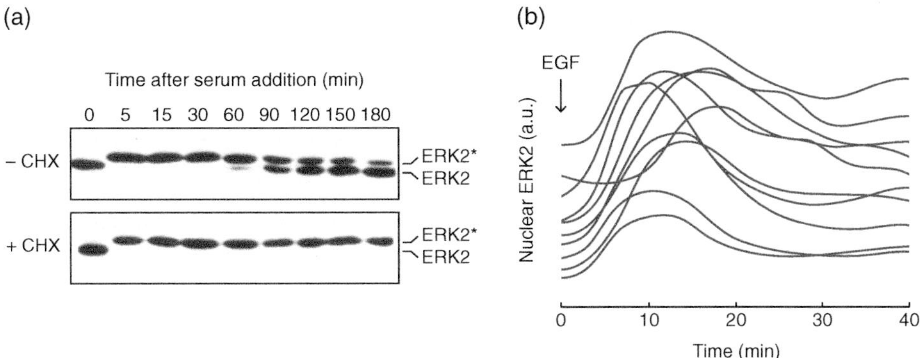

Figure 12.6 The time course of ERK2 activation and inactivation mammalian cells. (a) Bulk data from immunoblotting of ERK2* phosphorylation in NIH 3T3 cells, a mouse fibroblast cell line. ERK2* denotes the bis-phosphorylated, active form of ERK2, and CHX stands for cycloheximide. The mitogen was serum. (Based on data in Sun et al., Cell. 1993 and used with permission) (b) Single-cell dynamics from live cell imaging experiments in H1299 cells, a human lung cancer cell line. Here the mitogen was EGF, and the translocation of fluorescently tagged ERK2 to the nucleus, which accompanies ERK2 activation, was assessed. Tracings from ten individual cells are shown. (Adapted from Cohen-Saidon et al., *Mol Cell*. 2009 and used with permission.)

that this is the case came from comparing responses to mitogenic signals in NIH 3T3 cells, a commonly studied mouse cell line, in the presence and absence of the protein synthesis inhibitor cycloheximide. In the absence of cycloheximide, cells responded to mitogens with a pulse of ERK2 activation, as expected, but in the presence of cycloheximide, the ERK2 activation went up and remained high (**Figure 12.6a**). This suggests that perhaps some protein whose mRNA is induced by mitogens is a negative regulator of ERK2 and that blocking its synthesis interfered with the normal deactivation of ERK2. Sure enough, one of the proteins upregulated by mitogens turned out to be a dual-specificity phosphatase that can dephosphorylate the activating threonine and tyrosine phosphorylations on the ERK2 proteins. This phosphatase, usually called MKP1 (for MAP kinase phosphatase 1) or DUSP1 (for dual-specificity phosphatase-1), is thought to play an important role terminating receptor tyrosine kinase signaling in NIH 3T3 cells.

Here we will see how well a simple model of MKP1 induction can account for the kinetics of ERK2 activation and inactivation shown in **Figure 12.6**.

12.5 DELAYED NEGATIVE FEEDBACK CAN YIELD NEAR-PERFECT ADAPTATION

We begin by considering the activation and inactivation of ERK2. ERK2 activation involves the phosphorylation of two sites, a threonine and a tyrosine residue, by the kinase MEK. But for simplicity we will assume that activation is a one-step, mass action process, and that there are only two ERK2 species. Likewise, we will assume that the inactivation of ERK2 is a one-step, mass action process catalyzed by MKP1. If we let x denote inactive ERK2 and x^* active ERK2, and y denote MKP1, then the rate equation for x^* would be:

$$\frac{dx^*}{dt} = k_1 MEK \cdot (1 - x^*) - k_{-1} y \cdot x^*. \tag{12.17}$$

Note we have chosen units for x such that the total $x+x^*$ equals 1.

To write a rate equation for the phosphatase y, we assume that its synthesis rate is proportional to x^*, and for the moment assume its degradation is described by mass action kinetics:

$$\frac{dy}{dt} = k_2 x^* - k_{-2} y. \tag{12.18}$$

Equations 12.17 and 12.18 constitute our two-ODE model for the negative feedback regulation of ERK2 by MKP1 production.

We want to examine how the concentrations of x^* and y evolve with time for some choice of initial conditions—say, $x^*[0]=0$ and $y[0]=0$—and parameters (the four rate constants and the assumed concentration of active MEK). We turn to numerical solution of the system of two ODEs, plot the results, and adjust the parameter values until we obtain a reasonable-looking pulse of x^*. One such result is shown in **Figure 12.7**. In response to 1 unit of MEK, the ERK2 activity (x^*) rises quickly to a peak of about 0.38, and then slowly drops, approaching a steady-state value of about 0.1. The percent adaptation in this case works out to be 75%—not perfect adaptation, but pretty good. The next increments of MEK, though, result in poorer adaptation. We can do better.

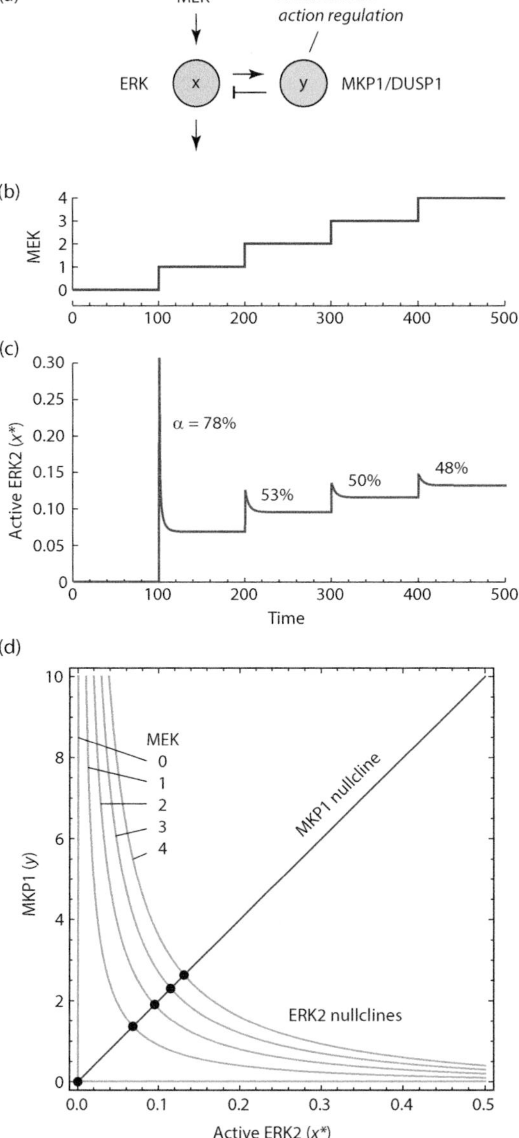

Figure 12.7 Partial adaptation in a model of the ERK2/MKP1 system that assumes mass action kinetics in the negative feedback. (a) Schematic of the adaptation circuit. b) Step changes in input (MEK). (c) Resulting changes in ERK2 activity. (d) Nullclines in the phase plane. The closed circles denote the steady states. The parameters chosen were $k_1=1$; $k_{-1}=10$; $k_2=2$; and $k_{-2}=0.1$, and MEK was stepped from 0 to 4.

12.6 ULTRASENSITIVITY IN THE FEEDBACK LOOP IMPROVES THE SYSTEM'S ADAPTATION

Because the ERK2/MKP1 model contains only two time-dependent variables (x^* and y), it is a candidate for phase plane analysis, which will let us see why the adaptation is reasonably good but not perfect, and get an idea of how to make it perfect. In a system with perfect adaptation, the x^* coordinates of the steady states should all be identical, irrespective of the amount of stimulus (MEK). So we start by plotting the nullclines in the phase plane and examine the steady states.

The nullclines are obtained, as usual, by setting the time derivatives in the rate equations (Eqs. 12.17 and 12.18) equal to zero. The x^* nullcline is defined by:

$$k_1 MEK \cdot (1 - x^*) - k_{-1} y \cdot x^* = 0 \tag{12.19}$$

$$y = \frac{k_1}{k_{-1}} MEK \frac{1 - x^*}{x^*}. \tag{12.20}$$

The y nullcline is:

$$k_2 x^* - k_{-2} y = 0. \tag{12.21}$$

$$y = \frac{k_2}{k_{-2}} x^*. \tag{12.22}$$

These nullclines are plotted in **Figure 12.7a**, using the same kinetic parameters that were (fairly arbitrarily) chosen for **Figure 12.6**. The x^* nullclines (in green) correspond to five different values of MEK (0, 1, 2, 3, and 4), and the single y nullcline is shown in blue. The plot does not show us anything about the initial response of x^* to MEK—the upward portion of the time course—but it does show where the system ultimately settles: the places where the x^* and y nullclines intersect. And the steady-state values of x^* increase with MEK, which is why the adaptation is not perfect.

There are two easy ways to make the blue y nullcline vertical or near vertical and thus make the adaptation perfect or near perfect. The first would be to make k_{-2} small. The result would be near perfect adaptation to an increase in MEK at the cost of an extremely slow decay in MKP1 concentration after MEK was turned off.

The second way would be to saturate the reaction that degrades y. Recall from Section 6.3 that saturating a degradation reaction results in a steady-state response curve that approaches vertical as the synthesis rate approaches the maximal degradation rate. If the *EC50* for the saturable degradation reaction is very small, the whole curve approaches a vertical line—that is, the rate of degradation is independent of the concentration of y. Note that this is essentially the same trick as was used to make the Barkai model of chemotaxis adapt perfectly—zero-order kinetics in the negative feedback loop.

Figure 12.8 shows the dynamics of the ERK2/MKP1 model assuming zero-order degradation of MKP1. The model now adapts perfectly, with the steady-state level of ERK2 activation always returning to

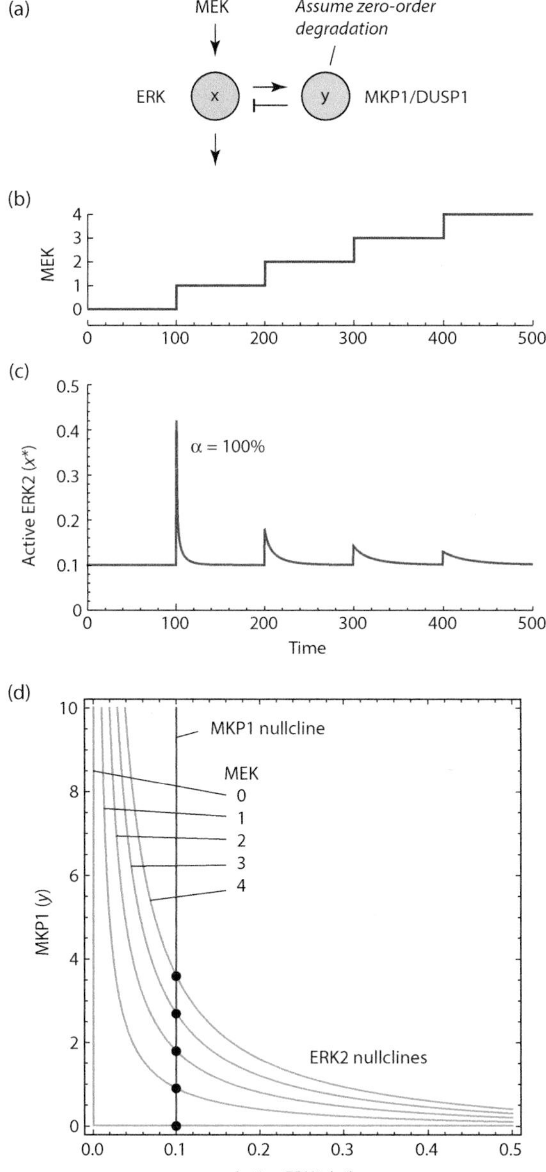

Figure 12.8 Perfect adaptation in a model of the ERK2/MKP1 system that assumes zero-order kinetics in the degradation of MKP1. (a) Schematic of the adaptation circuit. (b) Step changes in input (MEK). (c) Resulting changes in ERK2 activity. (d) Nullclines in the phase plane. The closed circles denote the steady states. The parameters used in **Figure 12.7** were also used here: $k_1=1$; $k_{-1}=10$; $k_2=2$; and $k_{-2}=0.1$, and MEK was stepped from 0 to 4.

$x^*=k_{-2}/k_2$, which equals 0.1 for the parameters chosen here. Any other variation on the negative feedback that produces a vertical or near-vertical y nullcline would yield perfect or near-perfect adaptation. For example, a negative feedback mediated by highly cooperative multisite phosphorylation, or by a high-affinity stoichiometric inhibitor that was only effective once its concentration exceeded that of a higher-affinity binding protein, would work. Other things being equal, adding ultrasensitivity, from any source, to the negative feedback improves the quality of the adaptation, just as it did in the Barkai model of chemotaxis.

12.7 INDUCTION OF IMMEDIATE-EARLY GENE PRODUCTS IS NOT REQUIRED FOR ERK INACTIVATION IN MANY CELL TYPES

So does the ERK2/MKP1 negative feedback loop account for the termination of EGFR signals? It may in NIH 3T3 cells, but in many other cell types, it does not. For example, in PC12 cells (a rat pheochromocytoma cell line), 3T3-L1 cells (mouse fibroblasts that can be induced to

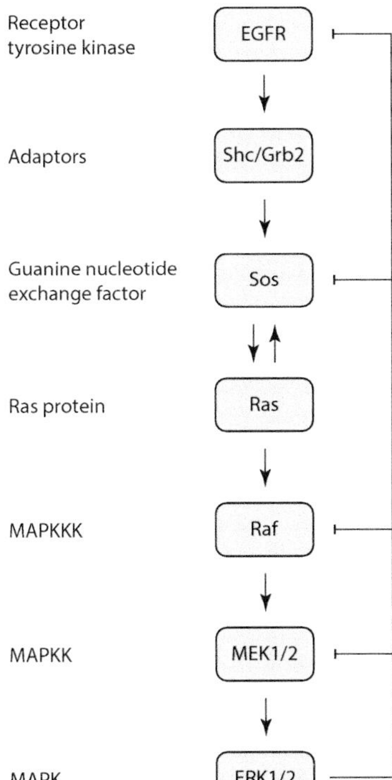

Figure 12.9 Additional negative feedback loops in MAPK signaling.

transdifferentiate into fat cells), and porcine aortic endothelial (PAE) cells, one can block the translation of all immediate-early response genes without significantly affecting the pulsatile nature of the ERK response.

There are other possible negative feedback loops that do not involve transcription or translation that could be responsible for the adaptation in these cases. For example, active ERK can phosphorylate and inactivate the upstream Sos protein, and there is evidence for negative feedback to the EGFR, Raf, and MEK as well (**Figure 12.9**). But there is another possibility as well—that adaptation is carried out not through negative feedback but through incoherent feedforward regulation. This is generally considered to be a different class of signaling motif, and we will explore this mechanism for adaptation in Chapter 12.

SUMMARY

Here we have explored two examples of adaptation: the adaptation of *E. coli* to the presence of chemoattractants, which allows the organism to search out food sources through a biased random walk, and the adaptation of mammalian cell lines to the presence of a mitogenic activator of the MAP kinase cascade. In both cases, models of the process can be made to adapt by including a negative feedback loop, and in both cases, saturating one or more regulatory reactions in the negative feedback loop allows the adaptation to be perfect. So would any other mechanism that makes the negative feedback mediator be produced or activated with very high ultrasensitivity. Negative feedback loops do not always yield adaptation—we saw that in Chapter 11, and we will see it again in Chapter 15—and they are not the only way of generating a transient response from a sustained stimulus. But negative feedback is so widespread that it probably does qualify as one of the most important mechanisms for adaptation in cell signaling.

FURTHER READING

E. COLI CHEMOTAXIS

Alon U, Surette MG, Barkai N, Leibler S. Robustness in bacterial chemotaxis. *Nature*. 1999 Jan 14;397(6715):168–71.

Barkai N, Leibler S. Robustness in simple biochemical networks. *Nature*. 1997 Jun 26;387(6636):913–17.

Berg HC. Bacterial microprocessing. *Cold Spring Harb Symp Quant Biol*. 1990;55:539–45.

Berg HC, Brown DA. Chemotaxis in Escherichia coli analysed by three-dimensional tracking. *Nature*. 1972 Oct 27;239(5374):500–504.

Endres RG. *Physical Principles in Sensing and Signaling: With an Introduction to Modeling in Biology*. Oxford University Press, Oxford, 2013.

Tu Y. Quantitative modeling of bacterial chemotaxis: signal amplification and accurate adaptation. Annu Rev Biophys. 2013;42:337–59.

ERK2 INACTIVATION AND MAP KINASE PHOSPHATASES

Alessi DR, Gomez N, Moorhead G, Lewis T, Keyse SM, Cohen P. Inactivation of p42 MAP kinase by protein phosphatase 2A and a protein tyrosine phosphatase, but not CL100, in various cell lines. *Curr Biol*. 1995 Mar 1;5(3):283–95.

Caunt CJ, Keyse SM. Dual-specificity MAP kinase phosphatases (MKPs): shaping the outcome of MAP kinase signalling. *FEBS J*. 2013 Jan;280(2):489–504.

Cohen-Saidon C, Cohen AA, Sigal A, Liron Y, Alon U. Dynamics and variability of ERK2 response to EGF in individual living cells. *Mol Cell*. 2009 Dec 11;36(5):885–93.

Sun H, Charles CH, Lau LF, Tonks NK. MKP-1 (3CH134), an immediate early gene product, is a dual specificity phosphatase that dephosphorylates MAP kinase in vivo. *Cell*. 1993 Nov 5;75(3):487–93.

NEGATIVE FEEDBACK LOOPS IN MAP KINASE SIGNALING

Lake D, Corrêa SA, Müller J. Negative feedback regulation of the ERK1/2 MAPK pathway. *Cell Mol Life Sci*. 2016 Dec;73(23):4397–413.

ADAPTATION 2

13

Incoherent Feedforward Regulation and State-Dependent Inactivation

IN THIS CHAPTER . . .

DOI: 10.1201/9781003124269-13

INTRODUCTION

Negative feedback is not the way of producing adaptation. There are at least two other simple adaptation mechanisms—incoherent feedforward regulation and **state-dependent inactivation**—and in this chapter we will work through biological examples of each.

13.1 RECEPTOR TYROSINE KINASE ACTIVATION IS FOLLOWED BY TRANSITORY RAS ACTIVATION

As mentioned in Chapter 1, an incoherent feedforward system is one where an upstream regulator regulates a downstream target in two different and opposite ways, with a time lag between the two. As an example, we turn again to receptor tyrosine kinase signaling, this time focusing on the regulation of Ras.

Receptor tyrosine kinases relay signals onward by recruiting specific proteins to autophosphorylated docking sites on the receptor. Curiously, two of the proteins recruited to various receptor tyrosine kinases, including the epidermal growth factor receptor (EGFR), are enzymes that have opposite effects on the downstream target Ras. The first is the guanine nucleotide exchange factor Sos, which translocates from the cytosol to the activated EGFR through the intermediacy of the Shc and/or Grb2 adaptor proteins. Since EGFRs are transmembrane proteins, this puts Sos in proximity of Ras, which is also membrane associated. The receptor-associated Sos then causes inactive GDP-bound Ras to drop its GDP, which allows it to pick up a GTP and flip into its activated conformation (**Figure 13.1**).

Note that Ras–GTP is not only activated by Sos but is also a stoichiometric activator of Sos (Figure 13.1), so there is positive feedback in the system. We will return to this positive feedback loop in Chapter 16, but for now we will ignore it and focus on how Ras activation is turned off.

The second protein to be recruited to the autophosphorylated EGFR is the GTPase-activating protein p120 GAP, which causes Ras to hydrolyze its bound GTP to GDP. This hydrolysis allows Ras to flip back to its inactive conformation (**Figure 13.1**). The actions of Sos and GAP on Ras constitute a cycle of activation and inactivation that is roughly analogous to a phosphorylation–dephosphorylation cycle, except that the "marks" that distinguish active from inactive Ras are non-covalently bound GTP vs. GDP molecules rather than covalently bound phosphate vs. hydroxyl groups on amino acid side chains.

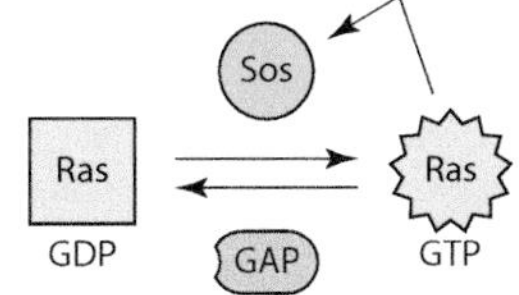

Figure 13.1 Schematic view of the activation of Ras by Sos and the inactivation of Ras by GAP. Note there is also positive feedback—active Ras stoichiometrically binds to and increases the activity of Sos.

Both the Sos and GAP proteins are recruited to the EGFR in EGF-treated cells fairly quickly, with both peaking about 1 min after addition of EGF for the experiment shown in **Figure 13.2a**. But there is a time lag of 30 s or so between when Sos begins to appear at the plasma membrane and when GAP does. This fits well with the time course of Ras activation: Ras–GTP binding rises for the first 30 sec and then falls over the next few minutes (**Figure 13.2a**, bottom). This sequential recruitment of antagonistic enzymes is a conceptually simple mechanism for generating a pulsatile response, and, as shown below, it has the potential to yield perfect adaptation.

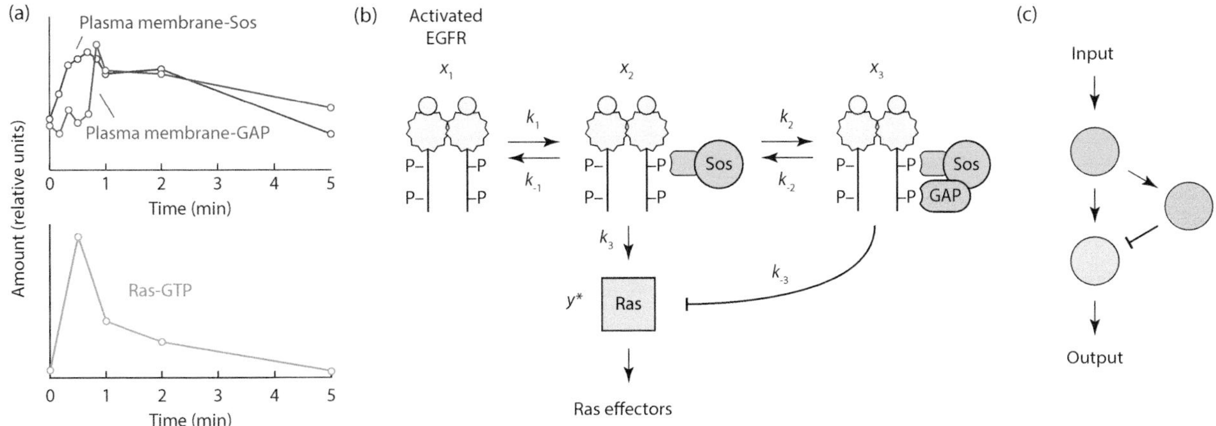

Figure 13.2 Activation of the EGFR results in a short pulse of Ras–GTP production. (a) Experimental data showing the recruitment of Sos (blue) and GAP (red) to the plasma membrane (top panel) and the resulting activation of Ras (green, bottom panel) in rat pheochromocytosis cells (PC12 cells) treated with EGF (adapted from Sasagawa et al., *Nat Cell Biol.* 2005 and used with permission). (b) Schematic view of the sequential binding of Sos and GAP to activated EGF receptors. (c) Even more schematic view of the logic of this incoherent feedforward circuit.

13.2 THE SEQUENTIAL RECRUITMENT OF SOS AND GAP TO THE EGFR CAN BE VIEWED AS INCOHERENT FEEDFORWARD REGULATION

Here we will formulate a simple mass action model of the sequential activation and inactivation of Ras. We begin with the activated and autophosphorylated EGFR, and we assume it can exist in three forms: with no recruited proteins, which we designate x_1; with the Sos protein recruited, which we call x_2; and with both Sos and GAP recruited, designated x_3 (**Figure 13.2b**). By assuming that GAP only binds to receptors that have already bound Sos, we ensure that the two proteins will be recruited sequentially and also guarantee that there will be a time lag, as seen in **Figure 13.2a**, between the recruitment of Sos and the recruitment of GAP. This system can be viewed as carrying out incoherent feedforward regulation (**Figure 13.2c**); the activation of the EGFR not only turns Ras on but also brings about its inactivation.

We can write down mass action rate equations for the three forms of the EGFR as follows:

$$\frac{dx_1}{dt} = -k_1 x_1 Sos + k_{-1} x_2 \tag{13.1}$$

$$\frac{dx_2}{dt} = k_1 x_1 Sos - k_{-1} x_2 - k_2 x_2 GAP + k_{-2} x_3 \tag{13.2}$$

$$\frac{dx_3}{dt} = k_2 x_2 GAP - k_{-2} x_3. \tag{13.3}$$

Note that these equations put only two constraints on the three time-dependent variables, since Eq. 13.2 is a combination of the other two. We obtain a third constraint from the conservation relationship:

$$x_{tot} = x_1 + x_2 + x_3. \tag{13.4}$$

For simplicity, we will assume that the concentrations of free Sos and GAP are sufficiently high relative to the total concentration of activated EGFRs (x_{tot}) that they are not changed appreciably when the proteins dock. We can then simply lump the approximately constant concentrations of free Sos and GAP into the rate constants.

Using Eq. 13.4 to eliminate one variable (x_1), we reduce the model to:

$$\frac{dx_2}{dt} = k_1(x_{tot} - x_2 - x_3) - k_{-1}x_2 - k_2x_2 + k_{-2}x_3 \quad (13.5)$$

$$\frac{dx_3}{dt} = k_2x_2 - k_{-2}x_3. \quad (13.6)$$

Once these two variables are solved for, the conservation equation (Eq. 13.4) can be used to calculate the third (x_1).

Finally, we can write a rate equation for the activation of Ras (y) by EGFR–Sos (x_2) and the inactivation of Ras by EGFR–Sos–GAP (x_3):

$$\frac{dy^*}{dt} = k_3x_2(y_{tot} - y^*) - k_{-3}x_3 \cdot y^*, \quad (13.7)$$

where y^* denotes active Ras, y denotes inactive Ras, and we have used the conservation relationship $y_{tot} = y + y^*$ to eliminate y from Eq. 13.7. Equations 13.5–13.7 constitute our three-ODE model of Ras regulation.

13.3 INCOHERENT FEEDFORWARD SYSTEMS CAN YIELD PERFECT ADAPTATION

We can solve these equations numerically for some choice of rate constants and initial conditions. The results are shown in **Figure 13.3**.

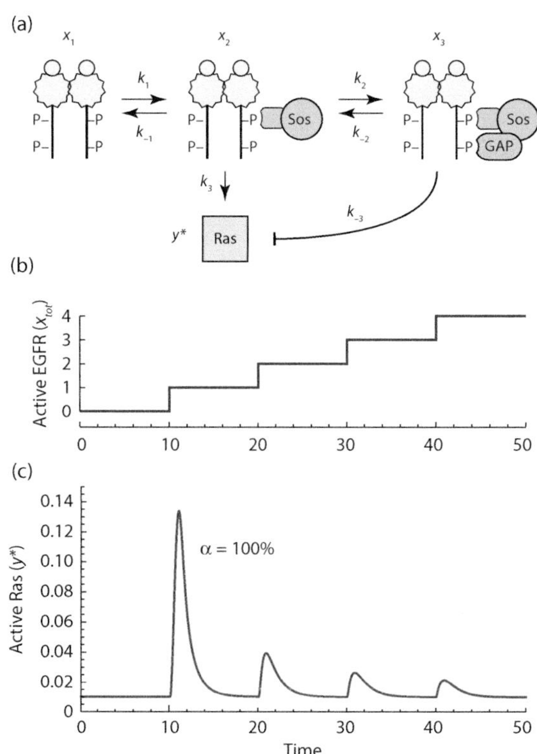

Figure 13.3 Perfect adaptation in Ras regulation. (a) Schematic view of the model, shown again (cf. **Figure 13.2**) for reference here. (b) Steps in the concentration of active EGFR (x_{tot}). (c) The resulting dynamics of Ras activation. Although the pulses of activity differ in height and shape, the adaptation is always perfect ($\alpha = 100\%$). We have assumed $k_1 = k_2 = k_3 = 1$; $k_{-1} = k_{-2} = 0.1$; $k_{-3} = 10$; and $y_{tot} = 1$.

Each successive step in the concentration of active EGFR (x_{tot}) produces a pulse of Ras activity, and the system returns to the same baseline level of Ras activity irrespective of the value of x_{tot}—perfect adaptation.

To see why the system adapts perfectly, we return to the rate equations. At steady state, all three derivatives must equal zero, which means:

$$0 = k_1(x_{tot} - x_2 - x_3) - k_{-1}x_2 - k_2x_2 + k_{-2}x_3 \quad (13.8)$$

$$0 = k_2x_2 - k_{-2}x_3 \quad (13.9)$$

$$0 = k_3x_2(y_{tot} - y^*) - k_{-3}x_3 \cdot y^*. \quad (13.10)$$

We can solve these three equations simultaneously to yield expression for x_2, x_3, and y^* at steady state:

$$(x_2)_{ss} = \frac{K_2x_{tot}}{1 + K_2 + K_1K_2} \quad (13.11)$$

$$(x_3)_{ss} = \frac{x_{tot}}{1 + K_2 + K_1K_2} \quad (13.12)$$

$$(y^*)_{ss} = \frac{K_2y_{tot}}{K_2 + K_3}. \quad (13.13)$$

An expression for the steady-state level of x_1 follows quickly from Eqs. 13.11, 13.12, and the conservation relationship (Eq. 13.4):

$$(x_1)_{ss} = \frac{x_{tot}K_1K_2}{1 + K_2 + K_1K_2}. \quad (13.14)$$

As usual, we have taken $K_1=k_{-1}/k_1$, $K_2=k_{-2}/k_2$, and $K_3=k_{-3}/k_3$.

Equation 13.13 shows that the steady-state level of y^* is independent of x_{tot}. Thus, the system always returns to the same level of y^*—the adaptation is perfect. The steady-state values of the two antagonistic species x_2 and x_3 (Eqs. 13.11 and 13.12) do depend on x_{tot}, but they scale the same way, with both being directly proportional to x_{tot}.

13.4 STRICT ORDERING OF SOS AND GAP BINDING TO THE EGFR IS NOT REQUIRED FOR PERFECT ADAPTATION

So far we have assumed that GAP only binds EGFRs that have already recruited an Sos molecule. This assumption ensures that the two opposing factors will be recruited sequentially, and it also simplifies the model by making it so that we need not consider the formation or dissociation of EGFR–GAP complexes. What if we relax the assumption, making GAP binding slower than Sos binding but not dependent on it?

We now have four distinct complexes of the activated EGFR, as shown in **Figure 13.4a**. We have kept the same numbering scheme as before for EGFR (x_1), EGFR–Sos (x_2), and EGFR–Sos–GAP (x_3), and have added a species (x_4) for EGFR–GAP. The rate constants k_4 and k_{-4} describe the binding and dissociation of GAP to EGFR, and k_5 and k_{-5} the binding and dissociation of Sos to EGFR–GAP. We can write rate equations for the four EGFR species:

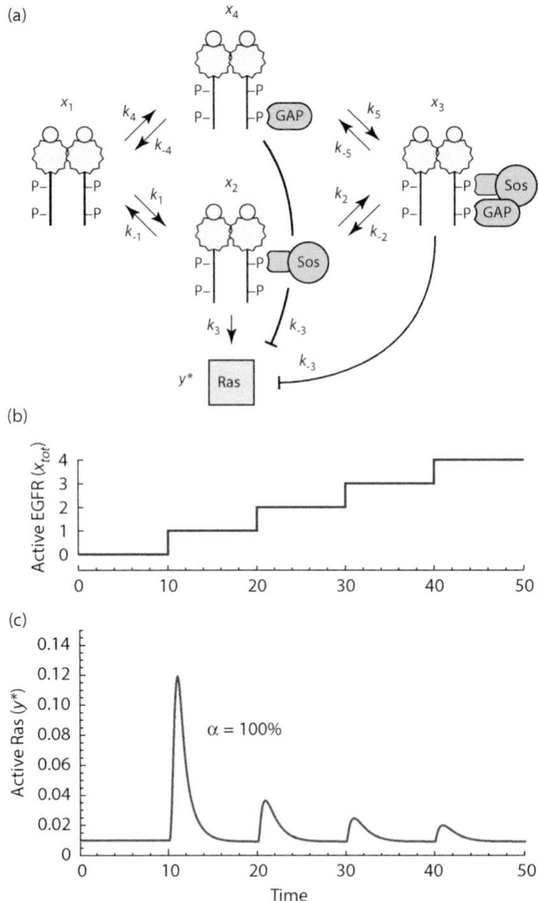

Figure 13.4 Perfect adaptation in Ras regulation in the absence of strict ordering in the recruitment of Sos and GAP to the active EGFR. (a) Schematic view of the model. (b) Steps in the pulses of activity differ in height and shape, the adaptation is always perfect ($\alpha = 100\%$). We have assumed $k_1=k_2=k_3=k_5=1$; $k_4=0.1$; $k_{-1}=k_{-2}=k_{-4}=k_{-5}=0.1$; $k_{-3}=10$; and $y_{tot}=1$. Since we have assumed k_4 is 10× smaller than k_1, EGFR binding to GAP is slower than to Sos, but both binding reactions do occur.

$$\frac{dx_1}{dt} = -k_1x_1 + k_{-1}x_2 - k_4x_1 + k_{-4}x_4 \tag{13.15}$$

$$\frac{dx_2}{dt} = k_1x_1 - k_{-1}x_2 - k_2x_2 + k_{-2}x_3 \tag{13.16}$$

$$\frac{dx_3}{dt} = k_2x_2 - k_{-2}x_3 \tag{13.17}$$

$$\frac{dx_4}{dt} = k_4x_1 - k_{-4}x_4 - k_5x_4 + k_{-5}x_3. \tag{13.18}$$

As before, we have assumed that the concentrations of Sos and GAP are approximately constant and have lumped these quantities into the rate constants. We can then use the conservation law $x_{tot} = x_1 + x_2 + x_3 + x_4$ to eliminate x_1 from Eqs. 13.16 and 13.18 and reduce the system to three ODEs:

$$\frac{dx_2}{dt} = k_1(x_{tot} - x_2 - x_3 - x_4) - k_{-1}x_2 - k_2x_2 + k_{-2}x_3 \tag{13.19}$$

$$\frac{dx_3}{dt} = k_2x_2 - k_{-2}x_3 \tag{13.20}$$

$$\frac{dx_4}{dt} = k_4\left(x_{tot} - x_2 - x_3 - x_4\right) - k_{-4}x_4 - k_5x_4 + k_{-5}x_3. \tag{13.21}$$

Finally, we write a rate equation for Ras activation and inactivation. For simplicity we will assume that EGFR–GAP (x_4) and EGFR–Sos–GAP (x_3) are equally active in catalyzing the conversion of Ras–GTP to Ras–GDP. This yields:

$$\frac{dy^*}{dt} = k_3x_2\left(y_{tot} - y^*\right) - k_{-3}\left(x_3 + x_4\right)y^*. \tag{13.22}$$

Equations 13.18–13.22 constitute a 4 ODE model of Ras activation and inactivation. We can solve this numerically; as a first go, we assume the same rate constants as we did for the strictly ordered model, and that the binding of EGFR to GAP is 10× slower than the binding to Sos. The results are shown in **Figure 13.4c**. The system still responds to each step in x_{tot} and appears to adapt perfectly, irrespective of the level of stimulus (x_{tot}).

To prove that this is in fact the case, we set each of the rate equations equal to zero—the requirement for the system to be in steady state—and derive algebraic equations for each of the species at steady state. The equation for y^* at steady state is more complicated than it was in the strictly sequential model:

$$\left(y^*\right)_{ss} = \frac{k_1k_3k_{-2}\left(k_5 + k_{-4}\right)y_{tot}}{k_4k_{-1}k_{-2}k_{-3} + k_1k_3k_{-2}\left(k_5 + k_{-4}\right) + k_1k_2k_{-3}\left(k_5 + k_{-4} + k_{-5}\right)}, \tag{13.23}$$

but, again, the steady-state level of Ras activation does not depend on the stimulus, x_{tot}. This is why the adaptation is perfect.

On the other hand, the assumption that the concentrations of Sos and GAP are much higher than the concentrations of the EGFR–Sos and EGFR–Sos–GAP complexes does turn out to be critical for perfect adaptation. Thus, the wiring of the circuit is not sufficient to guarantee perfect adaptation; the parameters of the system, in this case the relative concentrations of the proteins, are important too. Note that the same was true of the Barkai model for bacterial chemotaxis analyzed in Chapter 11, where a simple negative feedback circuit yielded perfect adaptation only if it was assumed that the receptor concentration was high enough that the methylase and demethylase that acted upon it were operating in their zero-order regimes.

13.5 THE VOLTAGE-SENSITIVE SODIUM CHANNEL ALSO UNDERGOES SEQUENTIAL ACTIVATION AND INACTIVATION

The most famous pulse generator in biology is probably the voltage-sensitive sodium channel, the protein at the heart of the action potential in nerves and heart muscle cells. In response to a depolarization of the plasma membrane (where depolarization means that the cytoplasmic side of the plasma membrane becomes less negatively charged relative to the extracellular side), the voltage-sensitive sodium channel undergoes a conformation change that allows millions of sodium ions to flow through it into the cell (**Figure 13.5a**). But then after a few milliseconds, the protein undergoes a second conformation change that plugs the sodium pore (**Figure 13.5a**), allowing the Na^+–K^+ ATPase pump to restore the intracellular Na^+ to its normal low concentration and the membrane potential to its initial negative

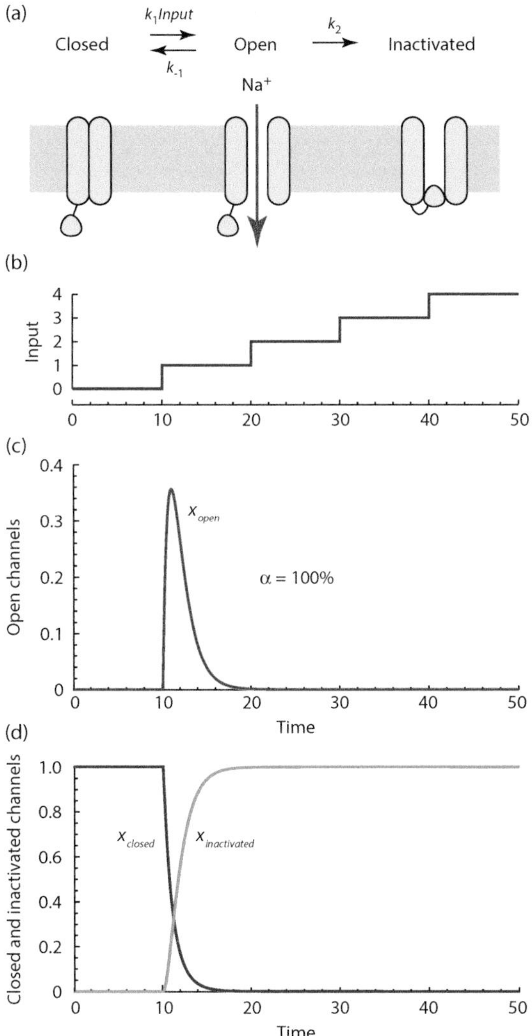

Figure 13.5 The voltage-dependent sodium channel. (a) Schematic view of the conversion of a closed channel to an open channel in response to membrane depolarization, followed by channel inactivation. The inactivated channel then slowly returns to the closed state; here we assume that the time scale of this return is too long to be relevant to our modeling. In addition, we are ignoring one of the hallmarks of the channel—the fact that the influx of sodium through an open channel further depolarizes the membrane and causes other closed channels to open—in the interest of focusing on the recovery (adaptation) phase of the response. (b–d) The fraction of the channels that are closed (d), open (c), and inactivated (d) in response to steps of depolarization (b). Note the toilet flush phenomenon. We have assumed $k_1 = k_2 = 1$ and $k_{-1} = 0.1$.

value. After this repolarization, the inactivated receptor slowly returns to the initial closed state. Only then is the receptor able to be opened again in response to a depolarization.

Just as there was in the case of Ras activation, there is a positive feedback loop built into the system: when sodium rushes in through the open channel, the membrane becomes further depolarized, leading to activation of more channels. This provides the action potential with its bursty, all-or-none (or almost all-or-none) character. But there is no obvious negative feedback loop, as there was in the bacterial chemotaxis system, and no clear feedforward regulation à la EGFR–Sos–GAP (at least not according to the usual definitions). Thus the central motifs we have previously seen in adapting circuits are not present here.

To model the system, let us call the three states of the sodium channel x_{closed}, x_{open}, and $x_{inactivated}$ (**Figure 13.5a**), and assume that the recycling of the channel from the inactivated state to the closed state is too slow to be relevant to our model. Let us also ignore the positive feedback; once again it is an interesting feature but is not necessary for adaptation. We write three rate equations for these three time-dependent species, assuming mass action kinetics for each of the interconversions:

$$\frac{dx_{closed}}{dt} = -k_1 Input \cdot x_{closed} + k_{-1}x_{open} \tag{13.24}$$

$$\frac{dx_{open}}{dt} = k_1 Input \cdot x_{closed} - k_{-1}x_{open} - k_2 x_{open} + k_{-2}x_{inactivated} \tag{13.25}$$

$$\frac{dx_{inactivated}}{dt} = k_2 x_{open} - k_{-2}x_{inactivated}. \tag{13.26}$$

In addition, we have the conservation equation $x_{tot} = x_{closed} + x_{open} + x_{inactivated}$. By setting the derivatives all equal to zero, we can solve for the steady-state levels of all three channel species, and for the open channel we get:

$$\left(x_{open}\right)_{ss} = \frac{k_1 \cdot k_{-2} Input \cdot x_{tot}}{k_1 k_2 Input + k_1 \cdot k_{-2} Input + k_{-1} k_{-2}}. \tag{13.27}$$

In general the steady-state value of x_{open} will depend on *Input* since it appears in the numerator and in some, but not all, of the terms in the denominator (Eq. 13.27). However, if either k_{-1} or k_{-2}, or both, are equal to zero, then:

$$\left(x_{open}\right)_{ss} = \frac{k_1 \cdot k_{-2} x_{tot}}{k_1 k_2 + k_1 \cdot k_{-2}}, \tag{13.28}$$

and we have perfect adaptation. As long as one or the other or both of the two steps that lead from the closed receptor to the inactivated receptor are irreversible, the steady-state activity of the receptor is unaffected by the input.

The response of the model to a series of step increases in the *Input* depolarization, calculated by numerical integration of the rate equations (Eqs. 13.24–13.26), is shown in **Figure 13.5**. We have assumed that the second step is irreversible ($k_{-2}=0$), which means that the steady-state level of x_{open} should be zero. In response to the first increment of *Input* there is a brisk increase in x_{open}, followed by a return toward $x_{open}=0$ (**Figure 13.5c**). But the response to the next increment is virtually imperceptible. This is because there are hardly any closed receptors left to be opened; essentially all of the receptors are in the inactivated state. Thus the system functions like a toilet. In response to a stimulus (pushing down on the flusher) there is a response (the toilet flushes), and then the toilet is refractory to a second stimulus until the toilet tank slowly refills. Many of us find it kind of nice that our nervous systems and our toilets operate by similar principles.

Circuits like these, that adapt without explicit negative feedback or incoherent feedforward regulation, have been dubbed state-dependent inactivation systems, and although they are not always foremost in the minds of students of adaptation, they appear to be commonplace in biology.

And what about the positive feedback that we have ignored here? It is not critical for a discussion of adaptation, but it is certainly an important aspect of neuronal signaling. We will return to neuronal signaling and add positive feedback back into the model in Chapter 15, when we explore the FitzHugh–Nagumo model, and in Chapter 16 when we examine excitable monostable systems.

13.6 EGFR INTERNALIZATION CAN BE VIEWED AS STATE-DEPENDENT INACTIVATION

A second example of state-dependent inactivation is provided by the internalization of the EGF receptor. This is shown schematically in **Figure 13.6a**. First the plasma membrane-bound receptor is activated as a result of ligand binding, and then the activated receptor becomes internalized. The internalized receptor is either degraded or recycled to the plasma membrane. This process is substantially slower than the incoherent feedforward regulation of Ras, occurring over tens of minutes; the downregulation of signaling in the EGFR/MAPK system appears to operate at multiple points in the pathway and over different time scales.

The essence of this process is sequential activation-inactivation, just as it was with the voltage-dependent sodium channel, although the time scales are different (tens of minutes vs. milliseconds) and the mechanisms of activation and inactivation are different as well (dimerization and trans-autophosphorylation followed by clathrin-dependent and clathrin-independent internalization vs. two conformation changes).

We can model this process by writing rate equations for the three EGFR species shown in in **Figure 13.6a**: the inactive receptor ($EGFR_{off}$), the

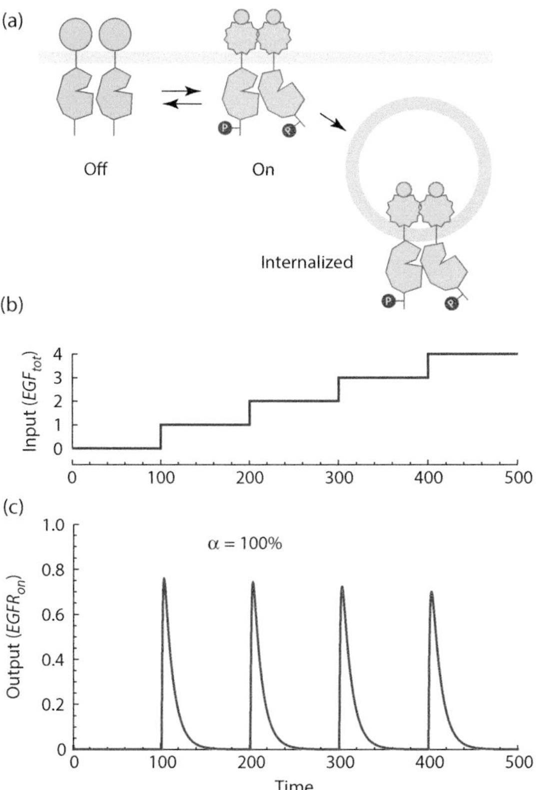

Figure 13.6 Perfect adaptation in the internalization of the EGF receptor. (a) Schematic of the receptor as a three-state system. We assume that the return of the internalized receptor to the plasma occurs on a longer time scale. (b, c) EGFR activity (c) in response to step changes in EGF (b).

EGF-bound active receptor ($EGFR_{on}$), and the internalized EGF–EGFR complex ($EGFR_{in}$):

$$\frac{dEGFR_{off}}{dt} = -k_1 EGF_{free} . EGFR_{off} + k_{-1} EGFR_{on} \quad (13.29)$$

$$\frac{dEGFR_{on}}{dt} = k_1 EGF_{free} \cdot EGFR_{off} - k_{-1} EGFR_{on} - k_2 EGFR_{on} \quad (13.30)$$

$$\frac{dEGFR_{in}}{dt} = k_2 EGFR_{on}. \quad (13.31)$$

Note that we are assuming that on the time scale of interest, the return of the internalized EGF–EGFR complex to the plasma membrane is insignificant. Next we write conservation equations for EGF and EGFR, bearing in mind that in vivo the concentration of EGF is not likely to greatly exceed that of the EGFR; in fact, as discussed in Chapter 4.4, proteomics data indicate that there are probably more EGFR than EGF molecules present. The two conservation equations are:

$$EGF_{tot} = EGF_{free} + EGFR_{on} + EGFR_{in} \quad (13.32)$$

$$EGFR_{tot} = EGFR_{off} + EGFR_{on} + EGFR_{in}. \quad (13.33)$$

We can use Eqs. 13.32 and 13.33 to reduce the model to two rate equations with two time-dependent variables, $EGFR_{on}$ and $EGFR_{in}$:

$$\frac{dEGFR_{on}}{dt} = k_1 (EGF_{tot} - EGFR_{on} - EGFR_{in}) \cdot (EGFR_{tot} - EGFR_{on} - EGFR_{in}) - k_{-1} EGFR_{on} - k_2 EGFR_{on} \quad (13.30)$$

$$\frac{dEGFR_{in}}{dt} = k_2 EGFR_{on}. \quad (13.31)$$

By setting the derivatives equal to zero we can see that at steady state, the output $EGFR_{on}=0$, irrespective of the input (EGF_{tot}). And by examining the time course in response to steps of EGF_{tot}, we see that the system responds with a pulse of output followed by perfect adaptation (**Figure 13.6**). Note that we did not get the same toilet flush effect that we did with the voltage-sensitive sodium channel model, because we assumed that both EGF and EGFR are taken out of play by internalization after the first increase in EGF_{tot}, some EGFR will be left on the cell surface to respond to the next increment.

13.7 GPCR SIGNALING IS SWITCHED FROM G-PROTEINS TO β-ARRESTIN VIA A MECHANISM AKIN TO STATE-DEPENDENT INACTIVATION

One final variation on state-dependent inactivation is provided by GPCR signaling, and we will take the well-studied β-adrenergic receptor as an example. As we mentioned in Chapter 2, the binding of an agonist ligand to the receptor shifts the receptor between a conformation where it is inactive as a guanine nucleotide exchange factor and one where it is active (**Figure 13.7**). The active receptor is a substrate for βARK and other protein kinases, and the kinases phosphorylate multiple sites in the receptor's C-terminus and cytoplasmic loops. The phosphorylations allow the β-arrestin protein to bind to the active receptor, and stoichiometrically prevent it from activating G-proteins. So far this is just like

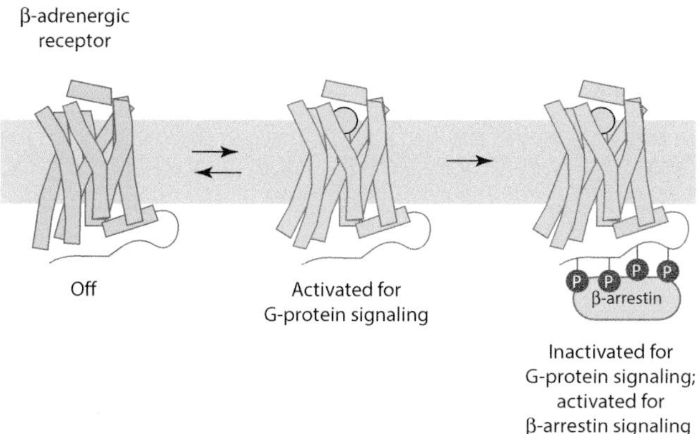

Figure 13.7 A G-protein-coupled receptor as a three-state system.

the conversion of the voltage-sensitive sodium channel from off to on to inactivated. Although the β-arrestin-bound receptor is unable to active G-proteins, it is able to activate a different pathway, the ERK MAPK cascade. Both pathways ultimately contribute to the receptor's output. Thus, this variation on state-dependent inactivation yields a pulse of G-protein activation followed, after a time lag, by ERK signaling.

The same is true of the μ-type opioid receptor, and in this case the two pathways contribute to different aspects of opioid pharmacology. Based on experiments in mice, G-protein signaling seems to be the main pathway for opioid-induced analgesia, whereas β-arrestin mediates respiratory depression and also appears to be important for making opioids addictive. Moreover, agonists have been identified that are biased toward activation of one pathway or the other. This raises the hope for selective pathway activators as analgesics with less potential than standard opioids for great harm.

SUMMARY

In Chapter 12 we showed how negative feedback can allow a cell signaling circuit to adapt to the sustained presence of a pathway input and we worked through examples where particular assumptions about the response functions in the pathway made the adaptation perfect. Here we have examined two other mechanisms that can yield perfect adaptation: incoherent feedforward regulation and state-dependent inactivation. All three of these mechanisms appear to be in common use in cell signaling. Indeed, EGFR receptor signaling appears to use all three—multiple negative feedback loops (e.g. ERK-induced MKP1/DUSP1 expression and negative regulation of various upstream proteins by ERK phosphorylation), incoherent feedforward regulation (e.g. the sequential recruitment of Sos and GAP to the activated receptor), and state-dependent inactivation (e.g. receptor internalization). This emphasizes that nature can use multiple signaling motifs to accomplish adaptation, even in a single pathway. And it probably uses these multiple mechanisms because turning off signaling is important.

Conversely, a single motif may sometimes be able to generate more than one systems-level behavior. This is well-illustrated by the various types of response that can come from negative feedback systems. Negative feedback can speed and stabilize a response without necessarily making the response pulsatile (Chapter 11), and it can give rise to adaptation (Chapter 12). Moreover, in the next chapter, we will see how negative feedback can destabilize a steady state and how that destabilization can result in a system that never settles into a steady state at all and instead exhibits sustained oscillations.

FURTHER READING

RAS REGULATION

Gaul U, Mardon G, Rubin GM. A putative Ras GTPase activating protein acts as a negative regulator of signaling by the Sevenless receptor tyrosine kinase. *Cell*. 1992 Mar 20;68(6):1007–19.

Jadwin JA, Oh D, Curran TG, Ogiue-Ikeda M, Jia L, White FM, Machida K, Yu J, Mayer BJ. Time-resolved multimodal analysis of Src Homology 2 (SH2) domain binding in signaling by receptor tyrosine kinases. *Elife*. 2016 Apr;12(5):e11835.

Sasagawa S, Ozaki Y, Fujita K, Kuroda S. Prediction and validation of the distinct dynamics of transient and sustained ERK activation. *Nat Cell Biol*. 2005 Apr;7(4):365–73.

Simanshu DK, Nissley DV, McCormick F. RAS proteins and their regulators in human disease. *Cell*. 2017 Jun 29;170(1):17–33.

Simon MA, Dodson GS, Rubin GM. An SH3-SH2-SH3 protein is required for p21Ras1 activation and binds to sevenless and Sos proteins in vitro. *Cell*. 1993 Apr 9;73(1):169–77.

Trahey M, McCormick F. A cytoplasmic protein stimulates normal N-ras p21 GTPase, but does not affect oncogenic mutants. *Science*. 1987 Oct 23;238(4826):542–5.

VOLTAGE-DEPENDENT SODIUM CHANNELS

Catterall WA, Lenaeus MJ, Gamal El-Din TM. Structure and pharmacology of voltage-gated sodium and calcium channels. *Annu Rev Pharmacol Toxicol*. 2020 Jan 6;60:133–154.

EGFR INTERNALIZATION

Tomas A, Futter CE, Eden ER. EGF receptor trafficking: consequences for signaling and cancer. *Trends Cell Biol*. 2014 Jan;24(1):26–34.

Wiley HS. Trafficking of the ErbB receptors and its influence on signaling. *Exp Cell Res*. 2003 Mar 10;284(1):78–88.

G-PROTEIN COUPLED RECEPTORS AND β-ARRESTIN

DeWire SM, Ahn S, Lefkowitz RJ, Shenoy SK. Beta-arrestins and cell signaling. *Annu Rev Physiol*. 2007;69:483–510.

Manglik A, Lin H, Aryal DK, McCorvy JD, Dengler D, Corder G, Levit A, Kling RC, Bernat V, Hübner H, Huang XP, Sassano MF, Giguère PM, Löber S, Da Duan, Scherrer G, Kobilka BK, Gmeiner P, Roth BL, Shoichet BK. Structure-based discovery of opioid analgesics with reduced side effects. *Nature*. 2016 Sep 8;537(7619):185–190.

McDonald PH, Lefkowitz RJ. Beta-arrestins: new roles in regulating heptahelical receptors' functions. *Cell Signal*. 2001 Oct;13(10):683–9.

STATE-DEPENDENT INACTIVATION

Friedlander T, Brenner N. Adaptive response by state-dependent inactivation. *Proc Natl Acad Sci USA*. 2009 Dec 29;106(52):22558–63.

OVERVIEWS OF ADAPTATION AND PULSE GENERATORS

Ferrell JE Jr. Perfect and near-perfect adaptation in cell signaling. *Cell Syst*. 2016 Feb 24;2(2):62–7.

Ma W, Trusina A, El-Samad H, Lim WA, Tang C. Defining network topologies that can achieve biochemical adaptation. *Cell*. 2009 Aug 21;138(4):760–73.

Tyson JJ, Chen KC, Novak B. Sniffers, buzzers, toggles and blinkers: dynamics of regulatory and signaling pathways in the cell. *Curr Opin Cell Biol*. 2003;15(2):221–231.

NEGATIVE FEEDBACK 3

Oscillations

14

IN THIS CHAPTER . . .

INTRODUCTION

One of the conclusions of Chapter 11 was that negative feedback can stabilize a steady state. Curiously, negative feedback can also destabilize a steady state, if the negative feedback loop is long enough. It is this destabilization that allows negative feedback loops to sometimes function as biochemical oscillators, and we will explore a classic example of a negative feedback oscillator, the Goodwin oscillator, in this chapter.

DOI: 10.1201/9781003124269-14

TABLE 14.1 **Biological Oscillations**

Rhythm	Period
Human brain waves	0.03 to 0.3 s
Human heartbeat	0.5 to 1 s
Calcium oscillations in astrocytes	30 s
Xenopus embryonic cell cycle	25 min (1.5×10^3 s)
Somitogenesis	30 min (1.8×10^3 s) in zebrafish 90 min (5.4×10^3 s) in chicks
p53, NF-κB, ERK oscillations in cell culture	A few hours (~10^4 s)
Circadian oscillations	1 day (8.6×10^4 s)
Human menstrual cycle and other circalunar rhythms	28 days (2.4×10^6 s)
Circannual rhythms	365 days (3.1×10^7 s)
Cicada life cycle	13 or 17 years (4.1 or 5.3×10^8 s)
Bamboo flowering cycle	3 to 150 years (9.4×10^7 to 4.7×10^9 s)

14.1 BIOLOGICAL OSCILLATIONS CONTROL MYRIAD ASPECTS OF LIFE AND OPERATE OVER A TEN-BILLION-FOLD RANGE OF TIME SCALES

Anyone who has seen a living, beating heart or watched a fertilized *Xenopus* egg divide, knows how compelling biological oscillations can be. Biological oscillations range in time scale from periods of 0.03–0.3 s for the various classes of cortical brain waves to 150 years, or ~5×10^9 s, for the flowering cycle of some species of bamboo (**TABLE 14.1**). That is a span of about 11 orders of magnitude, a truly remarkable range.

If you were to attend a recent conference on biological oscillations, you probably would hear mostly about circadian oscillations, cell cycles, somite formation, and p53 and NF-κB oscillations—all fascinating processes where good progress has been made. But a lot remains to be learned, especially with the slower biological clocks. For example, how do those cicadas figure out if this is the right year to emerge and mate? How do those bamboo plants know when they turn 150?

In the next two chapters we will examine two basic classes of biological oscillator circuit, starting with the simplest: oscillators constructed from a single negative feedback loop. The classic example is the Goodwin oscillator, and it consists of a three-tier transcription-translation cascade with negative feedback connecting the bottom to the top.

14.2 THE GOODWIN OSCILLATOR IS BUILT UPON A THREE-TIER CASCADE WITH HIGHLY ULTRASENSITIVE NEGATIVE FEEDBACK

A one-variable negative feedback model, like that considered in Chapter 11, cannot generate oscillations; one-variable systems always go monotonically toward or away from a steady state, so there is no possibility of up-and-down oscillatory outputs. Two-variable negative feedback models can generate pulses, as explored in Chapter 12, and damped oscillations, but they cannot continue to oscillate indefinitely. However, three-variable negative feedback models can oscillate, and that is where we will begin here.

Brian Goodwin proposed the negative feedback oscillator model that bears his name in the 1960s. It is not a model of some particular known biological oscillator; it is more an example of a "toy" oscillator that is relatively easy to analyze and understand. However, it has proven enormously useful, and it has been invoked, sometimes with modifications, to model a wide variety of real oscillatory biological phenomena, from circadian oscillations to the periodic gliding motions of myxobacteria.

The Goodwin oscillator is built out of a cascade of three synthesis-destruction processes. At the top level there is transcription, producing an mRNA we will call x. Next comes the translation of x to produce an enzyme y. Then there is the synthesis of some small molecule z by y. Finally, there is feedback, with z inhibiting the production of x, perhaps through some unmentioned ligand-regulated repressor protein. This leads to a decrease in x, followed by a decrease in y and then a drop in z, and with z gone, the production of x can resume. Intuitively it seems like this logic might result in sustained cycles of production and destruction of x, y, and z.

To see if this is the case, we want to construct a model of this process and examine the resulting dynamics. We start by writing rate equations for x, y, and z:

$$\frac{dx}{dt} = k_1 \frac{K^n}{K^n + z^n} - k_{-1}x, \tag{14.1}$$

$$\frac{dy}{dt} = k_2 x - k_{-2}y, \tag{14.2}$$

$$\frac{dz}{dt} = k_3 y - k_{-3}z. \tag{14.3}$$

The equation for the production and destruction of mRNA x (Eq. 14.1) assumes that the rate of transcription–the first term–will be maximal, and equal to k_1, when the concentration of z is zero, and that the rate of transcription decreases as z increases. It also assumes that the function that describes the decrease is an inhibitory Hill function and that the inhibition of transcription is an ultrasensitive function of z. The reason for this assumption was not that Goodwin knew of some experimental system where transcription had been shown to be inhibited in an ultrasensitive fashion. Rather, it was because the model did not yield sustained oscillations if simple stoichiometric inhibition was assumed. For a three-variable negative feedback loop with a single ultrasensitive step built in, one must assume a very high degree of ultrasensitivity ($n > 8$) to generate sustained oscillations. When Goodwin was working, this assumption was thought to be highly implausible, although now many examples of regulatory reactions with high degrees of ultrasensitivity are known (Table 5.1).

The parameter K determines the concentration of z required to half-maximally inhibit the production of x. K is inversely proportional to the strength of the negative feedback; as K approaches infinity, the strength of the feedback approaches zero, and as K approaches zero, the strength of the feedback becomes maximal.

The rest of the model is more straightforward. The degradation terms for all three species are simple mass action expressions, and the synthesis terms for y and z are mass action expressions as well.

We can explore the dynamics of the model by choosing arbitrary values for the rate constants (let us take them all equal to 1), for n ($n = 10$), and for K ($K = 0.1$ to start), as well as initial values for the three variables ($x[0] = y[0] = z[0] = 0$). Numerical solution of the rate equations yields the time courses shown in **Figure 14.1b**. Species x grows right

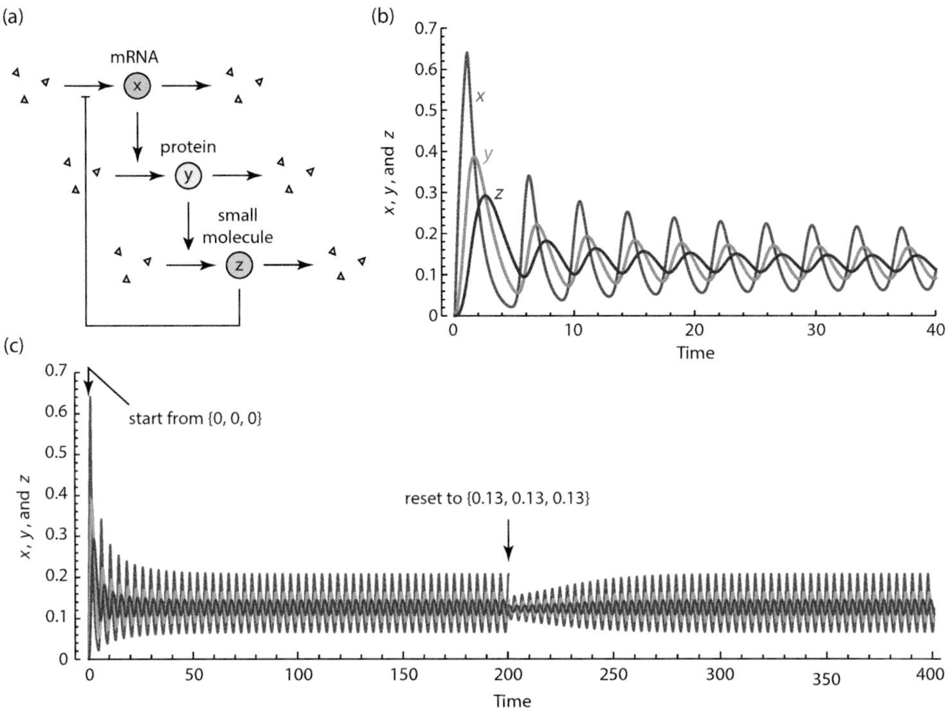

Figure 14.1 The Goodwin oscillator. (a) Schematic diagram of the model. (b,c) Time courses. For both panels we have assumed that all of the rate constants are equal to 1, $K=1$, and $n=10$, and that the initial values of **x**, **y**, and **z** are {0, 0, 0}.

from the start, followed by y and then z. The rising concentration of z then slows production of x to the point where x begins to fall, followed by decreases in y and z. As time goes on, the oscillations in all three species dampen, but eventually they appear to settle into a constant amplitude rhythm. It is perhaps easier to appreciate that the amplitude really does stop dampening by plotting a longer time course (**Figure 14.1c**). If we reset the circuit so that $x=y=z=0.13$, which we do when $t=200$ in Figure 14.1c, the system works its way back to the same "groove," with the same amplitudes and phase relationships for the three variables, except that this time the oscillations grow into the final rhythm rather than shrinking down to it.

Another helpful way to portray the oscillations, and to see how the oscillations approach this groove, is to plot the trajectories in phase space. Since there are three time-dependent variables, phase space is three-dimensional, and we can plot the oscillations in this three-dimensional space (**Figure 14.2b**). However, often two-dimensional projections of phase space, like that in **Figure 14.2c** where the trajectories are projected into the x-y plane, are invoked because they are easier to apprehend. In both representations, it is clear that one sample trajectory (the blue one that starts at $x[0]=y[0]=z[0]=0$) spirals counter-clockwise inward, approaching an avocado-shaped closed curve (the dashed white curve), and the other sample trajectory (red) spirals outward, approaching the same closed curve from the other side. This closed curve is called a limit cycle. A limit cycle is analogous to a steady state, but rather than being a single point that trajectories approach (if it is a stable steady state) or diverge from (if it is an unstable steady state), the limit cycle is a closed curve that attracts trajectories (if it is a stable limit cycle, as this one is) or repels them (in the case of unstable limit cycles, which we will not encounter in our modeling). A stable limit cycle is an attractor, just as a stable steady state is. Moreover, any orbit that starts exactly on a limit cycle, stable or unstable, will remain there forever.

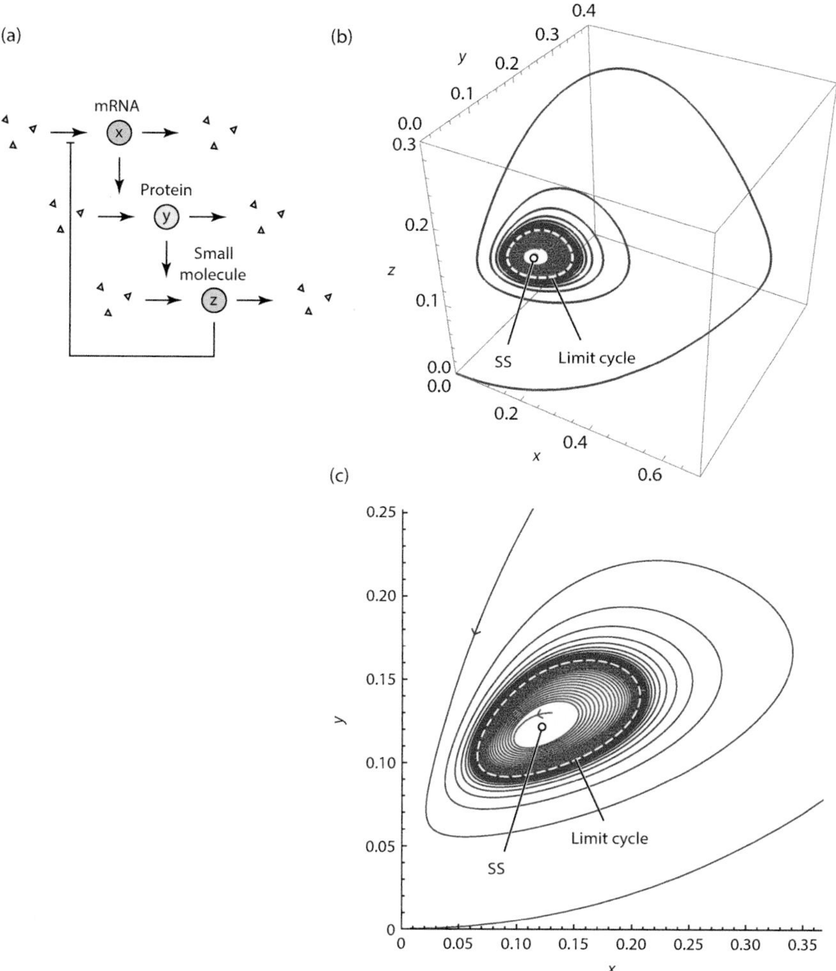

Figure 14.2 Limit cycle oscillations in phase space and projected onto the *x*–*y* phase plane for the Goodwin oscillator. (a) Schematic. (b) Trajectories in phase space. The blue curve corresponds to initial conditions of **x**, **y**, and **z**={0, 0, 0}. The red curve starts at **x**, **y**, and **z**={0.13, 0.13, 0.13}. The dashed white curve is the limit cycle, and the hollow circle is the steady state (SS). (c) The same trajectories projected onto the *x*-*y* plane.

14.3 LINEAR STABILITY ANALYSIS YIELDS A PAIR OF COMPLEX EIGENVALUES

Does an oscillatory system like this even have a steady state? After all, the system never comes to rest. To answer this question, we start with the rate equations and set them all equal to zero, as required for the system to be in steady state:

$$0 = k_1 \frac{K^n}{K^n + z^n} - k_{-1}x, \tag{14.4}$$

$$0 = k_2 x - k_{-2} y, \tag{14.5}$$

$$0 = k_3 y - k_{-3} z. \tag{14.6}$$

We cannot solve this set of equations in closed form but can solve it numerically, and for our choice of parameters, the solution is:

$$x_{ss} = y_{ss} = z_{ss} \approx 0.122. \tag{14.7}$$

Thus the system has a single steady state. Note that its coordinates are why we started one of our trajectories at $x[0]=y[0]=z[0]=0.13$; we were starting close to the steady state to see what would happen, and what happened was that the trajectory spiraled out away from the steady state, ultimately approaching the limit cycle.

So is this single steady state stable or unstable? To address this we need to carry out linear stability analysis and calculate the eigenvalues and eigenvectors of the three-variable system at the steady state. The procedure is the same as what we did in Chapter 9.3, except here we have three variables rather than two.

First, we define functions f, g, and h:

$$f = k_1 \frac{K^n}{K^n + z^n} - k_{-1}x, \tag{14.8}$$

$$g = k_2 x - k_{-2} y, \tag{14.9}$$

$$h = k_3 y - k_{-3} z. \tag{14.10}$$

Next, we calculate the partial derivatives and arrange them in a Jacobian matrix:

$$jacobian = \begin{bmatrix} \frac{\partial f}{\partial x} & \frac{\partial f}{\partial y} & \frac{\partial f}{\partial z} \\ \frac{\partial g}{\partial x} & \frac{\partial g}{\partial y} & \frac{\partial g}{\partial z} \\ \frac{\partial h}{\partial x} & \frac{\partial h}{\partial y} & \frac{\partial h}{\partial z} \end{bmatrix} = \begin{bmatrix} -k_{-1} & 0 & -\frac{k_1 K^n n z^{n-1}}{\left(K^n + z^n\right)^2} \\ k_2 & -k_{-2} & 0 \\ 0 & k_3 & -k_{-3} \end{bmatrix}. \tag{14.11}$$

Finally, we plug in the assumed values for the rate constants, K, and n, and the z-coordinate of the steady state, and calculate the eigenvectors and eigenvalues. The three eigenvalues (to three significant figures) are:

$$\begin{aligned} \lambda_1 &= -3.06 \\ \lambda_2 &= 0.0316 + 1.79i \\ \lambda_3 &= 0.0316 - 1.79i \end{aligned} \tag{14.12}$$

and the corresponding eigenvectors are {–0.880, 0.427, –0.207}, {0.880, 0.213–0.379i, –0.103–0.179i}, and {0.880, 0.213+0.379i, –0.103+0.179i}.

The first eigenvector represents a special direction in real phase space, and it corresponds to a negative eigenvalue, so trajectories that begin from this general direction are initially attracted to the steady state. However, as they get close to the steady state, they are repelled out to the limit cycle. Thus the steady state shares some of the dynamical character of the saddles we saw in two-variable bistable systems (Chapter 9), first attracting and then repelling. We can plot the one-dimensional stable manifold of the steady state by making all of the rate constants in the model negative, picking a starting point close to the steady state along the eigenvector and then numerically calculating the trajectory of this time-reversed model. The result is shown in **Figure 14.3b**, with this stable manifold, the limit cycle, and one sample trajectory all projected onto the x–y plane.

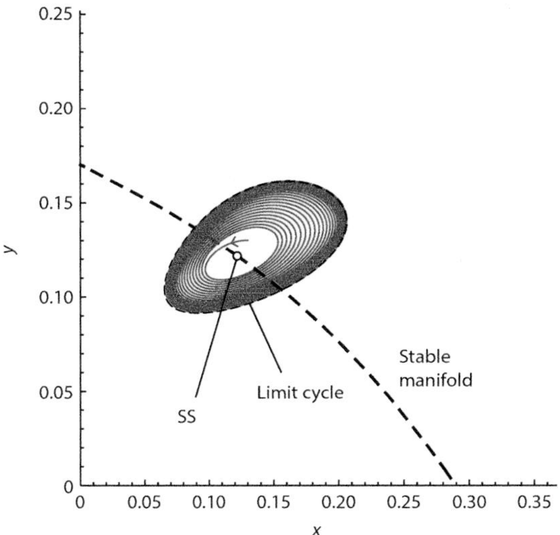

Figure 14.3 The stable manifold of the steady state of the Goodwin oscillator. The stable manifold (dashed thick black curve) is shown projected in the *x*–*y* plane, together with the steady state (open circle), the limit cycle (dashed closed curve), and one trajectory (red curve).

The other two eigenvectors and eigenvalues are complex numbers rather than real numbers—pairs of complex conjugates. To see what the complex eigenvalues mean physically, recall that exponential functions with imaginary exponents can be related to trigonometric functions through Euler's equation:

$$e^{iy} = \cos y + i\sin y, \tag{14.13}$$

and so:

$$e^{x+iy} = e^{x}(\cos y + i\sin y). \tag{14.14}$$

Thus, the response to an infinitesimal perturbation corresponding to the second and third eigenvalues is proportional to:

$$\begin{aligned} e^{(0.0316+1.79i)t} &= e^{0.0316t}\left(\cos(1.79t)+i\sin(1.79t)\right) \\ e^{(0.0316-1.79i)t} &= e^{0.0316t}\left(\cos(1.79t)-i\sin(1.79t)\right). \end{aligned} \tag{14.15}$$

There is a periodic component to the trajectories, the real part of which is the cosine function, and there is a real exponential component, with a positive exponent. The cosine explains why the trajectories circle the steady state in the phase plane, and the positive exponential explains why they spiral outward from the steady state rather than inward. From Eq. 14.15 it follows that the period of the oscillations when the trajectory is very close to the steady state is $2\pi / 1.79$ time units (~3.51), and by numerical simulation the period is not too different from this even as the trajectory approaches the limit cycle (~3.70).

Thus, the three-variable Goodwin oscillator model possesses a single steady state. Linear stability analysis of the model at this steady state yields one negative real eigenvalue, which means that from one special direction the steady state attracts trajectories. But there are also two complex eigenvalues. The imaginary parts of these eigenvalues show that the trajectories have an inherent periodicity, and the magnitude of the imaginary parts determines the period of the oscillations close to the steady state. The real parts determine whether the oscillations will spiral out from the steady state (in the phase plane) or spiral in toward it. In our case the real parts were positive, which means that the oscillations spiral out. A steady state like this is often called an unstable spiral point.

It is interesting that even though biological oscillations are time-dependent, dynamical behaviors, analysis of the steady state—the point in phase space where the model's behavior is *not* time-dependent—yields a great deal of insight into the model's dynamics.

There are other oscillator models whose steady states have real rather than complex eigenvalues and we will see examples of this in the next chapter. But a pair of complex eigenvalues ensure that a model will exhibit at least damped oscillations, and complex eigenvalues with a positive real part ensure that the oscillations will not be damped.

14.4 OSCILLATIONS ARE BORN AND EXTINGUISHED AT HOPF BIFURCATIONS

We can make the negative feedback weaker by increasing the assumed value of K, and see what happens to the oscillations and to the stability of the steady state. Figure 14.4b shows the model with K set at 0.1. **Figure 14.4a** assumes stronger negative feedback (K=0.01), and the result is a loss of amplitude. **Figures 14.4c,d** assume weaker (C, K=1) and absent (D, K=∞) negative feedback, and the results are damped oscillations (C) and no oscillations whatsoever (D).

By taking a finer range of K values, we can carry out a one-dimensional bifurcation analysis of the system. As shown in **Figure 14.4e**, oscillations are present when K is less than about 0.175. Above this value of K, the real portions of the complex conjugate of eigenvalues, Re[λ_2] and Re[λ_3], are negative—the steady state is a stable spiral point—and above it they are positive, so the spiral point becomes unstable, and a stable limit cycle is born. This transition in the stability of the steady state and the dynamics of the system is termed a Hopf bifurcation.

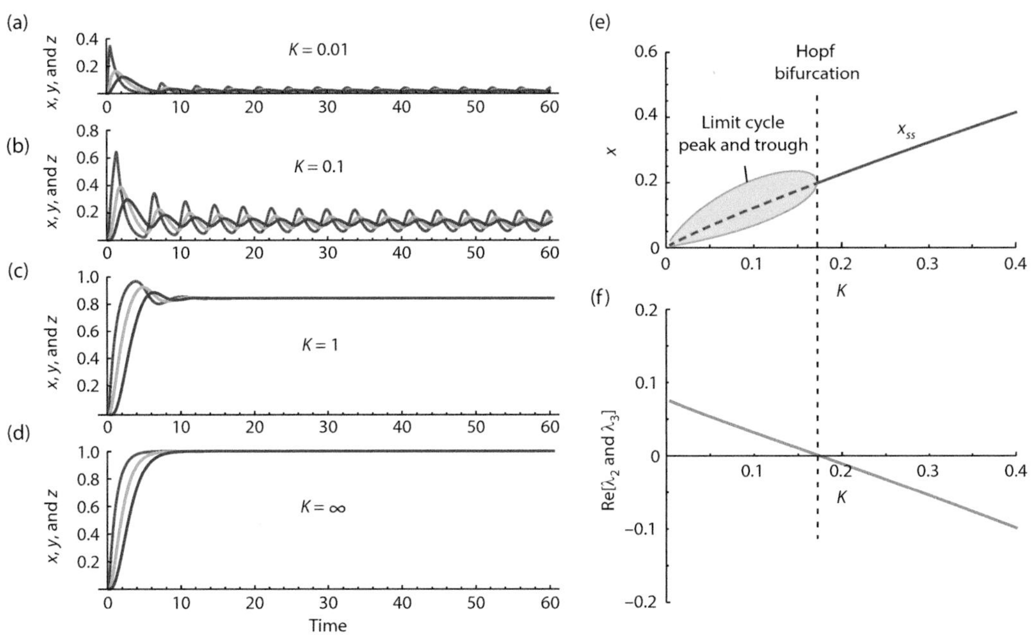

Figure 14.4 Behavior of the Goodwin oscillator with various strengths of negative feedback. (a–d) The time course of the system with K=0.01 (a), 0.1 (b), 1 (c), and ∞ (d). The larger the value of K is, the weaker the feedback. The red curves are x, the green curves y, and the blue curves z. (e) One-dimensional bifurcation diagram, showing the steady-value of x as a function of K. Below $K \approx 0.175$, the steady state becomes unstable and sustained oscillations commence. The point where the oscillations commence (if K is going down) or cease (if K is going up) is a Hopf bifurcation. The stable steady states are shown as the solid red curve, and the unstable steady states as the dashed red curve. The peak and trough levels for the limit cycle oscillations are shown as the thinner red curves with pink fill. (f) The real portions of the second and third eigenvalues as a function of K. The Hopf bifurcation occurs when the real portions of the eigenvalues change from positive to negative.

Hopf bifurcations are the most common way that oscillations are born and extinguished in models of biological oscillations.

The amplitude of the oscillations is a sensitive function of K, rising from 0 at the Hopf bifurcation to a maximum when K is close to our original value of 0.1, and then falling again as K decreases and the negative feedback strength increases further (Figure 14.5). The period of the limit cycle is much less sensitive to the assumed negative feedback strength, varying by less than 10% over the oscillatory range of K values. These qualities—a variable amplitude and a less variable period—are commonly found in negative feedback oscillators.

14.5 SIMPLE HARMONIC OSCILLATORS ARE NOT LIMIT CYCLE OSCILLATORS

Most manmade clocks, as opposed to biological clocks, are harmonic oscillators. Are harmonic oscillators similar to the Goodwin oscillator in terms of their dynamics, or are they a different kind of thing?

For a mass on an ideal spring in the absence of friction (Figure 14.5a), the equation of motion is:

$$m\frac{d^2x}{dt^2} = -kx \tag{14.16}$$

or:

$$\frac{d^2x}{dt^2} = -\frac{k}{m}x, \tag{14.17}$$

where m is the mass, k is the spring constant, and x is the displacement of the mass from its resting position. This is a second-order differential equation, but we can convert it to two first-order differential equations by noting that the mass's velocity v is $\frac{dx}{dt}$, which means that:

$$\frac{dx}{dt} = v \tag{14.18}$$

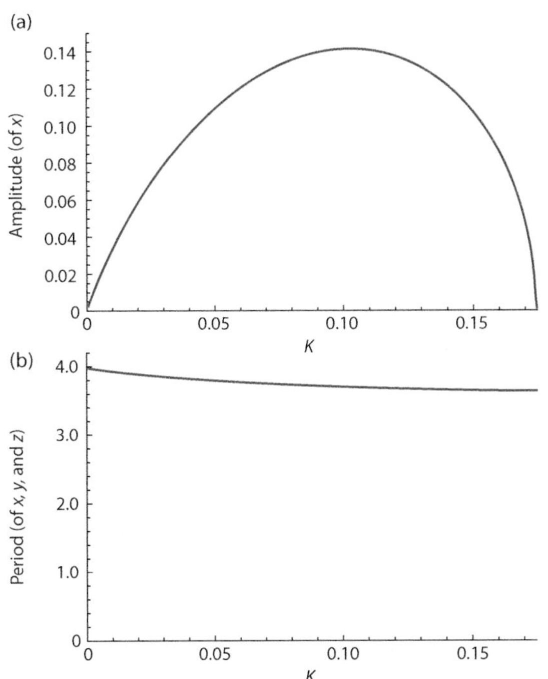

Figure 14.5 Amplitude (a) and frequency (b) of the limit cycle oscillations in the Goodwin oscillator as a function of K.

$$\frac{dv}{dt} = -\frac{k}{m}x. \tag{14.19}$$

Equations 14.18 and 14.19 constitute our two-ODE model of a harmonic oscillator.

We can solve these ODEs analytically:

$$x[t] = x[0]\cos(\omega t) - \omega v[0]\sin(\omega t) \tag{14.20}$$

$$v[t] = v[0]\cos(\omega t) - \omega x[0]\sin(\omega t), \tag{14.21}$$

where the frequency $\omega = \sqrt{\frac{k}{m}}$. We can then plot these solutions either as time courses (Figure 14.5b), or in the x–v phase plane (**Figure 14.6b**), for some choice of the initial conditions (say $x[0]=1$, $v[0]=0$) and ω (say $\omega = 1$). The trajectories are, as expected, sine and cosine waves in the time course plots, and a closed circle in the phase plane (**Figure 14.5b,c**). One might be tempted to call this circle a limit cycle. However, unlike a limit cycle, no trajectories spiral into or away from it. And the system does not eventually approach the same "groove"

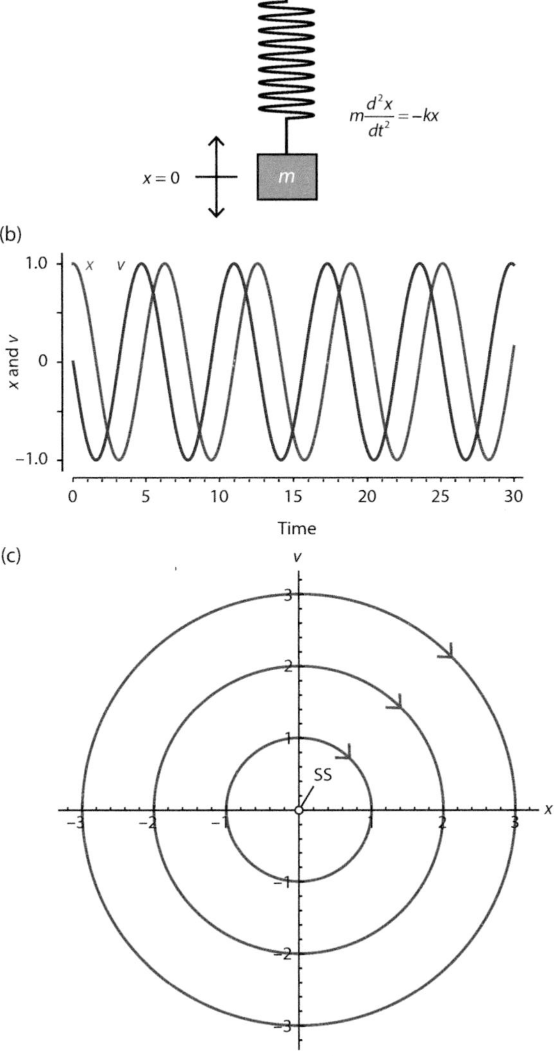

Figure 14.6 The harmonic oscillator. (a) Schematic of a mass on an ideal spring. (b) Time course of the position (x) and velocity (v) as a function of time. The parameters are $m=k=1$, and the initial conditions are $x[0]=1$, $v[0]=0$. (c) Trajectories in phase space. Parameters and initial conditions are as in (b) except that $x[0]=1$, 2, or 3.

irrespective of its starting conditions, the way it did for the Goodwin oscillator. Instead, if two different starting conditions put you on different circles, they will oscillate with different amplitudes (but with identical periods).

What about the steady state of the harmonic oscillator system? By setting Eqs. 14.18 and 14.19 equal to zero, it quickly follows that the nullclines are the two axes in the phase plane, and that there is a single steady state for the system at {0, 0}. And as to the stability of the steady state, what would happen if you perturbed the system by some small amount δ in the x direction (or in any direction, actually)? The answer is that the trajectory would not spiral into the steady state, nor away from it. It would just orbit the steady state at a constant distance of δ. Thus the steady state is neither a stable spiral point nor an unstable spiral point.

And, finally, what does linear stability analysis say about the steady state? The Jacobian matrix is:

$$jacobian = \begin{bmatrix} \frac{\partial f}{\partial x} & \frac{\partial f}{\partial v} \\ \frac{\partial g}{\partial x} & \frac{\partial g}{\partial v} \end{bmatrix} = \begin{bmatrix} 0 & 1 \\ -\omega & 0 \end{bmatrix}, \quad (14.22)$$

and the eigenvalues of this matrix are $-i\sqrt{\omega}$ and $i\sqrt{\omega}$—two purely imaginary numbers. Since the real parts of the eigenvalues are neither positive nor negative, the steady state is neither unstable nor stable. A steady state of this sort, which in our case sits in the center of all the orbits, is called a center.

To sum it up: the harmonic oscillator is a wonderful, simple system to study, but it is not a limit cycle oscillator and does not behave like the oscillators encountered in cell signaling.

SUMMARY

Here we have examined the Goodwin oscillator model, in which a negative feedback loop of sufficient length (at least three variables) and with a substantial amount of ultrasensitivity (n>8) generates limit cycle oscillations. The system has a single steady state with one negative eigenvalue and two complex ones, and when the negative feedback is dialed up to the point where it is strong enough to allow sustained oscillations to occur, the real parts of the complex eigenvalues switch from negative to positive. The point where the real parts go through zero and a stable limit cycle is born is termed a Hopf bifurcation. The Goodwin oscillator always approaches the same "groove"—the stable limit cycle—irrespective of the system's initial conditions. This distinguishes it from the familiar harmonic oscillator, where the initial position and velocity of the oscillator determines its amplitude.

Goodwin oscillators, sometimes with phosphorylation/dephosphorylation reactions in place of the synthesis/destruction reactions and with zero-order ultrasensitivity in place of the cooperative inhibition, have been used to model a wide variety of biological clocks and timers. However, biological oscillator circuits often possess an additional feature, a positive feedback loop that functions as a bistable trigger, and in Chapter 15 we will see how this can change the basic character of the oscillator.

FURTHER READING

BIOLOGICAL OSCILLATIONS

Elowitz MB, Leibler S. A synthetic oscillatory network of transcriptional regulators. *Nature*. 2000;403(6767):335–8.

Ferrell JE Jr, Tsai TY, Yang Q. Modeling the cell cycle: why do certain circuits oscillate? *Cell*. 2011;144(6):874–85.

Goldbeter A. *Biochemical Oscillations and Cellular Rhythms: The Molecular Bases of Periodic and Chaotic Behaviour*. Cambridge University Press, Cambridge, 1996.

Goldbeter A. *La Vie Oscillatoire: Au Coeur des Rythmes du Vivant*. Odile Jacob Sciences, 2010.

Igoshin OA, Goldbeter A, Kaiser D, Oster G. A biochemical oscillator explains several aspects of Myxococcus xanthus behavior during development. *Proc Natl Acad Sci USA*. 2004;101(44):15760–65.

Potvin-Trottier L, Lord ND, Vinnicombe G, Paulsson J. Synchronous long-term oscillations in a synthetic gene circuit. *Nature*. 2016;538(7626):514–17.

Raible F, Takekata H, Tessmar-Raible K. An overview of monthly rhythms and clocks. *Front Neurol*. 2017;8:189.

Strogatz SH. *Nonlinear Dynamics and Chaos: With Applications to Physics, Biology, Chemistry, and Engineering*. Westview Press, Cambridge, MA, 1994.

Zheng X, Lin S, Fu H, Wan Y, Ding Y. The bamboo flowering cycle sheds light on flowering diversity. *Front Plant Sci*. 2020;11:381.

THE GOODWIN OSCILLATOR

Goodwin BC, ed. *Oscillatory Behavior in Enzymatic Control Processes*. Permagon Press, Oxford, 1965.

Griffith JS. Mathematics of cellular control processes. I. Negative feedback to one gene. J Theor Biol. 1968;20(2):202–208.

RELAXATION OSCILLATORS

15

IN THIS CHAPTER . . .

DOI: 10.1201/9781003124269-15

INTRODUCTION

Although a simple negative feedback loop of sufficient length, like the Goodwin oscillator, can give rise to limit cycle oscillations, many biological oscillators possess an additional circuit element: a bistable trigger. The trigger fires once per cycle, and its presence changes the character of the oscillations. This is why many biological oscillations look more like a succession of spikes than a sinusoidal (harmonic oscillator) or almost-sinusoidal (Goodwin oscillator) ebb and flow. Oscillator circuits with a fast positive feedback trigger and a slower negative feedback loop are termed **relaxation oscillators**. The pacemaker cells of the cardiac sinoatrial node, which give rise to periodic spikes in membrane potential that spreads throughout the heart to make the heart contract, represent a classic example of a relaxation oscillator (**Figure 15.1a**). Another good example is the embryonic cell cycle, which is driven by regular spikes of Cdk1 activity (**Figure 15.1b**). Although the time scales of these oscillations are quite different and the proteins involved are completely unrelated, the character of the oscillations is remarkably similar (**Figure 15.1**).

Here we will begin with an analysis of the cell cycle oscillator, followed by the Fitzhugh–Nagumo model, which is a favorite of physicists that can be used to model neuronal action potentials and cardiac pacemaker oscillations. We will finish with a depletion-based oscillator model recently proposed to account for oscillations in the RhoA GTPase cycle.

15.1 THE *XENOPUS* EMBRYONIC CELL CYCLE IS DRIVEN BY A RELIABLE BIOCHEMICAL OSCILLATOR

Many model organisms have contributed to our understanding of the cell cycle oscillator, including fission yeast (*Schizosaccharomyces pombe*), budding yeast (*Saccharomyces cerevisiae*), *Drosophila melanogaster* embryos, and various mammalian cell lines. But the most detailed quantitative picture arguably comes from the eggs and embryos of the South African clawed-toed frog, *Xenopus laevis*.

Unfertilized *Xenopus* eggs are huge cells (~1 μL in volume, compared with ~2–3 pL for typical somatic cells) that are arrested in metaphase of meiosis II. After fertilization, they complete meiosis II, expelling the tiny second polar body that contains half of the mother's sister chromatids, and then they enter interphase of the first mitotic cell cycle. The remaining sister chromatids from the egg (in the female pronucleus) and the sister chromatids from the sperm (in the male pronucleus) are replicated, the pronuclei move toward each other, and they enter mitosis. At the end of mitosis the replicated sister

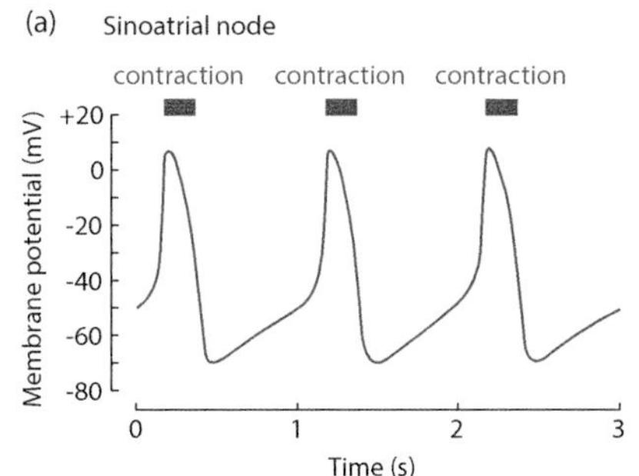

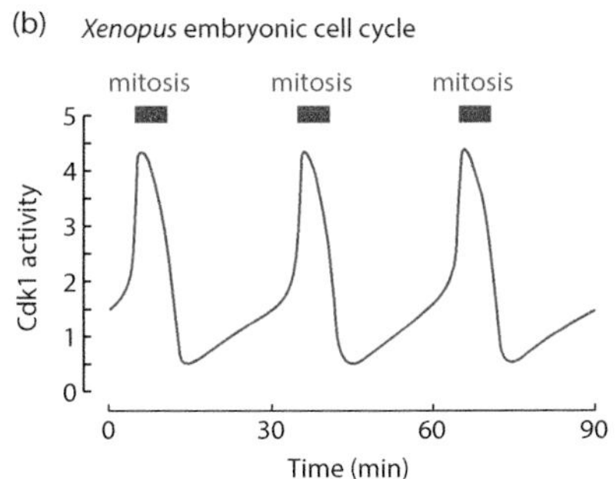

Figure 15.1 Two relaxation oscillators. (a) Electrical oscillations in the sinoatrial node of the heart. (b) Cdk1 activity (in arbitrary units) in the *Xenopus laevis* embryonic cell cycle.

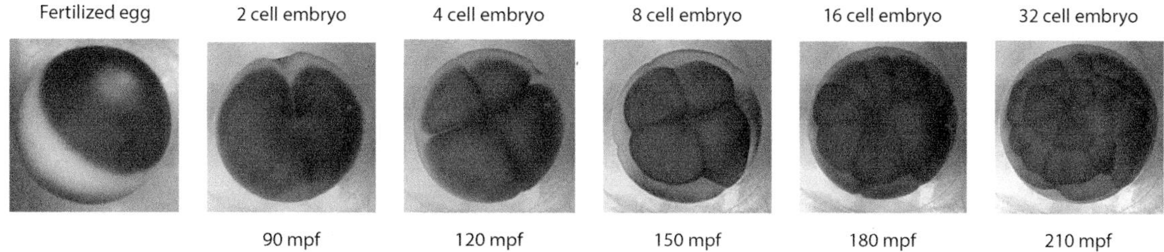

Figure 15.2 The rapid, nearly synchronous divisions of the early embryonic cell cycle in *Xenopus laevis*. Photos: Graham Anderson. Here mpf stands for minutes postfertilization.

chromatids are pulled apart and two daughter cells are formed; the spherical egg is split longitudinally, from the dark brown animal pole down to the cream-colored vegetal pole (**Figure 15.2**). In total, this process takes about 90 min. This is followed by a succession of cell cycles that consist of a ~15-min interphase, during which DNA replication occurs, and a ~15-min M-phase. No cell growth takes place during these cell cycles; the fertilized egg simply divides in half, and then in quarters, and so on, until after 12 cell cycles the embryo is composed of about 4,000 cells (**Figure 15.2**).

The embryonic cell cycle is streamlined—there are no G1 or G2 phases and no S-phase or M-phase checkpoints. This simplicity, plus the rapidity of the cycles (~30 min vs. ~24 h for typical mammalian cells in culture), has made the *Xenopus* embryo a workhorse system for studies of cell cycle regulation. In addition, the regularity of the cycles—the period varies only a few percent between cells within a cleaving embryo or even between embryos—makes the system appealing to those interested in biological clocks and oscillators. Something this precise and "physics-like" seems likely to be understandable. And finally, it turns out that undiluted, crudely-fractionated extracts from *Xenopus* eggs can carry out the cell cycle in vitro (**Figure 15.3**). These cell cycles can be monitored in various ways. For example, if sperm chromatin is added to the extract, it will turn into an intact nucleus during interphase that undergoes nuclear envelope breakdown whenever the extract goes into mitosis, and then nuclear envelope reformation when the extract progresses into the next interphase, and these events can be followed by microscopy. Microtubule polymerization can also be used to read out cell cycle progression (**Figure 15.3b**); during interphase, the cytoplasm is filled with a network of microtubules, whereas during M-phase most of these microtubules are depolymerized, and only the spindle microtubules remain. Cycling extracts can also be sampled repeatedly for biochemical assays. **Figure 15.3c** shows experimental measurements of the activity of cyclin B-Cdk1 during a cycle: it starts low, rises gradually, and then spikes upward right when mitosis begins. It then plummets back to its original level of activity, at which point the extract exits mitosis and enters the next interphase, ready to carry out the cycle again.

Extracts can be manipulated in ways that would be impossible with intact cells. A good example of these sorts of extract-only approaches came during the work on cyclin function in the mid-1980s. It was already known that the cyclin proteins accumulate during interphase and then become degraded prior to division in the embryonic cell cycle, but it was not certain that these oscillations drove the cell cycle. Minshull, Blow, and Hunt inhibited cyclin synthesis in cycling *Xenopus* extracts by adding in antisense oligonucleotide, and found that the mitotic entry was delayed or blocked. At about the same time, Murray and Kirschner treated cycling *Xenopus* extracts with ribonuclease

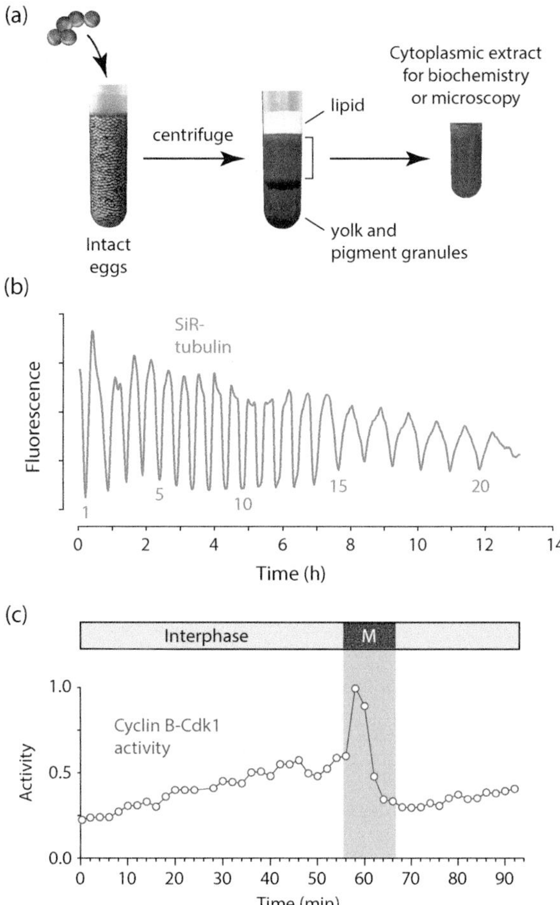

Figure 15.3 Cell cycles in cell-free *Xenopus* egg extracts. (a) Preparation of a cycling extract by centrifugation of packed eggs. (Adapted from Cheng and Ferrell, *Science*. 2019.) (b) Oscillations in tubulin polymerization in a cycling extract. Tubulin polymerization is quantified using a compound whose fluorescence increases when bound to microtubules (a silicon rhodamine taxane derivative, SiR-tubulin). This extract carried out 20 cell cycles, with the troughs of SiR-tubulin fluorescence corresponding to mitosis. (Adapted from Afanzar et al., *Elife*. 2020.) (c) Cdk1 activity as a function of time in a cycling extract. (Adapted from Kamenz et al., *Curr Biol*. 2020.)

(RNase), which degraded mRNAs and blocked mitotic entry, then neutralized the RNase and added back a cyclin mRNA to the extract, and found that this cyclin mRNA restored the extract's ability to cycle. These discoveries placed cyclins at the core of the cell cycle oscillator.

15.2 THE CELL CYCLE OSCILLATOR INCLUDES A NEGATIVE FEEDBACK LOOP AND A BISTABLE TRIGGER

The current view of regulatory circuit that drives the cell cycle is shown schematically in **Figure 15.4**. Oscillations are now know to be driven by the synthesis of several related mitotic cyclin proteins (cyclins B1α, B1β, B2, B4, B5, and A1, here collectively referred to as cyclin B). The cyclin B binds with high affinity to Cdk1, which is present in modest excess, and when the cyclin B–Cdk1 complex is in the right phosphorylation state, it is active as a protein kinase, phosphorylating hundreds of substrate proteins at many hundreds of phosphorylation sites. The collective effect of these phosphorylations is the dramatic cellular changes of mitotic entry, including chromatin condensation, nuclear envelope breakdown, vesiculation of the golgi, endoplasmic reticulum, and mitochondria, and reorganization of the microtubules into a football-shaped spindle.

Four particular regulatory proteins that are targets of active Cdk1 are of special importance for this regulatory circuit. The first is the Greatwall kinase (Gwl), which is activated by Cdk1 and then inactivates the phosphatase PP2A-B55 by phosphorylating and activating a pair of stoichiometric inhibitors of PP2A-B55 (ENSA and ARPP19).

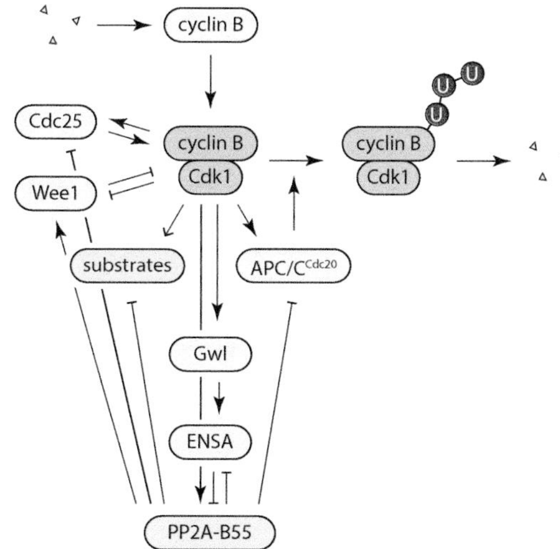

Figure 15.4 The regulatory circuit that drives mitotic entry and exit. Mitosis is driven by the collective effects of the phosphorylation of hundreds of substrate proteins (blue), which is in turn determined by the activity of the master mitotic kinase cyclin B–Cdk1 (pink) and the phosphatase PP2A-B55 (green). The cyclical changes in the activities of cyclin B–Cdk1 and PP2A-B55 are ensured by a negative feedback loop: Cdk1 activates APC/C^{Cdc20}, which then polyubiquitylates cyclin B, leading to its degradation by the proteasome. There are also several interlinked positive and double-negative feedback loops (e.g., Cdk1 activates Cdc25C and Cdc25C activates Cdk1) that make Cdk1 activation and PP2A-B55 inactivation explosive in character.

PP2A-B55 is the phosphatase responsible for dephosphorylating many of the substrates that Cdk1 phosphorylates, so the inactivation of PP2A-B55 potentiates the effect of Cdk1 activation.

The second key regulatory target of Cdk1 is the anaphase-promoting complex or cyclosome (APC/C), which in the embryonic cell cycle is bound to the Cdc20 activator protein (as opposed to the Cdh1 activator protein, which is important in somatic cells but not in the embryonic cell cycle). APC/C^{Cdc20} phosphorylation leads to activation. Active APC/C^{Cdc20} polyubiquitylates the N-terminus of cyclin B, leading to its rapid destruction by the proteasome. This constitutes a negative feedback loop: cyclin B-Cdk1 activates APC/C^{Cdc20}, which feeds back to destroy cyclin and thereby inactivate Cdk1. Thus, like the Goodwin oscillator, the cell cycle oscillator possesses a negative feedback loop. This loop is essential for oscillations; engineered cyclin proteins that lack their N-termini and cannot be polyubiquitylated by APC/C^{Cdc20} will drive mitotic entry normally, but then arrest in mitosis because Cdk1 cannot be inactivated. It has been conjectured that all biochemical oscillators must possess a negative feedback loop, or the equivalent, and whether this is or is not true in general, it is true for the embryonic cell cycle.

The remaining key targets of Cdk1 are a pair of enzymes with opposite effects on Cdk1, Wee1 and Cdc25C. These enzymes are responsible for determining whether the cyclin B–Cdk1 complex is in the right phosphorylation state to be fully active. Wee1 is a protein kinase, and it phosphorylates Cdk1 at a residue in its catalytic cleft and renders Cdk1 unable to position the γ-phosphate in ATP properly for phosphotransfer. Cdc25C is the phosphatase that dephosphorylates the Wee1-phosphorylated site and restores full activity to cyclin B1–Cdk1. Moreover, cyclin B1–Cdk1 activates it activator, Cdc25C, through multisite phosphorylation, and inactivates Wee1, also through multisite phosphorylation. These interlinked feedback loops—one a positive feedback loop and the other a double-negative feedback loop—function as the bistable mitotic trigger. The loops are also reinforced by additional feedback. For example, Cdc25C is inactivated by PP2A-B55, which is inactivated by cyclin B1–Cdk1, which means that cyclin B1–Cdk1 –| PP2A–B55 –| Cdc25C –> cyclin-B1–Cdk1, a double-negative feedback loop. And Wee1 is activated by PP2A-B55, which means that cyclin B1–Cdk1 –| PP2A–B55 –> Wee1 –| cyclin-B1–Cdk1, another double-negative feedback loop. Multiple interlinked feedback loops appear to be common in bistable regulatory systems.

In summary, cyclin B1-Cdk1 activation is driven by cyclin synthesis during interphase; boosted to mitotic levels by a bistable trigger consisting of interlinked positive and double-negative feedback loops; and then inactivated through a negative feedback loop where activated APC/C^{Cdc20} feeds back to bring about the proteolysis of cyclin.

15.3 A SIMPLIFIED MODEL CAPTURES THE BASIC DYNAMICS OF THE CELL CYCLE OSCILLATOR

To a hardcore student of the cell cycle, every step shown in **Figure 15.4** is interesting and important. But the system is too complicated to be useful as an introduction to relaxation oscillators. Fortunately, it can be pared down to a much simpler model, which is relatively easy to learn from and which fairly faithfully reproduces the dynamics of cyclin B-Cdk1 oscillations.

We focus on three key time-dependent species. The first is cyclin B, whose concentration is regulated by synthesis and degradation. We will call the total concentration of cyclin B x_{tot}. The second is the cyclin B-Cdk1 complex, whose activity is regulated by phosphorylation and dephosphorylation. We will call the active cyclin B-Cdk1 species $y*$ and the inactive form y. The third is APC/C^{Cdc20}, which is also regulated by phosphorylation, and which we will call $z*$ (active) or z (inactive). This is shown schematically in **Figure 15.5a**.

We assume that cyclin B is translated at a constant rate (k_1), and that it rapidly binds to the excess Cdk1 with high affinity, so that the production of cyclin B-Cdk1 complexes also occurs at a constant rate of k_1. We assume that the polyubiquitylation of cyclin B (in active or inactive complexes, or free) leads immediately to its destruction, and hence the destruction of x_{tot}, by the proteasome, and we assume that the process is described by mass action kinetics, and so is proportional to

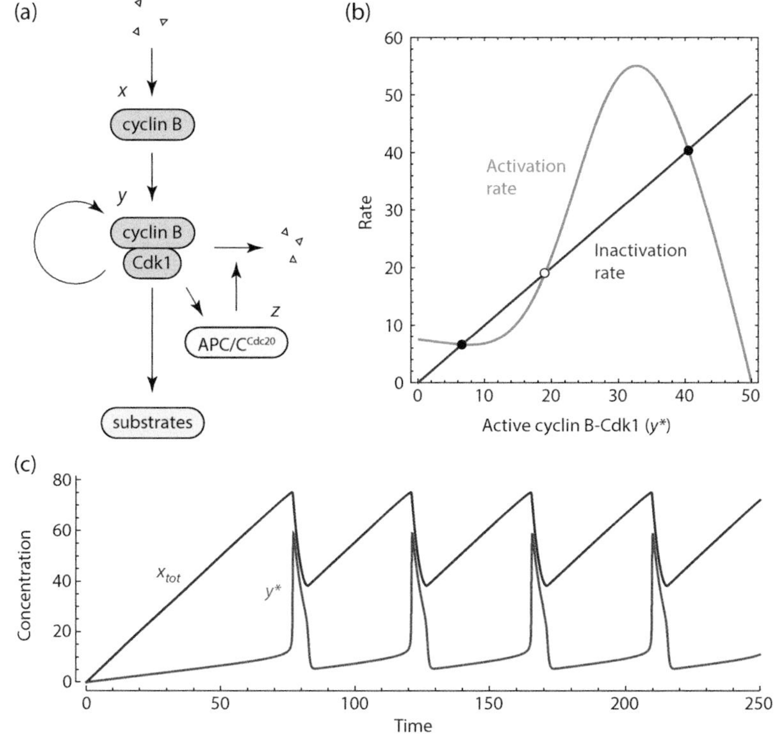

Figure 15.5 A simplified model of the cell cycle oscillator yields sustained oscillations in cyclin B abundance (x_{tot}) and cyclin B–Cdk1 activity (y*). (a) Schematic of the oscillator circuit. (b) Rate-balance analysis for the mitotic trigger, showing the rate of y activation and inactivation as a function of $y*$ for one value of x_{tot} ($x_{tot}=50$). (c) Time courses. For panels b and c, the parameters were: $k_1=1$; $k_{-1}=0.2$; $K_1=30$; $n_1=5$; $a_2=0.03$; $k_2=5$; $k_{-2}=1$; $K_2=30$; and $n_2=5$.

x_{tot} and to the concentration of active APC/C^{Cdc20} (z^*). The resulting rate equation is:

$$\frac{dx_{tot}}{dt} = k_1 - k_{-1}x_{tot}z^*. \tag{15.1}$$

Next we consider the activity of the cyclin B–Cdk1 complexes. The circuits that regulate cyclin B–Cdk1 activity include multiple interlinked positive and double-negative feedback loops, which collectively make the steady-state response of cyclin B–Cdk1 activity as a function of the cyclin B concentration hysteretic and bistable. For simplicity we will reduce the interlinked loops to a single loop, with inactivated cyclin B–Cdk1 (x) activated by the phosphatase Cdc25, which is itself activated by active cyclin B–Cdk1 (y^*). We will ignore the double-negative Wee1 loop and assume that y^* is inactivated by a constitutive kinase with mass action kinetics. We will make use of the fact that the binding of cyclin B to Cdk1 is fast, so that that amount of free cyclin B is negligible and $x_{tot} = y + y^*$. This leaves us with the following rate equation:

$$\frac{dy^*}{dt} = k_2 Cdc25 \cdot (x_{tot} - y^*) - k_{-2}y^*. \tag{15.2}$$

The steady-state activity of Cdc25 has been shown experimentally to be a switch-like function of y^*, with a fairly high basal activity. Thus:

$$Cdc25_{ss} \propto a_2 + \frac{(y^*)^{n2}}{K_2^{n2} + (y^*)^{n2}}. \tag{15.3}$$

Since Cdc25 is regulated by phosphorylation and dephosphorylation, which occur quickly relative to protein synthesis and destruction, we will assume that for any given value of y^*, the activity of Cdc25 is approximately the steady-state activity given by Eq. 15.3. This allows us to reduce Eq. 15.2 to one with a single variable that changes slowly (x_{tot}) and a single variable that changes more rapidly (y^*):

$$\frac{dy^*}{dt} = k_2\left(a_2 + \frac{(y^*)^{n2}}{K_2^{n2} + (y^*)^{n2}}\right)(x_{tot} - y^*) - k_{-2}y^*. \tag{15.4}$$

Note that this is similar to what we did in Chapter 8 to model the bistable Mos/MAPK switch that drives *Xenopus* oocyte maturation: we combined a simple positive feedback loop with a nonlinear Hill function for the mediators of the feedback and produced a bistable system—a system where, for some values of parameters and x_{tot}, the rates of activation and inactivation of y^* balance at three steady states, two of which are stable (**Figure 15.5b**). In the present case, though, the feedback promotes the activation of a species (y^*) rather than the production of a species (Mos).

Finally, we need to implement the negative feedback that degrades cyclin B after mitotic entry. The activation of APC/C^{Cdc20} (z) is the result of phosphorylation by cyclin B–Cdk1, and it is known to be switch-like in character. Since z^* is regulated by phosphorylation and dephosphorylation, we will assume that z^* is at the steady-state level determined by whatever y^* is.[1]

1 Note that this is a fiction—it has been experimentally shown that z^* lags significantly behind y^*, and it has been hypothesized that this time lag contributes to the robustness of the cell cycle oscillator. Nevertheless, making this simplification allows us to have a model with only two time-dependent variables, x_{tot} and y^*, and this allows us to make use of 2D phase plane analysis to probe the workings of the model. So the fiction is useful for the purposes of teaching.

Since the activation of APC/C^{Cdc20} is switch-like, we use a Hill function for the steady-state activity of $z*$, in fractional terms:

$$z* = \frac{y*^{n1}}{K_1^{n1} + y*^{n1}}. \tag{15.5}$$

Substituting this into Eq. 15.1 yields:

$$\frac{dx_{tot}}{dt} = k_1 - k_{-1}x_{tot}\frac{y*^{n1}}{K_1^{n1} + y*^{n1}}. \tag{15.6}$$

Equations 15.4 and 15.6 constitute our two-ODE model of cell cycle oscillations: one equation for the relatively slow synthesis and destruction of cyclin (x_{tot}) (Eq. 15.6) and one for the relatively rapid activation and inactivation of cyclin B–Cdk1 (y) (Eq. 15.4).

Figure 15.5c shows the time course of cyclin abundance (x_{tot}) and cyclin B–Cdk1 activity ($y*$) for this model, using judiciously chosen parameters ($k_1=1$, $k_{-1}=0.2$, $K_1=30$, $n_1=5$, $k_2=5$, $k_{-2}=1$, $a_2=0.03$, $K_2=30$, and $n_2=5$) and initial conditions of $x_{tot}[0]=y*[0]=0$. The model yields sustained sawtooth waves of cyclin abundance and periodic spikes of cyclin B–Cdk1 activity. These are very similar in character to the oscillations in these quantities measured experimentally, and they are quite different from the sinusoidal oscillations seen in the harmonic oscillator, or the almost-sinusoidal oscillations seen in the Goodwin oscillator. Periodic spikes and sawtooth waves are common in biological oscillations.

15.4 THE CELL CYCLE MODEL HAS A SINGLE UNSTABLE STEADY STATE

To better understand the dynamics of the system, we can carry out local stability analysis of the steady state. We plot the two nullclines of the two-ODE system in the phase plane, determine the position of the steady state, and calculate the eigenvalues of the Jacobian matrix at the steady state.

An equation for the x_{tot}-nullcline can be obtained by setting the time derivative in Eq. 15.6 equal to zero:

$$0 = k_1 - k_{-1}x_{tot}\frac{y*^{n1}}{K_1^{n1} + y*^{n1}}. \tag{15.7}$$

Equation 15.7 can be rearranged to yield an expression for x_{tot} as a function of $y*$:

$$x_{tot} = \frac{k_1}{k_{-1}}\left(\frac{K_1^{n1} + y*^{n1}}{y*^{n1}}\right). \tag{15.8}$$

This x_{tot}-nullcline is shown as the blue curve in **Figure 15.6b**.

For the $y*$-nullcline, we set the time derivative equal to zero in Eq. 15.4:

$$0 = k_2\left(a_2 + \frac{(y*)^{n2}}{K_2^{n2} + (y*)^{n2}}\right)(x_{tot} - y*) - k_{-2}y*. \tag{15.9}$$

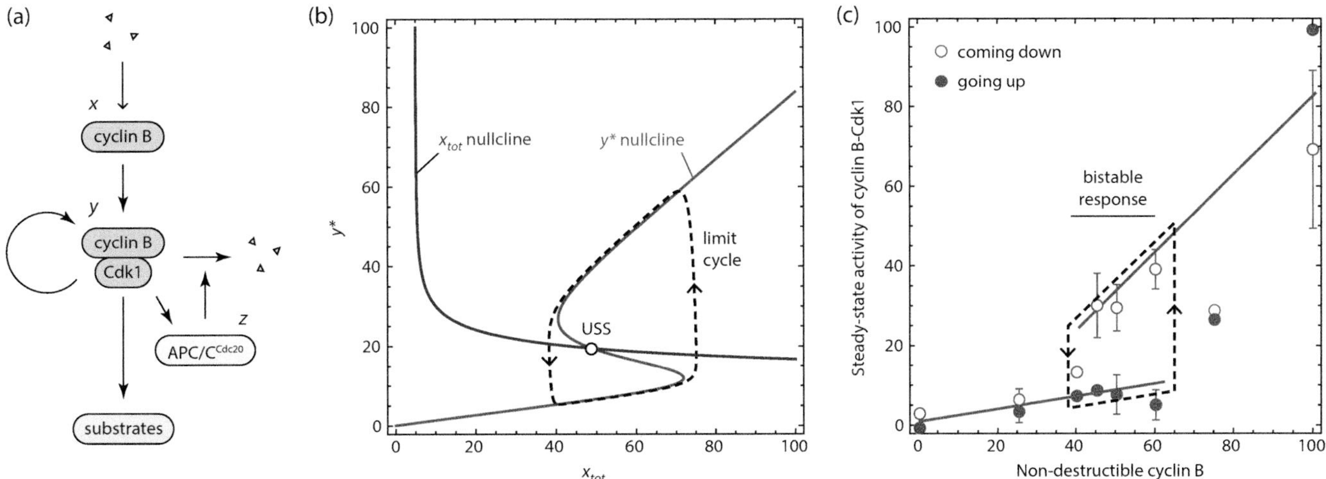

Figure 15.6 Nullclines and the limit cycle for the two-variable cell cycle model. (a) Schematic of the circuit, repeated from Figure 15.5a for reference. (b) Nullclines and the limit cycle. The x_{tot}-nullcline is shown in blue; the y^*-nullcline in red. USS denotes the unstable steady state, and the black dashed line is the limit cycle. (c) Experimental data on the steady-state activity of cyclin B–Cdk1 (which corresponds to y^* in our model) as a function of the total concentration of nondestructible cyclin (which corresponds to x_{tot}). The y^*-nullcline in panel b should correspond to this steady-state response, and the model agrees reasonably well with the experimental data. The limit cycle in panel b is approximated by a walk around the hysteretic response curve as shown in panel c. (Adapted from Pomerening et al., *Nat Cell Biol.* 2003.)

Solving for x_{tot} yields:

$$x_{tot} = \frac{\left(k_{-2} + k_2\left(a_2 + \frac{(y^*)^{n2}}{K_2^{n2} + (y^*)^{n2}}\right)\right)}{k_2\left(a_2 + \frac{(y^*)^{n2}}{K_2^{n2} + (y^*)^{n2}}\right)} y^*. \tag{15.10}$$

This is shown as the red curve in **Figure 15.6b**. Note that when plotted this way, with x_{tot} on the x-axis and y^* on the y-axis, the y^*-nullcline is an S-shaped, multivalued function.

The y^*-nullcline represents the steady-state response of y^* to x_{tot} in the absence of any negative feedback. This relationship has been measured experimentally and, as shown in **Figure 15.6c**, the experimental and modeled curves are similar. Both show a monostable, low value of y^* for low x_{tot} concentrations, a monostable, high value of y^* for high concentrations, and bistability when x_{tot} is between about 40 and 70 nM (note that we have chosen parameters for the model such that the concentrations end up in nM units and time ends up in minutes). In the bistable region, the portion of the S-shaped curve with negative slope corresponds to the unstable steady state for the no-negative-feedback system.

The two nullclines intersect at a single point, with $x_{tot} \approx 48.8$ nM and $y^* \approx 19.4$ nM, and we can carry out linear stability analysis for the system at this steady state. If we let:

$$f = k_1 - k_{-1} x_{tot} \frac{y^{*n1}}{K_1^{n1} + y^{*n1}}, \text{ and} \tag{15.11}$$

$$g = k_2\left(a_2 + \frac{(y^*)^{n2}}{K_2^{n2} + (y^*)^{n2}}\right)(x_{tot} - y^*) - k_{-2} y^*, \tag{15.12}$$

then the Jacobian matrix is:

$$J=\begin{pmatrix}\dfrac{\partial f}{\partial x_{tot}} & \dfrac{\partial f}{\partial y^*}\\ \dfrac{\partial g}{\partial x_{tot}} & \dfrac{\partial g}{\partial y^*}\end{pmatrix}=\begin{pmatrix} -k_{-1}\dfrac{y^{*n1}}{K_1^{n1}+y^{*n1}} & k_{-1}n_1x_{tot}\left(\dfrac{y^{*2n1-1}}{\left(K_1^{n1}+y^{*n1}\right)^2}-\dfrac{y^{*n1-1}}{K_1^{n1}+y^{*n1}}\right)\\ k_2\left(a_2+\dfrac{y^{*n2}}{K_2^{n2}+y^{*n2}}\right) & -k_{-2}+k_2n_2\left(x_{tot}-y^*\right)\left(-\dfrac{y^{*2n2-1}}{\left(K_2^{n2}+y^{*n2}\right)^2}+\dfrac{y^{*n2-1}}{K_2^{n2}+y^{*n2}}\right)-k_2\left(a_2+\dfrac{y^{*n2}}{K_2^{n2}+y^{*n2}}\right)\end{pmatrix} \quad (15.13)$$

Plugging in the parameters and the steady-state values of x_{tot} and y^*:

$$J=\begin{pmatrix} -0.021 & -0.231\\ 0.663 & 1.809\end{pmatrix}. \quad (15.14)$$

The eigenvectors of this matrix are {0.131, −0.991} and {−0.935, 0.356} and the corresponding eigenvalues are $\lambda_1 \approx 1.721$ and $\lambda_2 \approx 0.067$. This means that the single steady state is unstable—it repels in all directions. Note that for this set of parameters, the eigenvalues are a pair of positive real numbers rather than the pair of complex conjugates we obtained in the case of the Goodwin oscillator, which means that very close to the steady state, there is no intrinsic periodicity to the trajectories. Nevertheless, they acquire a counterclockwise rotation as they move away from the steady state and eventually converge upon a stable limit cycle (**Figure 15.6b**, black dashed line).

To further explore the dynamics of the system, suppose that we start a trajectory from the phase plane origin. This is right on the y^*-nullcline, so $\frac{dy^*}{dt}=0$, but because APC/C^{Cdc20} is inactive, cyclin will accumulate and the trajectory will crawl to the right. This results in the initial linear increase in x_{tot} and the gradual increase in y^* in the time course (**Figure 15.5c**). Eventually though, the trajectory reaches the end of this part of the y^*-nullcline—the sharp knee that would be a saddle-node bifurcation if this were a one-variable, bistable system with no negative feedback. At this point, the trajectory turns nearly straight upward, because the reactions that determine the activity of Cdk1 are rapid, and it approaches the upper part of the S-shaped nullcline (**Figure 15.6b**). This produces the spike in y^* activation in the time course (**Figure 15.5c**). By the time the trajectory reaches this leg, the high level of y^* has turned APC/C^{Cdc20} on, which means that cyclin B destruction is now faster than synthesis. This makes both the cyclin level (x_{tot}) and the Cdk1 activity (y^*) begin to drop (**Figure 15.5c**), and the trajectory crawls down the upper leg of the y^*-nullcline (**Figure 15.6b**). This continues until the trajectory reaches the end of this part of the y^*-nullcline and so turns nearly straight down toward the lower leg of the nullcline. Cyclin B levels plummet, Cdk1 activity drops back to basal levels, APC/C^{Cdc20} turns back off, and the system is ready to begin another cycle of cyclin accumulation, followed by Cdk1 activation, cyclin destruction, and Cdk1 inactivation.

This is how a typical relaxation oscillator works. It is built from a hysteretic bistable switch, with the oscillations being a walk around the hysteretic loop. And the relaxations for which the oscillator is named? They are the quick bursts upward and downward, which can be thought of as releasing some "tension" that builds up when you get toward the end of the nullcline you are walking along. The quickness of these relaxations gives the oscillator its spiky, nonsinusoidal character.

15.5 TUNING THE OSCILLATOR CHANGES THE PERIOD MORE THAN THE AMPLITUDE

The cell cycle oscillator is driven by cyclin B synthesis. What happens if we increase or decrease the cyclin synthesis rate?

Figure 15.7a–f shows time courses for the oscillator with the synthesis rate, k_1, ranging from 0.1 to 3. Oscillations begin somewhere between k_1=0.1 and 0.2, and they persist until k_1 reaches a bit less than 3. The bifurcation diagram corresponding to these changes in k_1 is shown in **Figure 15.7g**. There are Hopf bifurcations at k_1=0.139 and 2.86, where the real portions of both eigenvalues (shown for λ_1 in **Figure 15.7h**) pass through zero.

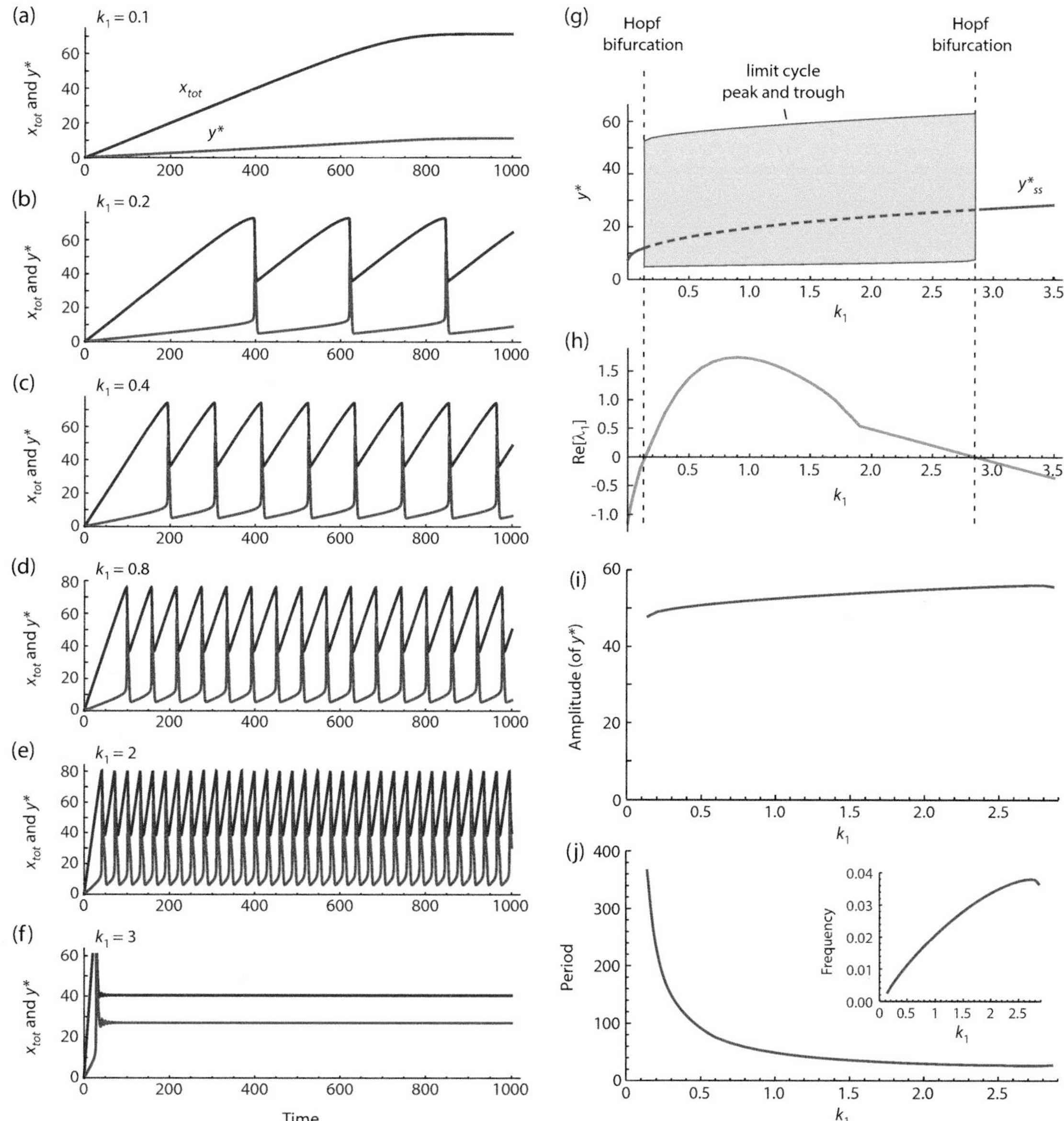

Figure 15.7 Varying the cyclin synthesis rate. (a–f) Time courses assuming cyclin synthesis rates (k_1) between 0.1 and 3 μM/min. In panels b through E the model yields limit cycle oscillations. (g, h) One-parameter bifurcation analysis, varying the synthesis rate k_1. (g) Steady-state values of Cdk1 activity (y^*) as a function of k_1. Stable steady states are depicted by solid red curves; unstable, by the dashed red curve. In the unstable regime, the peak and trough of the limit cycle is also depicted. Oscillations are born and extinguished at Hopf bifurcations, at k_1=0.139 and 2.86. (h) The real portion of one of the eigenvalues, λ_1, as a function of k_1. The Hopf bifurcations occur when the eigenvalue equals zero. The same is true for the other eigenvalue, λ_2. (i, j) The amplitude (i) and period or frequency (j) of the limit cycle as a function of the cyclin synthesis rate k_1.

Note that the amplitude of the oscillations varies relatively little over almost the entire range (**Figure 15.7i**), whereas the period or frequency varies substantially (**Figure 15.7j**). This is the opposite of the behavior we saw with the Goodwin oscillator, where the period was nearly constant but the amplitude varied substantially (**Figure 14.5**). The constant amplitude and period tunability seen here for the cell cycle oscillator turns out to be a fairly general property of relaxation oscillators. This is probably why the sinoatrial node pacemaker, another classic biological relaxation oscillator, can be tuned over a substantial range of frequencies (~40 to 160 beats per min) without changing its amplitude much, so that at the extremes of frequency it is still able to perform its vital function, making the heart beat.

15.6 PHASE PLANE ANALYSIS SHOWS WHY THE HOPF BIFURCATIONS OCCUR WHERE THEY DO

Plotting the nullclines in the phase plane provides an explanation for why the system oscillates with our original choice of parameters (including the choice of $k_1 = 1$) and why the two Hopf bifurcations occur at $k_1 = 0.139$ and 2.86. When $k_1 = 1$, the x_{tot}-nullcline (**Figure 15.8**, blue curve) intersects the S-shaped y^*-nullcline (**Figure 15.8**, red curve) in the middle section of the S, that is, the segment where the slope of the y^*-nullcline is negative. This is also the section of the nullcline that would be unstable if this were a one-variable system, with the y^* ODE representing the rate of y^* production as a function of x_{tot} in the absence of feedback. In fact, whenever the blue curve intersects this middle portion of the red curve, there are sustained limit cycle oscillations, and whenever it intersects the top or bottom portion there are no oscillations.

We can see why this is the case through linear stability analysis, examining the signs of the four partial derivatives rather than their exact values. For the function f, which defines the x_{tot}-nullcline when $f=0$, the value of f is positive for the part of the phase plane to the left of the nullcline and negative to the right of the nullcline (**Figure 15.8b**). This means that irrespective of where the steady state is positioned on the x_{tot}-nullcline, $\frac{\partial f}{\partial x_{tot}}$ will be a negative number (since f decreases from left to right) and $\frac{\partial f}{\partial y^*}$ will be a negative number as well (since f decreases from bottom to top). This gives us the signs of two elements of the Jacobian matrix. For the function g, which defines the S-shaped y^*-nullcline when $g=0$, $\frac{\partial g}{\partial x_{tot}}$ will be always be a positive number, since g is positive to the right of the nullcline and negative to the left of it (**Figure 15.8c**). But the sign of $\frac{\partial g}{\partial y^*}$ depends on whether or not the steady state is on the middle part of the S-shaped nullcline. If it is on the middle part, then $\frac{\partial g}{\partial y^*}$ will be positive; otherwise, $\frac{\partial g}{\partial y^*}$ will be negative. This means the signs of the elements of the Jacobian matrix if the steady state is on the unstable part of S-shaped nullcline are:

$$\begin{pmatrix} neg & neg \\ pos & pos \end{pmatrix}, \tag{15.15}$$

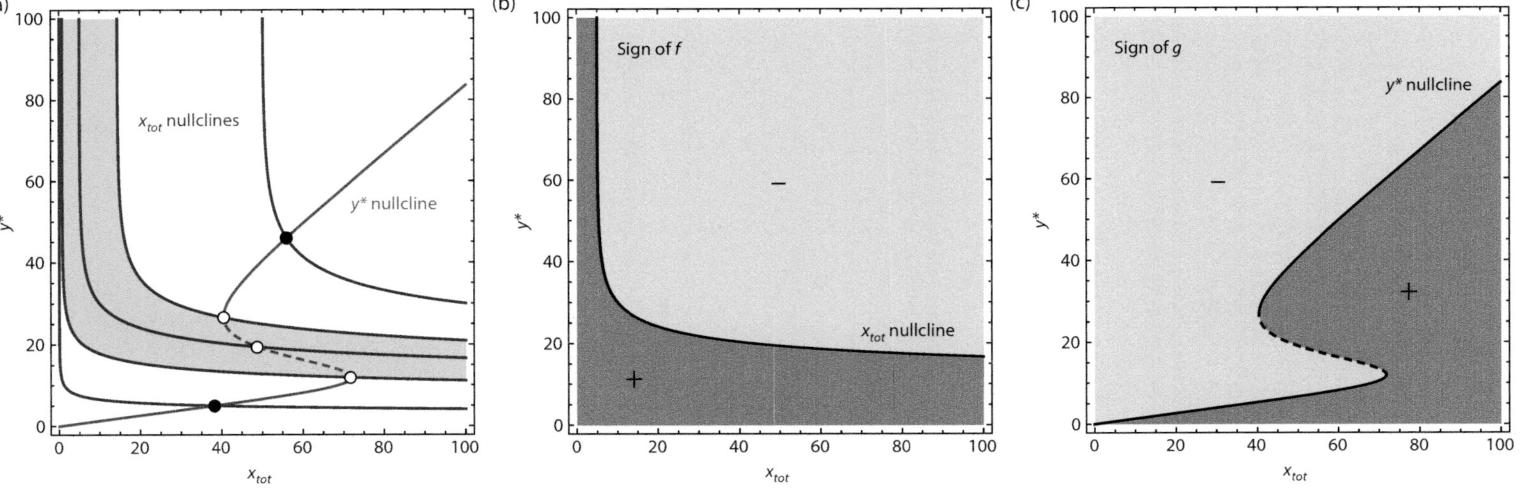

Figure 15.8 Bifurcation analysis in the phase plane. (a) Varying k_1. The $\boldsymbol{x_{tot}}$-nullclines for five values of k_1 (0.001, 0.139, 1, 2.86, and 10 μM/min, from bottom to top) are shown in blue. The y^*-nullcline is shown in red. The part of the nullcline that corresponds to unstable steady states for y^* as a function of $\boldsymbol{x_{tot}}$ in the absence of negative feedback is shown as a dashed segment. Hopf bifurcations occur when the intersection of the nullclines moves from the stable part of the y^*-nullcline (solid red curve) to the unstable part (dashed red curve) or the reverse. (b, c) The signs of f and g on either side of the nullclines.

and otherwise the signs are:

$$\begin{pmatrix} neg & neg \\ pos & neg \end{pmatrix}. \tag{15.16}$$

Next, we make use of the formula for the two eigenvalues of a 2×2 matrix:

$$\lambda_{1,2} = \frac{\tau \pm \sqrt{\tau^2 - 4\Delta}}{2} \tag{15.17}$$

where τ is the trace of the matrix (the sum of the diagonal elements) and Δ is the determinant of the matrix (the product of the diagonal elements minus the product of the off-diagonal elements). First let us assume that the steady state is on the top or bottom portions of the S-shaped nullcline, so that the sign pattern shown in Eq. 15.16 pertains. The trace will be:

$$\tau = neg + neg = neg \tag{15.18}$$

and the determinant will be

$$\Delta = neg \cdot neg - (pos \cdot neg) = pos \tag{15.19}$$

If the quantity $\tau^2 - 4\Delta \geq 0$, then $\frac{\tau - \sqrt{\tau^2 - 4\Delta}}{2}$ is a negative real number since both terms in the numerator are negative real numbers and $\frac{\tau + \sqrt{\tau^2 - 4\Delta}}{2}$ is a negative real number as well since the negative term in the numerator (τ) is larger in magnitude than the positive term ($\sqrt{\tau^2 - 4\Delta}$). If, on the other hand, the quantity $\tau^2 - 4\Delta < 0$, then $\frac{\tau + \sqrt{\tau^2 - 4\Delta}}{2}$ and $\frac{\tau - \sqrt{\tau^2 - 4\Delta}}{2}$ will be complex conjugates with negative real parts equal to $\tau/2$. The eigenvalues will either be two negative real numbers or will have negative real parts, and the steady state will be stable. Thus, theory says the system will not oscillate if the blue (x_{tot}) nullcline intersects the red (y^*) nullcline on the top or bottom portions, and this is what we found in **Figure 15.8a**.

So what if the x_{tot}-nullcline intersects the y^*-nullcline on its middle section? In this case, the trace will be:

$$\tau = neg + pos. \tag{15.20}$$

This means that the trace could be either a postive or a negative number, depending on whether $\frac{\partial f}{\partial x_{tot}}$ or $\frac{\partial g}{\partial y^*}$ is larger. Thus the eigenvalues might have positive real parts, but they might not. So the system might oscillate, if the parameters are right, but it will not necessarily oscillate.

We can show that this is in fact the case for the cell cycle oscillator. If we slow down the reactions of the bistable switch by dividing the values of the rate constants for the activation and inactivation of cyclin B–Cdk1 (k_2 and k_{-2}), by 100, so that the switch is no longer fast compared to the rates of cyclin synthesis and degradation, we are left with exactly the same nullclines as shown in **Figure 15.6b**, and exactly the same coordinates for the steady state (**Figure 15.9**). But the eigenvalues have become complex numbers with negative real

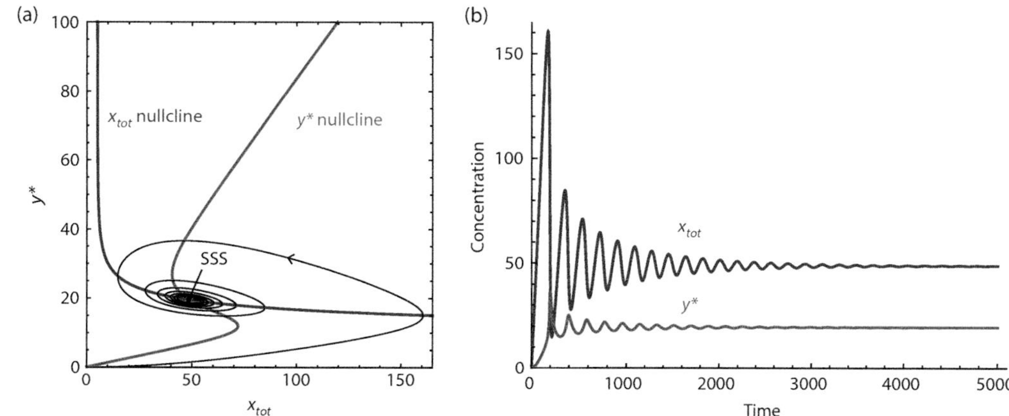

Figure 15.9 Slowing the bistable switch makes the oscillations become damped. (a) Phase plane plot of the nullclines, steady state, and one trajectory (starting at {0, 0}) for the oscillator model with k_2=0.05, k_{-2}=0.01, and all of the other parameters the same as in Figure 15.5. The trajectory spirals in toward the stable steady state or spiral point (SSS). (b) Time courses corresponding to the trajectory shown in panel a.

parts, the steady state is a stable spiral point, the oscillations have taken on more of a sinusoidal in character, and they peter out with time (**Figure 15.9**).

Thus the key elements of our cell cycle relaxation oscillator are: (1) a bistable trigger regulating the activation of cyclin B–Cdk1, built from a positive feedback loop: (2) negative feedback, which resets the system after its bistable trigger is fired; (3) a balance between the strengths of the positive and negative feedback that makes the negative feedback nullcline intersect the S-shaped positive feedback on its middle, unstable segment; and (4) quick kinetics in the bistable trigger relative to the synthesis and degradation of cyclin B.

15.7 INTERLINKED POSITIVE AND DOUBLE-NEGATIVE FEEDBACK LOOPS CAN MAKE THE MITOTIC TRIGGER MORE ALL-OR-NONE AND MORE ROBUST

Our simple model of the embryonic cell cycle included a single positive feedback loop in the mitotic trigger (**Figure 15.5**). However, we know that in reality the trigger possesses a double-negative feedback loop, through which cyclin B–Cdk1 inhibits Wee1 and Wee1 inhibits cyclin B-Cdk1, as well as the positive feedback loop. This is a conserved feature of the circuit, and such interlinked positive and/or double-negative feedback loops appear to be quite common in cell regulation. This raises the question of what it is such circuits might accomplish that a single loop system would not.

One possibility is that the circuits operate on different time scales, with the fast circuit providing a rapid response and the slower circuits providing irreversibility. This appears to be the case in fat cell differentiation in cells in culture, which is regulated by multiple interlinked positive and double-negative feedback loops with different speeds. However, in the case of Wee1 and Cdc25, the responses are essentially identical in speed.

Another possibility is suggested by rate-balance analysis of the trigger. **Figure 15.10a** shows the trigger circuit if the Wee1 loop is added, and **Figure 15.10b,c** shows the rate–balance plots assuming Wee1 is constitutively active (**Figure 15.10b**) or feedback-regulated (**Figure 15.10c**). With one loop (**Figure 15.10b**) it is pretty easy to

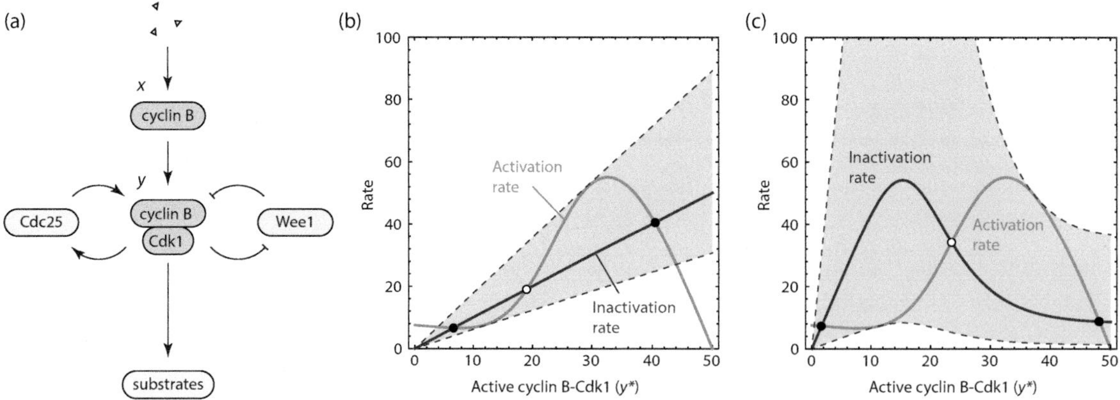

Figure 15.10 Adding a double-negative feedback loop to the mitotic trigger makes it more all-or-none in character and makes its bistability more robust with respect to parameter variation. (a) Schematic view of the mitotic trigger with the Wee1 double-negative feedback loop added. (b) Rate-balance analysis of the system without double-negative feedback. The off-state is at $y^*=6.6$ nM; the on-state is at $y^*=40.5$ nM; and the system yields bistability for a 2.6-fold range of inactivation rate constants (k_{-2}). (c) Rate-balance analysis with the double-negative feedback added. Now the off-state is at $y^*=1.6$ nM; the on-state is at $y^*=48.2$ nM; and the system yields bistability for a 27.8-fold range of inactivation rate constants (k_{-2}).

choose parameters that yield bistability, but the off-state is not completely off—6.6 nM of the 50 nM total cyclin B-Cdk1 complexes are active—and the on-state is not completely on—40.5 nM of the 50 nM complexes are active. The system can tolerate some perturbation in the parameter values and still remain bistable but not too much. For example, we can change the rate constant for cyclin B–Cdk1 inactivation (k_{-2}) or the assumed concentration of Wee1, which is built into that rate constant, by a factor of 2.6.

But if Wee1 is part of an ultrasensitive, double-negative feedback loop, as it really is, the situation is different. The rate curve for cyclin B–Cdk1 inactivation becomes essentially the mirror image of the activation rate curve (**Figure 15.10c**). Since Wee1 activity is low when Cdk1 activity is high, the curve dips down at high Cdk1 activities, which makes the on-state closer to fully-on (48.2 nM vs. 40.5 nM active cyclin B-Cdk1). The flux through the system at this on-state is much lower since the ATP-utilizing kinase Wee1 is so low in activity, which saves some metabolic energy. The off-state is, likewise, closer to being completely off (1.6 nM vs. 6.6 nM active cyclin B-Cdk1), because the blue curve is steeper at low cyclin B–Cdk1 activities than it was in the one-loop case. The range of k_{-2} (or [Wee1]) values over which the system is bistable becomes huge, increasing from 2.6-fold to 27.8-fold. Similar results are obtained if the Wee1 concentration is held constant and the Cdc25 concentration is varied.

Thus, adding double-negative feedback to the mitotic trigger can improve the performance of the trigger. It makes the switching more all-or-none in character, decreases the wasteful burning of ATP in the on-state, and increases the robustness of the system to changes in concentrations of its components. These improvements in performance may be one reason why interlinked feedback loops are commonly found in cell signaling systems.

15.8 THE FITZHUGH–NAGUMO MODEL ACCOUNTS FOR THE ELECTRICAL OSCILLATIONS OF THE SINOATRIAL NODE

One of the earliest relaxation oscillator models was proposed by the Dutch electrical engineer Balthasar van der Pol in 1920 in a paper in *Radio Review*. In fact it appears to be van der Pol who coined the term "relaxation oscillator." Van der Pol's model was motivated by the triode vacuum tube oscillator circuits he was building and studying, but as early as 1928 he and his collaborator Jan van der Mark hypothesized that something akin to this oscillator could plausibly be responsible for the heartbeat.

Our modern understanding of cellular electrophysiology began with the publication of five landmark papers by Alan Hodgkin and Andrew Huxley in 1952. The papers presented detailed, quantitative experiments on the responses of the giant squid axon to voltages[2]. The unusual size (up to 1.5 mm diameter) of this axon makes it easy to probe with microelectrodes, and as it turns out, the basic lessons obtained from this particular, peculiar axon apply pretty well to essentially all neuronal signaling, as well as the pacemaker rhythm of the modified muscle cells of the sinoatrial node. Hodgkin and Huxley accounted for their data through an experimentally inspired ODE model, which could be fitted to their experimental results beautifully. Hodgkin and Huxley's work arguably remains the most important triumph in mathematical biology.

One problem though is that the model is hard to understand; with four ODEs it is difficult to see why the model behaves the way it does. But in the early 1960s, Richard FitzHugh and Jin-Ichi Nagumo (and colleagues) came up with an electric circuit model that captures the essence of the Hodgkin–Huxley with only two time-dependent variables. This model, originally dubbed the Bonhoeffer–van der Pol model by FitzHugh but now almost universally called the FitzHugh–Nagumo model, has become probably the best-studied oscillator in nonlinear dynamics because of its relative simplicity and the rich behaviors it can produce.

Here we will work through the FitzHugh–Nagumo model and compare it to the cell cycle model analyzed in the first part of this chapter.

15.9 THE FITZHUGH–NAGUMO MODEL CONSISTS OF A QUICK BISTABLE SWITCH AND A SLOWER NEGATIVE FEEDBACK LOOP

The FitzHugh–Nagumo model imagines the neuron's signaling circuit to be composed of a bistable switch that operates on a relatively fast time scale that controls the potential across the nerve membrane (which we designate y), plus a slow negative feedback variable (which we designate x) that resets the neuron back to its resting potential after it fires. The two rate equations of the FitzHugh–Nagumo model are:

$$\frac{dy}{dt} = y - y^3 - x + k \tag{15.21}$$

$$\frac{dx}{dt} = \frac{1}{\tau}(y + k_2 - k_3 x). \tag{15.22}$$

There are three rate constants plus a variable (τ) that determines how slow the negative feedback is relative to the bistable switch. For the right choice of parameters (e.g., k_1=0.1, k_2=0.5, k_3=0.1, and τ=15),

2 Note that the giant squid axon is not an axon from a giant squid. Rather it is a really big axon from a squid of whatever size. The giant axons Hodgkin and Huxley used were from the longfin inshore squid, an animal ~30–50 cm in length. The function this axon performs is to tell the squid's jet propulsion system to fire, which is how the squid escapes from danger, and the large diameter of the axon allows it to conduct impulses quickly.

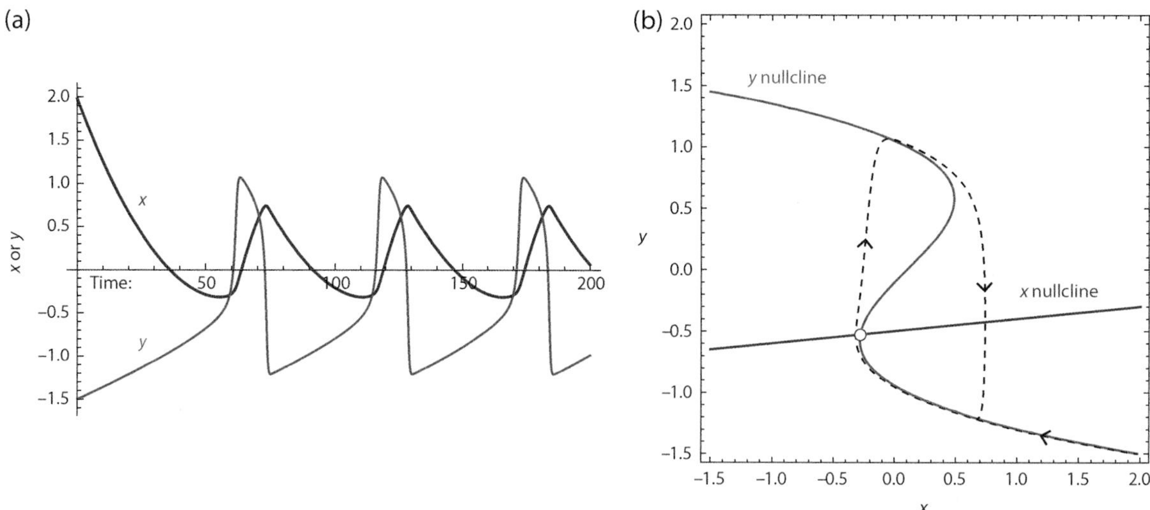

Figure 15.11 The FitzHugh–Nagumo oscillator. (a) The bistable output variable y and negative feedback variable x as a function of time. (b) Phase plane plot showing the nullclines (red and blue curves), the unstable steady state (open circle), and one trajectory (corresponding to the time course data in panel a, dashed black curve). The parameters chosen were $k_1=0.1$, $k_2=0.5$, $k_3=0.1$, and $\mu=15$.

voltage oscillations closely resembling those of the sinoatrial node can be obtained (**Figure 15.11a**). For that matter, they qualitatively resemble the oscillations of Cdk1 activity seen in the embryonic cell cycle, except for the fact that both of the variables take on negative as well as positive values.

These oscillations can be understood by examining the nullclines and trajectories in the x–y phase plane (**Figure 15.11b**). The y-nullcline is shaped like a backward S, so the response of y to a constant level of x would be hysteretic, with a bistable response at intermediate levels of x (from approximately $x=-0.28$ to 0.48). The x-nullcline is a straight line with positive slope that intersects the y-nullcline on its middle part. All told, the nullclines look a lot like the nullclines for the cell cycle oscillator (**Figure 15.6b**), except reversed left to right.

The eigenvalues of the Jacobian matrix evaluated at the steady state for our choice of parameters are $\lambda \approx 0.078 \pm 0.244i$. The imaginary parts mean that the steady state is a spiral point, and the positive real parts mean it is unstable. As was the case with the cell cycle oscillator, if we change some parameter in the FitzHugh–Nagumo model (say k_1) to change where the nullclines intersect, the Hopf bifurcations occur when the steady state changes from being on the middle part of the y-nullcline to the top or bottom part. And, finally, if we keep the nullclines in place but change the speed of the slow ODE (e.g., by changing τ; the smaller the value of τ, the faster the ODE is), we lose oscillations once the slow ODE gets too fast relative to the bistable switch. For the rate constants we chose, the oscillations are lost once τ falls below about 0.61. All of these behaviors correspond well to what we found for the cell cycle oscillator.

15.10 THE CELL CYCLE OSCILLATOR AND THE FITZHUGH–NAGUMO OSCILLATOR SHARE THE SAME SYSTEMS-LEVEL LOGIC

We have looked at two different two-variable relaxation oscillator models. One describes the embryonic cell cycle; the other, repetitive neuronal firing and the rhythm of the modified cardiomyocytes of the sinoatrial node. One is built from translation, proteolysis, and protein phosphorylation; the other from potentials and ion flows. The equations in the two models look pretty different.

Yet, in the most important ways the two models are very much alike, and this can be seen from the phase plots (**Figures 15.6b** and **15.11b**). Both models have an S-shaped or reverse S-shaped nullcline that can be viewed as a bistable switch. Both have a second nullcline for a slower recovery process, with a slope that is the same in sign as that of the middle part of the S-shaped nullcline. Both models require the two nullclines to intersect in the middle of the S to get oscillations. Both are parameterized so that the bistable switch is fast relative to the recovery process. The result is that both systems have a single unstable steady state, and both systems give rise to spiky relaxation oscillations, where the limit cycle can be thought of as a walk around a hysteretic stimulus/response loop.

15.11 DEPLETION CAN TAKE THE PLACE OF NEGATIVE FEEDBACK IN A RELAXATION OSCILLATOR

In Chapter 12 we examined how systems with negative feedback loops can function as pulse generators and can even adapt perfectly—return exactly to baseline after a pulse of output—if the system is set up properly. Then in Chapter 13 we examined systems that also generate pulses and adapt perfectly but do not, at least on the face of it, contain negative feedback loops. These included incoherent feedforward systems and three-state systems with state-dependent activation. Might such a system be able to substitute for negative feedback in the generation of sustained, limit cycle oscillations? The answer is yes, and a good example comes out of studies of cortical contraction in oocytes and eggs.

These spatial cortical waves are generated by the cortical actin cytoskeleton, and they arise through the activation of the Ras-like small GTPase RhoA. Wigbers, Tan, and their coworkers have presented a simple model for RhoA activation built on a three-state activation cycle. For our purposes, this is interesting because the model can, if the parameters are right, generate pulsatile oscillations in RhoA activity.

The RhoA cycle is shown schematically in **Figure 15.12a**. There are three time-dependent species. First, there is the cytoplasmic form of inactive, GDP-bound RhoA. We designate the fraction of the total RhoA in this state as y_1. Next there is the membrane-bound but still inactive form, y_2. And finally, there is the active, GTP-bound form, y_3. We assume that y_1 is taken away by binding to the membrane, yielding y_2, and is produced from y_3 by the simultaneous hydrolysis of its GTP to GDP and its dissociation from the membrane. This yields the first rate equation:

$$\frac{dy_1}{dt} = -k_1 y_1 + k_3 y_3. \tag{15.23}$$

Note that we could have added a term describing the production of y_1 by the dissociation of y_2 from the membrane, but for our purposes it is not necessary.

For the second rate equation, we assume that the conversion of inactive y_2 to active y_3, catalyzed by a guanine nucleotide exchange factor, takes place at a basal rate defined by k_2, but also that there is positive

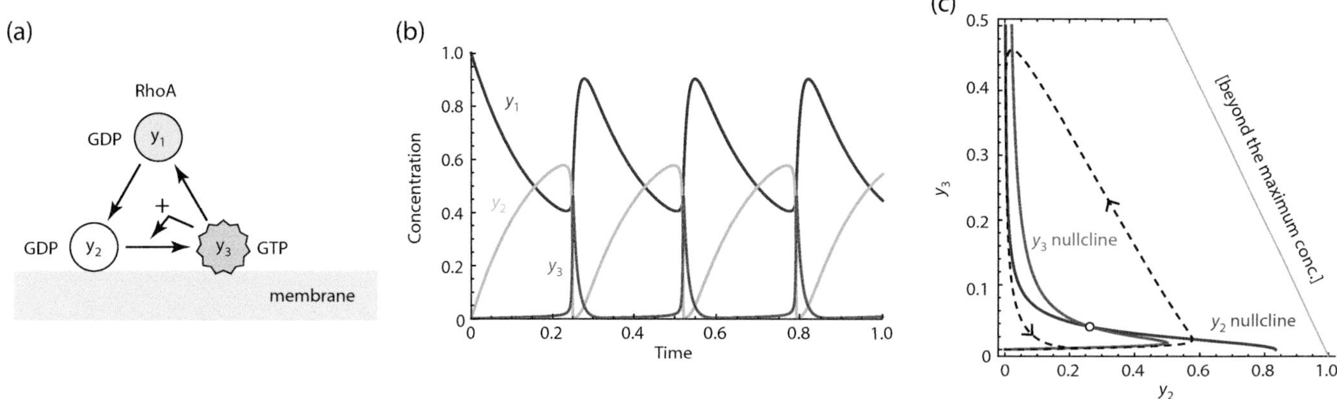

Figure 15.12 Oscillations in a three-state system with depletion. (a) Schematic depiction of the RhoA oscillator circuit. (b) Time courses for the three RhoA states showing spiky activation of RhoA. (c) Phase plane analysis showing the y_2-nullcline (blue), the y_3-nullcline (red), the unstable spiral point (open circle), and one trajectory starting from $y_1=1$, $y_2=0$, $y_3=0$ (dashed black line). The parameters were: $k_1=5$; $k_2=1$; $k_3=100$; and $k_{feedback}=10{,}000$.

feedback, with the feedback contribution to the rate being proportional to y_3^2. This introduces some nonlinearity that will be important for the behavior of the model. We could instead use a Hill equation here, but the simpler y_3^2 factor suffices.

$$\frac{dy_2}{dt} = k_1y_1 - k_2y_2 - k_{feedback}y_3^2y_2. \tag{15.24}$$

Finally, the rate equation for the active form y_3 is:

$$\frac{dy_3}{dt} = k_2y_2 + k_{feedback}y_3^2y_2 - k_3y_3. \tag{15.25}$$

These three ODEs constitute our three ODE model of RhoA activation and inactivation. Note that there is no explicit negative feedback in the model, just as there was no explicit negative feedback in the three-state model of sodium channel adaptation we analyzed in Chapter 13. But there is state-dependent inactivation, because we have assumed that only the active, membrane-bound form of RhoA (y_3) can be converted into the inactive, GDP-bound form.

Figure 15.12b shows the time courses of y_1, y_2, and y_3 for one choice of parameters ($k_1=5$; $k_2=1$; $k_3=100$; $k_{feedback}=10{,}000$), starting with $y_1=1$ and y_2 and $y_3=0$. Initially y_1 falls, and the inactive y_2 accumulates—the first phase of the oscillation. This also produces tiny amounts of y_3, which feeds back to increase the rate of y_3 production, and the concentration of y_3 rises exponentially—the second phase. The increase in y_3 occurs at the expense of y_2, and so eventually y_2 is depleted and no more y_3 can be produced. The concentration of y_3 plummets as it is converted to the inactive form y_1—the third phase—and the whole cycle starts over, with y_1 falling, y_2 rising, and ultimately y_3 spiking upward.

Note that the three ODEs imply that the total concentration of y must be constant, since:

$$\frac{dy_1}{dt} + \frac{dy_2}{dt} + \frac{dy_3}{dt} = 0 \tag{15.26}$$

Since we are measuring the three y species in fractional terms, it follows that:

$$1 = y_1 + y_2 + y_3 \tag{15.27}$$

We can use this relationship to reduce the model to two ODEs with two time-dependent variables (e.g., y_2 and y_3):

$$\frac{dy_2}{dt} = k_1(1 - y_2 - y_3) - k_2 y_2 - k_{feedback} y_3^2 y_2; \tag{15.24}$$

$$\frac{dy_3}{dt} = k_2 y_2 + k_{feedback} y_3^2 y_2 - k_3 y_3. \tag{15.25}$$

In this form, the system does appear to possess a negative feedback loop, since y_2 promotes the activation of y_3 and y_3 promotes the inactivation of y_2. With the system reduced to two time-dependent variables, we can carry out phase plane analysis of the model, plotting the y_2 and y_3-nullclines in the y_2–y_3 plane.

Figure 15.12c shows the two nullclines. The y_2-nullcline is a monotonic function and the nullcline for y_3-nullcline is something more complicated. It is not a full S-shaped nullcline like we saw with the cell cycle oscillator and the Fitzhugh–Nagumo oscillator; it is more like the bottom half of a bistable nullcline. If we start a trajectory from the origin, it crawls slowly up the y_3-nullcline until it runs out of nullcline. Then it bursts quickly upward, due to the rapid positive feedback, and to the left, due to the depletion of y_2 during the burst. Finally, once y_2 is essentially gone, the concentration of y_3 falls precipitously, the system approaches the y_3-nullcline, and the cycle begins again. The nullclines intersect at a single point, which represents an unstable steady state, and the limit cycle is stable, attracting trajectories from every part of the phase plane. Depletion oscillators like this one may be common in cell signaling—both the action potential (which we modeled with the Fitzhugh–Nagumo equations) and calcium oscillations can be viewed as depletion oscillators that cycle through three states.

SUMMARY

Here we have analyzed three relaxation oscillator models that produce periodic spikes of activity. The first was a two-ODE model inspired by the *Xenopus* embryonic cell cycle; the second, the two-ODE Fitzhugh–Nagumo model, which accounts for the neuronal action potential and for the periodic electrical pulses of cardiac pacemaker cells. Both of these models consist of slow negative feedback loops coupled to fast bistable triggers. The third model was developed to account for cortical contraction waves in oocytes. It uses state-dependent inactivation to reset the oscillator by depletion after each firing, although this depletion mechanism can be viewed as including a sort of implicit negative feedback. All of these models yield periodic bursts of activity in their time courses and limit cycle oscillations in the phase plane. All can be tuned over a range of frequencies without changing the amplitude of their outputs by much, and all have a single steady state that is either an unstable spiral point or a simple unstable steady state, for choices of parameters that yield sustained oscillations.

FURTHER READING

THE *XENOPUS LAEVIS* EMBRYONIC CELL CYCLE—DATA AND MODELS

Ferrell JE Jr. Feedback regulation of opposing enzymes generates robust, all-or-none bistable responses. *Curr Biol*. 2008 Mar 25;18(6):R244–5.

Mochida S, Hunt T. Protein phosphatases and their regulation in the control of mitosis. *EMBO Rep.* 2012 Mar;13(3):197–203.

Morgan DO. *The Cell Cycle: Principles of Control*. New Science Press, 2007.

Minshull J, Blow JJ, Hunt T. Translation of cyclin mRNA is necessary for extracts of activated xenopus eggs to enter mitosis. *Cell*. 1989 Mar 24;56(6):947–56.

Murray AW, Kirschner MW. Cyclin synthesis drives the early embryonic cell cycle. *Nature*. 1989 May 25;339(6222):275–80.

Murray AW, Hunt T. *The Cell Cycle: An Introduction*. W. H. Freeman & Co., 1993.

Novak B, Tyson JJ. Numerical analysis of a comprehensive model of M-phase control in Xenopus oocyte extracts and intact embryos. *J Cell Sci*. 1993 Dec;106 (Pt 4):1153–68.

Pomerening JR, Sontag ED, Ferrell JE Jr. Building a cell cycle oscillator: hysteresis and bistability in the activation of Cdc2. *Nat Cell Biol*. 2003 Apr;5(4):346–51.

Sha W, Moore J, Chen K, Lassaletta AD, Yi CS, Tyson JJ, Sible JC. Hysteresis drives cell-cycle transitions in Xenopus laevis egg extracts. *Proc Natl Acad Sci USA*. 2003 Feb 4;100(3):975–80.

Solomon MJ, Glotzer M, Lee TH, Philippe M, Kirschner MW. Cyclin activation of p34cdc2. *Cell*. 1990 Nov 30;63(5):1013–24.

Tsai TY, Choi YS, Ma W, Pomerening JR, Tang C, Ferrell JE Jr. Robust, tunable biological oscillations from interlinked positive and negative feedback loops. *Science*. 2008 Jul 4;321(5885):126–9.

Tyson JJ, Novak B. Bistability, oscillations, and traveling waves in frog egg extracts. *Bull Math Biol.* 2015 May;77(5):796–816.

THE VAN DER POL AND FITZHUGH–NAGUMO OSCILLATORS

FitzHugh R. Impulses and physiological states in models of nerve membrane. *Biophys J*. 1961;1:445–66.

Nagumo J, Arimoto S, Yoshizawa S. An active pulse transmission line simulating nerve axon. *Proc Inst Radio Engineers* 1962;50:2061–70.

van der Pol B. A theory of the amplitude of free and forced triode vibrations. Radio Review. 1920;1:701–710, 754–62.

van der Pol B. On "relaxation-oscillations". *The London, Edinburgh, and Dublin Philosophical Magazine and Journal of Science Ser 7*. 1926;2:978–92.

van der Pol B, van der Mark J. The heartbeat considered as a relaxation oscillation, and an electrical model of the heart. *The London, Edinburgh, and Dublin Philosophical Magazine and Journal of Science Ser 7*. 1928;6:763–75.

RhoA OSCILLATIONS

Wigbers M, Tan TH, Brauns F, Jinghui L, Swartz Z, Frey E, Fakhri N. A hierarchy of protein patterns robustly decodes cell shape information. *Nature Physics*, https://doi.org/10.1038/s41567-021-01164-9.

EXCITABILITY 16

IN THIS CHAPTER . . .

INTRODUCTION

Neurons quite commonly exhibit spontaneous, irregular spikes of output like those shown in **Figure 16.1a**, rather than the sustained relaxation oscillations analyzed in Chapter 15. Like relaxation oscillations, these spikes require positive feedback. Thus, the famous pufferfish poison tetrodotoxin, which blocks the voltage-dependent sodium channels responsible for positive feedback, extinguishes the calcium spikes seen in neuronal explants (**Figures 16.1a**).

Likewise, various mammalian cells exhibit spontaneous, irregular pulses of ERK1/2 activity (**Figure 16.1b**). These pulses are slower than neuronal calcium spikes, occurring on a time scale of hours, but otherwise are similar in character. Like neuronal spikes, there is generally a digital, all-or-none quality to the ERK1/2 spikes, although in both cases some submaximal spikes are seen. Adding epidermal growth factor (EGF) to the cells increases the average frequency of the ERK1/2 spikes without affecting their amplitude substantially (**Figure 16.1b**).

DOI: 10.1201/9781003124269-16

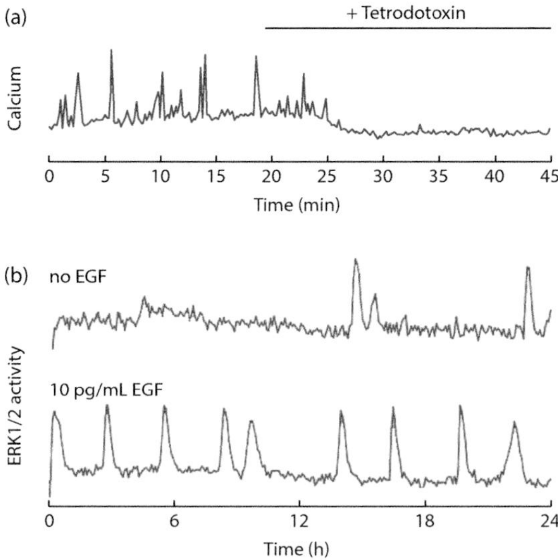

Figure 16.1 Excitability. (a) Irregular pulses of cytosolic calcium in neurons. The sodium channel blocker tetrodotoxin inhibits the calcium pulses. (Adapted from Dailey and Smith, *J Neurobiol.* 1994 with permission.) (b) Irregular pulses of ERK1/2 activation in MCF-10A cells, an immortal but nonmalignant human mammary epithelial cell line. (Adapted from Albeck et al., *Mol Cell.* 2013 with permission.)

These spikes of output, whether unprovoked or stimulated by some input, are termed excitable responses, and the cytoplasm or membrane that gives rise to the response is referred to as an excitable medium. The minimal circuits that produce excitable responses consist of a fast positive feedback loop and a slower negative feedback loop—exactly the ingredients required for a relaxation oscillator. The only difference is that the parameters of the circuit are such that there is a stable steady state rather than an unstable one.

Here we will examine the dynamics of a simple model of an excitable response, based on the receptor tyrosine kinase/MAP kinase cascade.

16.1 THE RECEPTOR TYROSINE KINASE/MAP KINASE SYSTEM INCLUDES MULTIPLE POSITIVE AND NEGATIVE FEEDBACK LOOPS

We start by examining the various feedback loops that could be important for these spikes of activity. Previous experimental and modeling studies have identified at least five plausible positive feedback loops (**Figure 16.2**), not including the Mos/MAPK loop that is important in oocyte maturation (Chapter 8) but thought not to be important in somatic cells.

First, activated EGF receptors can bring about the activation of other EGF receptors, possibly through double-negative feedback between the receptors and phosphotyrosine phosphatases. Second, active Ras-GTP can allosterically activate Sos, the upstream activator of Ras. Third, the distributive (or even semi-processive) dual phosphorylation of ERK1/2 by MEK can, in principle, yield a bistable response through an implicit positive feedback that arises from competition between the various ERK1/2 phosphoforms for MEK and the opposing phosphatases. Fourth, ERK1/2 activation can bring about release of fibroblast growth factor (FGF), which likewise can activate FGF receptors in the same cell or in neighboring cells. And finally, ERK1/2 activation can activate metalloproteases that release membrane-bound epidermal growth factor receptor (EGFR) ligands like epidermal growth factor (EGF) and TGF-α, freeing them up to activate more EGF receptors on either the same cell or neighboring cells. We do not currently know which of these loops is responsible for the pulses of ERK1/2 activation seen in MCF-10A cells. Somewhat arbitrarily, we will assume that it is the metalloprotease loop; the time course fits with the phenomenon, and there is evidence for its importance in some other cells and cell lines.

Likewise, as mentioned in Chapter 12, there are many choices for the relevant negative feedback (**Figure 16.2**), including induction of MAP kinase phosphatases (MKPs) and phosphorylation of the upstream regulators MEK, Raf, Sos, and the receptor tyrosine kinase itself. In addition there are various state-dependent inactivation systems, including the Ras cycle and EGFR internalization, with their implicit negative feedback. Somewhat arbitrarily, we will assume that MKP induction is the main relevant negative feedback here, and we will assume that it operates on a slower time scale than ERK-induced EGF release from the plasma membrane.

These are the two loops of our positive-plus-negative feedback system (**Figure 16.3a**). As was the case with our cell cycle oscillator model, and with the FitzHugh–Nagumo model, the model is simpler than the real system is known to be. But it is complicated enough to allow it to generate an excitable response and simple enough to be a good example to learn from.

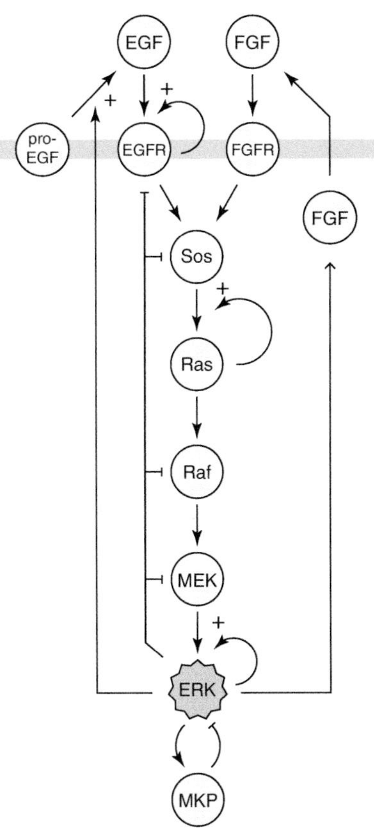

Figure 16.2 Feedback loops in the EGFR/MAP kinase cascade.

16.2 EXCITABLE RESPONSES CAN BE GENERATED BY A FAST POSITIVE FEEDBACK LOOP COUPLED TO A SLOW NEGATIVE FEEDBACK LOOP

We first formulate a rate equation for the positive feedback loop. For the no-feedback component of the activation of ERK1/2, by which we mean the activation in the absence of the ERK-induced EGF release, we will assume that that rate of ERK1/2 activation is proportional

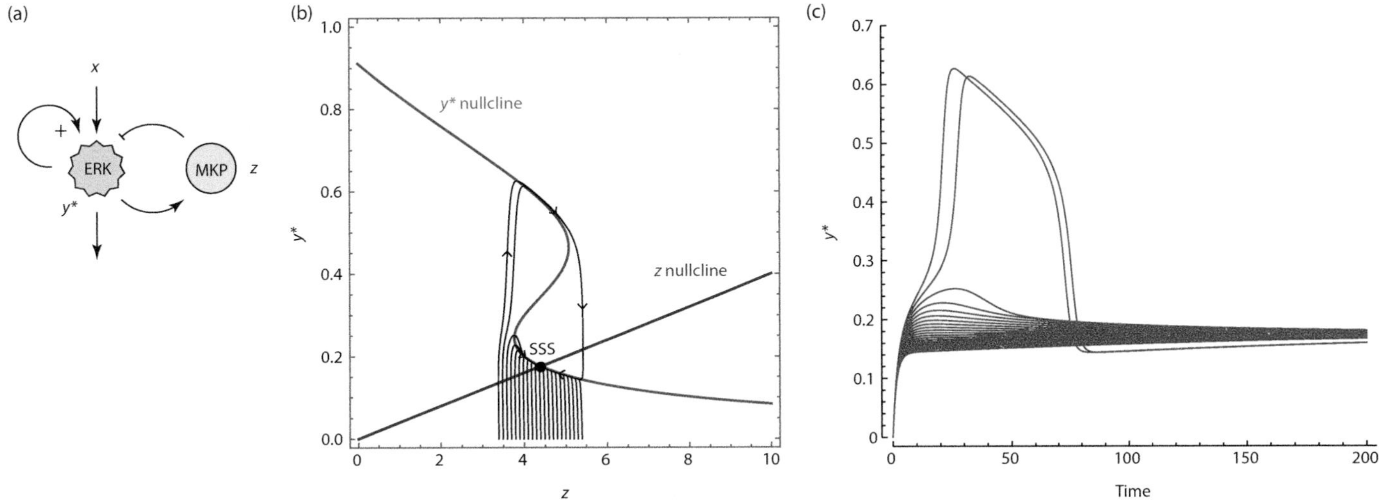

Figure 16.3 A minimal model of excitability in the MAPK cascade. (a) Schematic of the regulation of the ERK1/2 MAP kinase. (b) Nullclines (red and blue curves) and trajectories (black curves) in the phase plane. The concentration of MKP (z) is plotted on the x-axis, and the fraction of the ERK1/2 that is active (y^*) on the y-axis. (c) Time courses corresponding to the trajectories are shown in panel b.

to the concentration of added EGF (x) and the concentration of non-phosphorylated ERK1/2 ($1-y^*$), plus a term $k_0(1-y^*)$ that represents a basal rate of ERK1/2 activation (**Eq. 16.1**).

$$\textit{no feedback activation rate} = (k_0 + k_1 x)(1 - y^*). \tag{16.1}$$

For the feedback component of ERK1/2 activation, we will assume that the concentration of new EGF released from the membrane is proportional to a Hill function of the ERK1/2 activity and that the process is fast enough that we can assume the amount of EGF is always in steady state with the ERK1/2 activity. The Hill function allows us to obtain the S-shaped (or actually reverse S-shaped) nullcline that that we would need for either relaxation oscillations or excitability, and the assumption that the process is relatively fast allows us to end up with a model with only two time-dependent variables.

$$\textit{total activation rate} = (k_0 + k_1 x)(1 - y^*) + k_{pos}\frac{y^{*n}}{K_1^n + y^{*n}}(1 - y^*). \tag{16.2}$$

For the inactivation of ERK1/2, we will assume that there is a basal rate of inactivation that is proportional to y^*, the concentration of substrate to be dephosphorylated, and a rate of inactivation by the induced MKP protein that is proportional to both MKP (the enzyme, denoted z) and y^* (the substrate). This completes our rate equation for y^*:

$$\frac{dy^*}{dt} = (k_0 + k_1 x)(1 - y^*) + k_{pos}\frac{y^{*n}}{K_1^n + y^{*n}}(1 - y^*) - (k_{-1} + k_{neg} z)y^*. \tag{16.3}$$

In this scheme, the phosphatase z is the slowly changing variable. We will assume mass action kinetics for both its synthesis and degradation:

$$\frac{dz}{dt} = \frac{1}{\tau}(k_2 y^* - k_{-2} z). \tag{16.4}$$

Note that we have included a parameter τ that adjusts how fast z changes relative to y^*, just as we did in the FitzHugh–Nagumo model in Chapter 15.

Next we choose parameters to make the y^* nullcline have an inverted S shape, like the y nullcline in the FitzHugh–Nagumo model. For **Figure 16.3b** we have taken $k_0 = 0.1$, $k_1 = 1$, $x = 0$, $k_{-1} = 0.1$, $k_{pos} = 1$, $K_1 = 0.5$, and $k_{neg} = 0.1$.

Finally, we choose parameters to make the z nullcline intersect the y^* nullcline below the lower knee, so that the steady state sits on the bottom part of the y^* nullcline ($k_2 = 1$ and $k_{-2} = 0.4$ for **Figure 16.3b**). We choose a value of τ ($\tau = 10$) that makes the bistable switch (the y^* variable) change substantially faster than the negative feedback (the z variable). For these parameter choices, the system has a single steady state, at approximately $y^* = 0.176$ and $z = 4.394$, and the steady state is stable, with the eigenvalues of the Jacobian matrix being $\lambda_1 \approx -0.372$ and $\lambda_2 \approx -0.009$. As you might guess, the eigenvectors are nearly vertical and very nearly horizontal.

As expected, all of the trajectories eventually approach the stable steady state, but they do so by one of two qualitatively different routes. The black curves in **Figure 16.3b** show 21 such trajectories—10 starting to the left of the steady state, 10 starting to the right of the steady state, and one starting directly below the steady state. The trajectories to the right all approach the lower portion of the $y*$ nullcline, turn left, and then follow the nullcline up to the steady state (**Figure 16.3b**), resulting in time courses that monotonically approach the steady state (**Figure 16.3c**). The trajectories just to the left of the steady state also approach the lower portion of the $y*$ nullcline, and then they turn right and follow the nullcline down to the steady state (**Figure 16.3b**). However, the two trajectories furthest to the left do something altogether different. They miss the lower portion of the $y*$ nullcline, and then proceed up to the top portion, producing a spike in ERK activity. The trajectories then turn right and proceed down the top part of the $y*$ nullcline and continue until they run out of nullcline. They then fall down and finally head for the steady state (**Figure 16.3b**). The net result is a pulse of $y*$ in the time course (**Figure 16.3c**), or a walk around the hysteretic $y*$ curve in the phase plane (**Figure 16.3b**). This is much like a single cycle of a relaxation oscillator.

What if we were to start a trajectory such that it missed the bottom part of the $y*$ nullcline by the tiniest margin? The result is shown in **Figure 16.4**, with four trajectories that start very close together, too close to distinguish by eye. In the phase plane, the trajectories just pass the knee of the $y*$ nullcline, and then they follow the middle part, the *unstable* part, of the nullcline further upward. Eventually, and at different points, the trajectories all fall off the nullcline, either to the right or to the left, and make their way to the steady state. The resulting pulses vary substantially in amplitude and timing. Thus, even though the behavior of the system is, in principle, fully deterministic, near the pulse threshold it is so sensitive to initial conditions that it may as well be unpredictable.

We can see this as well by plotting the peak height—the maximum value of $y*[t]-y*[\infty]$) as a function of $z[0]$, taking the initial value of $y*$ in all cases to be 0. This is one measure of the output of the system,

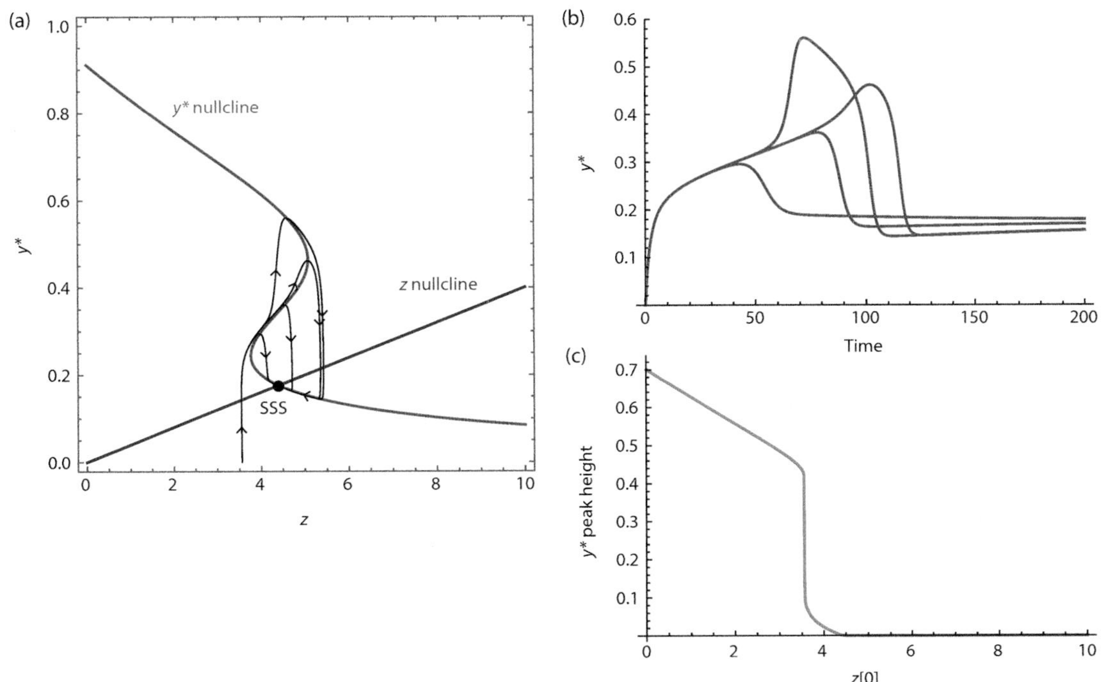

Figure 16.4 High sensitivity to initial conditions around the threshold. (a) Four trajectories starting at $y*[0]=0$ and $z[0]=4.394450$ to 4.394454. (b) Time courses of the same four trajectories. (c) Peak height (the maximum value of $y*[t]-y*[\infty]$) as a function of $z[0]$ for trajectories starting from $y*[0]=0$.

and perhaps a more important measure than the steady state output, since the steady state output will always be the same. As shown in **Figure 16.4c**, the resulting curve is continuous; there is no discontinuity the way there would be with a bistable system. But once the pulses of output begin to disappear (if you decrease $z[0]$) or appear (if you increase $z[0]$), the peak height changes enormously over a very small range of $z[0]$ values. The peak height is a highly ultrasensitive, but not discontinuous, function of the initial conditions.

16.3 NOISE CAN CAUSE AN EXCITABLE SYSTEM TO FIRE SPORADICALLY

We can imagine the irregular pulses of ERK activity as arising because even when the system is in steady state, there are fluctuations in something—say the rate of synthesis of z—that occasionally push the trajectory far enough to the left to allow it to produce a spike of y^*. We can model this by adding a noise term to the rate equation for z:

$$\frac{dz}{dt} = \frac{1}{\tau}(k_2 y^* - k_{-2} z) + \eta[t], \tag{16.5}$$

where the variable $\eta[t]$ fluctuates randomly with time. This is an example of a **Langevin equation**, a way of adding stochastic behavior to an otherwise deterministic ordinary differential equation. The results of such a simulation are shown in **Figure 16.5a,b**. For the parameters

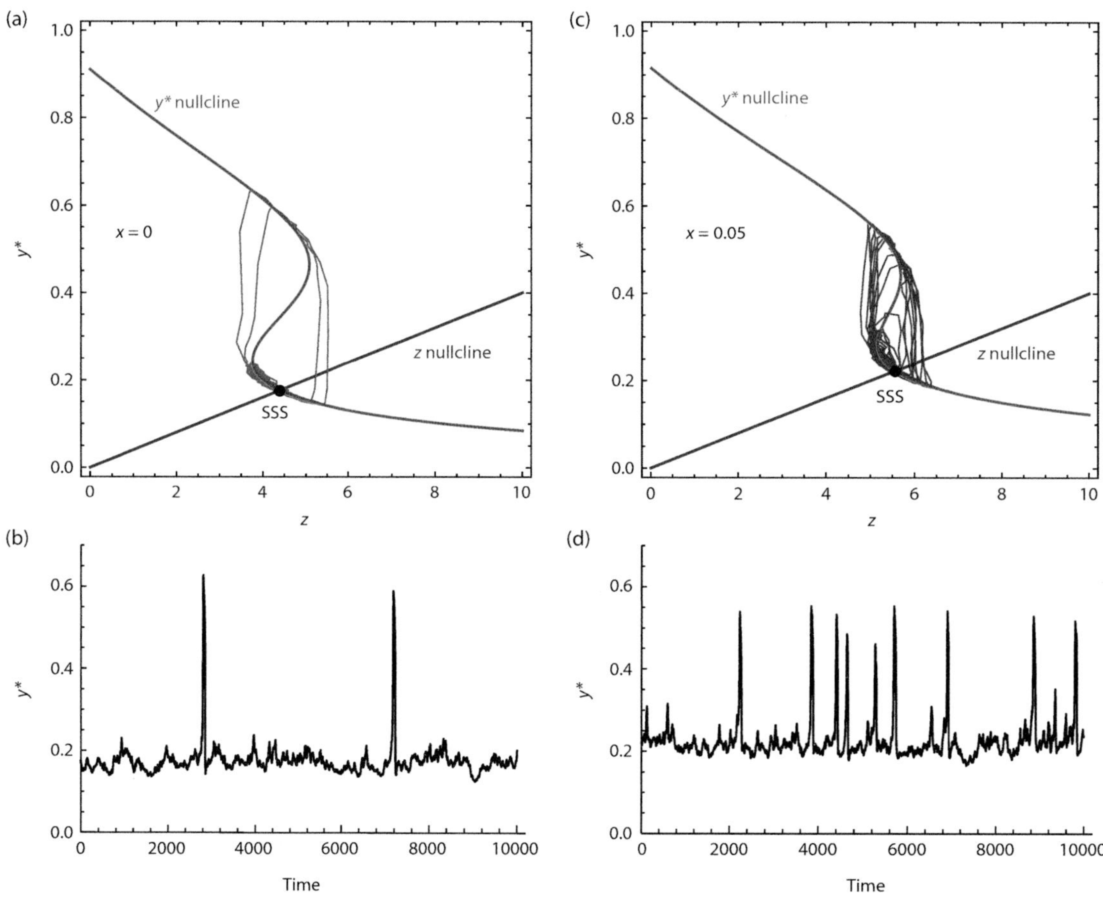

Figure 16.5 Sporadic bursts of ERK1/2 activity (y^*) driven by noise. (a, b) Phase plot (a) and the time course of y^* (b) for the model with no added EGF ($x=0$). (c, d) Phase plot (c) and the time course of y^* (d) for the model with 0.05 units of EGF ($x=0.05$). Note that the y^* nullcline is shifted to the right and so the steady state is closer to the excitation threshold.

chosen, and the noise function chosen (Gaussian white noise sampled once per time unit), we obtain infrequent, nearly all-or-none pulses of ERK1/2 activity. No two simulations are exactly the same; **Figure 16.5a,b** represents a typical one.

So far we have evaluated the model with x, the concentration of added EGF, equal to zero. If we add a small amount of x, the y^* nullcline shifts, and the steady state moves closer to the firing threshold. Intuitively it seems like this should make the random excursions of the system pass the firing threshold more frequently, and so the average frequency of the pulses should increase. This is in fact the case (**Figure 16.5c,d**). Thus, the model accounts for key aspects of the experimental data, including the sporadic pulses of ERK activity in the absence of EGF, the increased average frequency of the pulses when EGF is added, and the nearly (but not absolutely) all-or-none character of the pulses.

Very similar results would have been obtained using the FitzHugh–Nagumo model or the cell cycle oscillator model from Chapter 15 and adjusting the parameters to make the steady state sit on the bottom part of the S-shaped nullcline. The behaviors we found for the excitable ERK model can be found in a whole range of models.

SUMMARY

Here we have shown that interlinked positive and negative feedback loops can result in excitability, a phenomenon where some perturbations of the system from its steady state produce a transitory spike in output. The necessary ingredients are a bistable trigger that operates on a fast time scale, a negative feedback loop that operates on a slower time scale, and model parameters that make the steady state stable, but not too far in the phase plane from where an unstable steady state would be. When random fluctuations are imposed upon the system, the result is stochastic firing of the system's response. This is therefore an example of noise being not a hazard for the system but an essential element of the system's qualitative and quantitative behavior. Adding an external input that moves the steady state closer to the Hopf bifurcation where it would become unstable can increase the average frequency of these stochastic spikes of output.

FURTHER READING

EXAMPLES OF EXCITABILITY IN CELL SIGNALING

Albeck JG, Mills GB, Brugge JS. Frequency-modulated pulses of ERK activity transmit quantitative proliferation signals. *Mol Cell*. 2013 Jan 24;49(2):249–61.

Cai L, Dalal CK, Elowitz MB. Frequency-modulated nuclear localization bursts coordinate gene regulation. *Nature*. 2008 Sep 25;455(7212):485–90.

Dailey ME, Smith SJ. Spontaneous Ca2+ transients in developing hippocampal pyramidal cells. *J Neurobiol*. 1994 Mar;25(3):243–51.

Dalal CK, Cai L, Lin Y, Rahbar K, Elowitz MB. Pulsatile dynamics in the yeast proteome. *Curr Biol*. 2014 Sep 22;24(18):2189–94.

POSITIVE FEEDBACK LOOPS IN RECEPTOR TYROSINE KINASE/ERK SIGNALING

Baumdick M, Gelléri M, Uttamapinant C, Beránek V, Chin JW, Bastiaens PIH. A conformational sensor based on genetic code expansion reveals an autocatalytic component in EGFR activation. *Nat Commun*. 2018 Sep 21;9(1):3847.

Cox BD, De Simone A, Tornini VA, Singh SP, Di Talia S, Poss KD. In Toto Imaging of Dynamic Osteoblast Behaviors in Regenerating Skeletal Bone. *Curr Biol*. 2018 Dec 17;28(24):3937–3947.e4. doi:10.1016/j.cub.2018.10.052.

Das J, Ho M, Zikherman J, Govern C, Yang M, Weiss A, Chakraborty AK, Roose JP. Digital signaling and hysteresis characterize ras activation in lymphoid cells. *Cell*. 2009 Jan 23;136(2):337–51.

Hiratsuka T, Fujita Y, Naoki H, Aoki K, Kamioka Y, Matsuda M. Intercellular propagation of extracellular signal-regulated kinase activation revealed by in vivo imaging of mouse skin. *Elife*. 2015 Feb 10;4:e05178.

Markevich NI, Hoek JB, Kholodenko BN. Signaling switches and bistability arising from multisite phosphorylation in protein kinase cascades. *J Cell Biol*. 2004 Feb 2;164(3):353–9.

Reynolds AR, Tischer C, Verveer PJ, Rocks O, Bastiaens PI. EGFR activation coupled to inhibition of tyrosine phosphatases causes lateral signal propagation. *Nat Cell Biol*. 2003 May;5(5):447–53.

NEGATIVE FEEDBACK LOOPS IN MAP KINASE SIGNALING

Lake D, Corrêa SA, Müller J. Negative feedback regulation of the ERK1/2 MAPK pathway. *Cell Mol Life Sci*. 2016 Dec;73(23):4397–413.

WRAP-UP 17

IN THIS CHAPTER . . .

Let us finish now with a quick wrap-up.

17.1 THE BUILDING BLOCKS

We started by analyzing the building blocks of cell signaling: stoichiometric regulation (Chapters 2–4), which is how receptors work and how many downstream signaling processes work as well, activation/inactivation cycles (like phosphorylation, Chapter 5), and production/destruction cycles (Chapter 6). The simplest of these little systems respond to an input by exponentially approaching a new steady-state output, and their steady states are graded, Michaelian functions of the input. These are the basics.

We looked at how a graded, Michaelian response can be made more graded through negative cooperativity (Chapter 3) or more switch-like though positive cooperativity and other mechanisms for generating ultrasensitivity (Chapters 3 and 5). Ultrasensitivity is interesting in and of itself, and also because it makes a number of more complex behaviors, such as bistability and oscillations, easier to generate.

17.2 MOTIFS

Next we examined some simple circuit motifs—a bit bigger and more complicated than the basic building blocks but still simple enough to be understood. These included cascades (Chapter 7), positive feedback loops (Chapters 8–10), negative feedback loops (Chapter 11, 12, and 14), coherent and incoherent feedforward systems (Chapters 5 and 13), state-dependent inactivation systems (Chapter 13), and finally positive-plus-negative feedback circuits (Chapters 15 and 16).

DOI: 10.1201/9781003124269-17

We saw that sometimes a single motif can generate multiple qualitatively different responses, like stabilized monostable responses vs. pulses vs. oscillations from negative feedback, depending on the kinetic parameters and other details of the system.

17.3 SIGNAL PROCESSORS

We also saw how various types of signal processors can be constructed, including amplifiers (Chapter 7), doorbell switches (Chapter 5), toggle switches (Chapters 8 and 9), pulse generators (Chapters 12, 13, and 16), stabilizers (Chapter 11), and oscillators (Chapters 14 and 15). If you know how a system behaves through careful quantitative experiments, these chapters tell you what types of circuit are likely to be responsible.

17.4 NONLINEAR DYNAMICS

We learned various concepts and approaches from nonlinear dynamics. We examined sensitivity (Chapters 3, 5, and 7) and rate-balance analysis (Chapter 5). We used phase plane analysis, plotted nullclines, and made use of linear stability analysis (Chapter 9). We calculated eigenvalues and eigenvectors (Chapter 9) and classified steady states as stable, unstable, or saddle points (especially Chapters 9, 10, and 16), or as stable or unstable spiral points (Chapters 14 and 15). Finally, we examined various bifurcations: saddle-node (Chapters 8 and 9), pitchfork (Chapter 9), transcritical (Chapter 10), and Hopf (Chapters 14 and 15). These tools and concepts can provide insight into why systems behave the way they do and help conceptually unite all manner of biological phenomena.

* * *

There is of course more to explore, but these concepts, models, and approaches should allow the student of cell signaling to dig deep into many, many biological behaviors.

And that's all for now!

GLOSSARY

adaptation
In biology in general, adaptation is the process of adjusting to a change in the environment. In cell signaling it is a process that makes an output return to or toward baseline despite the presence of a sustained input. **Perfect adaptation** is achieved when the output returns exactly to where it was before the input was applied.

allosteric, allostery
A phenomenon where something that happens at one site in a macromolecule causes a change at a distant site. For example, the binding of a cyclin (an allosteric regulator) to a Cdk (the protein being regulated) causes a conformation change in the Cdk that aligns its catalytic residues and increases its activity.

antagonist
A substance, often a drug or other small molecule, that prevents another substance, often a hormone, from initiating a signal. Antagonists often work by binding to the hormone-binding site on a receptor, but they may also bind to the hormone itself or to an allosteric site on the receptor.

attractor
Another name for a stable steady state or a stable limit cycle.

bifurcation
In normal usage, a bifurcation is a fork, a splitting of one thing into two. In nonlinear dynamics it connotes a sudden change in the properties of a steady state in response to a gradual change in some parameter of the system. Common examples include the splitting of a steady state into two steady states (in the case of a **saddle-node bifurcation**), or into three (in the case of a **pitchfork bifurcation**), or the switching of a steady state from stable to unstable (as in a **Hopf bifurcation**).

bistable, bistability
A **bistable** system is one that has two stable steady states or equilibria.

coherent feedforward regulation
See **feedforward regulation.**

concerted model
Another name for the Monod–Wyman–Changeux model of **positive cooperativity** for multisubunit proteins. The key assumption is that all of the protein subunits flip in concert between two alternative conformations.

cooperativity
A phenomenon where one event (for example, ligand binding) makes it easier (positive cooperativity) or harder (negative cooperativity) for a second event to occur (e.g. binding of a second ligand molecule to the same protein or complex).

EC50
Effective concentration-50; the concentration of an input that produces half-maximal binding, phosphorylation, or other response.

effective Hill exponent
For a response that is not exactly described by a Hill equation, this is the Hill exponent that would yield a Hill curve with a similarly switch-like response. It is customarily defined as:

$$n = \frac{\mathrm{Log}_{10}[81]}{\mathrm{Log}_{10}[EC90 / EC10]}$$

where the *EC*90 is the concentration of input that yields a 90%-maximal response and the EC10 is the concentration that yields a 10%-maximal response.

eigenvalues and **eigenvectors**
Suppose you have a square matrix *J*, like an *n* x *n* Jacobian matrix of partial derivatives evaluated at a steady state. The eigenvectors **v** are *n*-dimensional vectors that satisfy the equation:

$$J\mathrm{v} = \lambda \mathrm{v},$$

where the scalars λ are the corresponding **eigenvalues**. If the eigenvalues and eigenvectors are real numbers, one can think of the eigenvectors as "special directions" such that if you perturb the system away from the steady state in this direction, it will come either straight back toward the steady state (if λ is negative) or straight away from the steady state (if λ is positive).

equilibrium constant

For an equilibrium reaction like the binding of a ligand to a receptor, the equilibrium constant K_{eq} is:

$$K_{eq} = k_{-1} / k_1,$$

where k_{-1} is the rate constant for the back (dissociation) reaction and k_1 is the rate constant for the forward (association) reaction. It is also equal to the concentration of free ligand at which the receptor's binding site is half-maximally occupied.

excitable systems, excitability

An excitable system is a **monostable** system where some perturbations from the steady state are amplified into explosive responses before the system settles back into the steady state. Nerve cells are the classic example; either spontaneously or in response to inputs, they may generate a spike of depolarization and of intracellular calcium.

feedforward regulation

In cell signaling, **feedforward regulation** means that an upstream input takes two distinct paths to produce a downstream output. **Coherent feedforward regulation** means that both paths act together rather than antagonistically. **Incoherent feed forward regulation** means that the two pathways are antagonistic.

full agonists

A substance, generally a hormone or drug, that activates a receptor maximally when it binds maximally.

G-protein-coupled receptors

The largest family of receptors in the human genome. **G-protein-coupled receptors** (GPCR) all have seven transmembrane segments, and generally (perhaps always) activate trimeric G-proteins when bound to agonist ligands.

Goldbeter–Koshland equation

An equation describing the steady-state response of a phosphorylation–dephosphorylation cycle if the rates of the forward and back reactions follow Michaelis–Menten kinetics. The equation can be written as:

$$y^*_{ss} = \frac{-K_{M2}kin - K_{M1}K_1pase + kin \cdot y_{tot} - K_1pase \cdot y_{tot} + \sqrt{4K_{M2}kin(kin - K_1pase)y_{tot} + (K_{M2}kin + K_{M1}K_1pase - kin \cdot y_{tot} + K_1pase \cdot y_{tot})^2}}{2(kin - K_1pase)}$$

where y^*_{SS} is the steady-state fraction of y that is phosphorylated, K_1 is the ratio of the k_{cat} values for the dephosphorylation and phosphorylation reactions, K_{M1} is the Michaelis constant for the kinase, K_{M2} is the Michaelis constant for the phosphatase, and *kin* and *pase* are the kinase and phosphatase concentrations.

Hill coefficient or Hill exponent

The number n in the **Hill equation.**

Hill equation

An equation of the form:

$$y = \frac{x^n}{K^n + x^n}$$

where x is an input and y is an output. The constant K is the ***EC50*** and the exponent n is the **Hill coefficient** or **Hill exponent**.

Hopf bifurcation

A bifurcation where, in response to a change in some parameter, a steady state changes from stable to unstable and limit cycle oscillations arise.

hyperbolic inhibition

An equation for a steady-state response y as a function of the concentration of an inhibitor x of the form:

$$y = \frac{K}{K + x}$$

where the constant K is the ***IC50***, the concentration of x where the response is half-maximally inhibited. This is also sometimes called **Michaelian inhibition**.

hyperbolic response

An equation for a steady-state response y as a function of the concentration of an inhibitor x of the form:

$$y = \frac{x}{K + x}$$

where the constant K is the ***EC50***, the concentration of x where the response is half-maximal. This is also sometimes called a **Michaelian response**.

hysteretic, hysteresis
Hysteresis is a phenomenon that occurs in systems being affected by a change in some parameter such that the state of the system depends not only on the value of the parameter, but also on the history of the system. In nonlinear dynamics the classic example of a **hysteretic** response occurs in bistable systems with positive feedback, where it may take more of a stimulus to push the system from the off-state to the on-state than it does to maintain the system in the on-state. If the feedback is strong enough to maintain the system in the on-state after the stimulus has been lowered to zero, the response is irreversible.

incoherent feedforward regulation
See feedforward regulation.

inverse agonists
A substance that decreases the basal activity of a receptor when bound.

Jacobian matrix
A matrix of partial derivatives. A 2×2 Jacobian matrix for the functions f and g is:

$$J = \begin{pmatrix} \frac{\partial f}{\partial x} & \frac{\partial f}{\partial y} \\ \frac{\partial g}{\partial x} & \frac{\partial g}{\partial y} \end{pmatrix}.$$

Koshland–Némethy–Filmer (KNF) model
A model for cooperativity in the binding of ligands to a multisubunit protein where it is assumed that the first binding event induces a conformation change in another subunit that either increases or decreases the affinity of that subunit for the ligand. It is sometimes referred to as the **sequential model**.

Langevin equation
An ordinary differential equation of the form:

$$\frac{dx}{dt} = F(x) + \eta(t),$$

where η is a time-dependent random noise term with a Gaussian probability distribution.

Langmuir equation
An equation for the equilibrium binding of a ligand x to a receptor y of the form:

$$\left(\frac{c_{xy}}{y_{tot}}\right)_{eq} = \frac{x}{K_{eq} + x},$$

where c_{xy} is the concentration of receptor–ligand complexes, y_{tot} is the total concentration of the receptor, and K_{eq} is the equilibrium constant.

limit cycle
A closed loop in the phase plane (or phase space) corresponding to sustained oscillations. If the limit cycle is stable, trajectories that start inside the limit cycle will spiral out toward it, and trajectories that start outside the limit cycle will spiral in toward it.

linear stability analysis
A procedure for analyzing the stability of a steady state that assumes that in some small neighborhood of the steady state, the rate of approaching or going away from the steady state $\left(\frac{dz}{dt}\right)$ is proportional to how far you are from the steady state (z):

$$\frac{dz}{dt} = \lambda z.$$

magnitude amplification
A phenomenon in a signaling cascade where the size of the output—the number or proportion of the molecules that are active—increases as you go from the top to the bottom of the cascade.

mass action kinetics or process
A reaction scheme that assumes the rate of the reaction is directly proportional to the concentrations of the reactants. In contrast to a **Michaelis–Menten** reaction, a mass action process is not saturable.

Michaelian inhibition
A synonym for **hyperbolic inhibition**.

Michaelian response
A synonym for **hyperbolic response**.

Michaelis–Menten equation
An equation for the rate of an enzyme-catalyzed reaction of the form:

$$V = V_{max}\frac{S}{K_m + S},$$

where V is the reaction rate, the constant V_{max} corresponds to the maximal rate of the reaction, and the parameter K_m corresponds to the substrate concentration at which the rate is half-maximal.

Monod–Wyman–Changeux (MWC) model
A model for positive cooperativity in the binding of a ligand to a multimeric receptor (or other macromolecular species); for example, the binding oxygen to hemoglobin or EGF to the EGFR. The model assumes that there are two conformations for the receptor, that the two conformations have different affinities for the ligands, and that the subunits of the receptor flip in concert between the two states. It is also sometimes called the **concerted model.**

monostable, monostability
A monostable system is one with a single stable steady state. This contrasts with a bistable system, which has two stable steady states, and an oscillatory system, which usually has a single unstable steady state or unstable spiral point.

negative cooperativity
A phenomenon where the binding of a ligand to a multimeric receptor makes it more difficult for a subsequent binding reaction to occur.

negative feedback
A phenomenon in biochemical regulation where a downstream protein negatively regulates its upstream activator or downstream protein positively regulates its upstream inactivator.

node
In nonlinear dynamics, **node** is another term for a steady state.

oocyte maturation
The process through which an immature oocyte re-enters meiosis and becomes ready for fertilization.

ordinary differential equation
A differential equation with a single independent variable, typically time (t), as opposed to a partial differential equation, where the derivative depends on both time and position.

partial agonists
A substance, generally a hormone or drug, that activates a receptor submaximally when it binds maximally.

phase plane
A plane or, often, a quadrant of a plane, where the two axes represent two variables, each of which may vary with time.

pitchfork bifurcation
A **bifurcation** where a single steady state splits into three as some parameter is varied.

positive cooperativity
A phenomenon where the binding of a ligand to a multimeric receptor directly or indirectly promotes subsequent binding reactions.

positive feedback
A phenomenon in biochemical regulation where a downstream protein positively regulates its upstream activator, or a downstream protein negatively regulates its upstream inactivator. The latter is sometimes called **double-negative feedback.**

protein kinase cascade
A succession of protein kinases where the first kinase phosphorylates and activates the second, and the second protein kinase phosphorylates and activates the third, and so on. The MAP kinase cascade, where Raf activates MEK and MEK activates MAPK, is a classic example.

rate-balance analysis, rate–balance plot
A graphical approach to the analysis of the steady states and the dynamics of systems with one time-dependent variable, where the forward and back reaction rates are plotted on one set of axes as a function of that variable, and the steady state(s) is/are deduced from the intersection points.

reciprocal regulation
A phenomenon where a regulator regulates a downstream target by increasing the rate of its activation *and* decreasing the rate of its inactivation (or the reverse).

relaxation oscillators
Oscillators composed of a positive feedback trigger operating on a fast time scale, and a slower negative feedback loop. Relaxation oscillators typically generate spiky oscillations like the repetitive action potentials of the sinoatrial node or the Cdk1 oscillations of the *Xenopus* embryonic cell cycle.

response regulator
In bacterial signal transduction, signals are often transmitted by **two-component systems**, where the first component is a histidine kinase receptor protein and the second component is a response regulator. Response regulators often regulate transcription.

saddle or saddle point
A steady state where linear stability analysis yields a positive and a negative eigenvalue.

saddle-node bifurcation
A bifurcation where a steady state appears out of thin air and then immediately splits into a saddle point (or, in one-variable systems, an unstable steady state) and a stable steady state, as some parameter is varied. Or, conversely, where a stable steady state and a saddle point (or unstable steady state) approach each other, annihilate each other, and disappear.

second messengers
Small molecules that are produced as a result of receptor activation and then regulate downstream effector proteins. Classic examples include cAMP, Ca^{2+}, diacylglycerol, and inositol trisphosphate.

sensitivity
In systems biology, **sensitivity** usually refers to how switch-like a response is. The two most common measures of sensitivity are the **effective Hill exponent** of a response, which is a global gauge of sensitivity, and the polynomial order of the response, which is a local measure.

sensitivity amplification
A phenomenon in a signaling cascade where the sensitivity of the response—the switch-like character of the response—increases as you go from the top to the bottom of the cascade.

separatrix
A boundary between two regions in the phase plane.

sequential model
Another name for the Koshland–Némethy–Filmer model of **cooperativity** for multisubunit proteins. The key assumption is that the binding of a ligand to one subunit induces a conformation change in other subunits that alters their ability to bind ligand.

signal transduction
The process of transferring a signal from one component to another in a cell or an organism. Alternatively, it can be taken to mean the process of converting a signal from one form (e.g. the free energy of a binding reaction) to another (e.g. a conformation change).

signaling cascade
A **protein kinase cascade**, or a succession of any of signaling proteins of any sort, where the main task of the first protein is to activate the second, the main task of the second protein is to activate the third, and so on.

state-dependent inactivation
A phenomenon where a protein can cycle between three activity states, like the voltage-dependent sodium channel does (going from off to on to inactivated).

steady state
A situation where the values of all of the time-dependent species are no longer changing with respect to time.

stoichiometric regulation
The type of regulation that occurs when one species affects another by binding to it. This contrasts to enzymatic regulation, where the regulator may affect an indeterminant number of downstream targets by, say, phosphorylating them.

subsensitive, subsensitivity
A response that is more graded than a benchmark Michaelian response, with an effective Hill exponent less than one.

transcritical bifurcation
A bifurcation where a steady state becomes unstable and a new stable steady state emerges as some parameter is varied. Alternatively, a bifurcation where a stable steady state and an unstable one converge, cross, and switch stabilities.

two-component system
The most common and best-studied type of bacterial signal transduction system, consisting in its simplest form of a receptor histidine kinase (the first component) and a **response regulator** (the second component) that is often a transcription factor.

two-state model
A model for receptor activation where the receptor interconverts between two discrete conformations, and ligand binding influences the equilibrium between the two.

ultrasensitive, ultrasensitivity
A response that is more switch-like than a benchmark Michaelian response, with an effective Hill exponent greater than one.

zero-order ultrasensitivity
A switch-like response that occurs in a phosphorylation–dephosphorylation cycle (or some analogous process) where the kinase, the phosphatase, or both enzymes are operating close to saturation.

INDEX

Note: **Bold** page numbers refer to tables and *italic* page numbers refer to figures.